T0320688

SPECIAL TECHNIQUES
FOR SOLVING INTEGRALS
Examples and Problems

SPECIAL TECHNIQUES FOR SOLVING INTEGRALS
Examples and Problems

Khristo N Boyadzhiev
Ohio Northern University, USA

World Scientific

EW JERSEY · LONDON · SINGAPORE · BEIJING · SHANGHAI · HONG KONG · TAIPEI · CHENNAI · TOKYO

Published by

World Scientific Publishing Co. Pte. Ltd.

5 Toh Tuck Link, Singapore 596224

USA office: 27 Warren Street, Suite 401-402, Hackensack, NJ 07601

UK office: 57 Shelton Street, Covent Garden, London WC2H 9HE

Library of Congress Cataloging-in-Publication Data
Names: Boyadzhiev, Khristo N., author.
Title: Special techniques for solving integrals : examples and problems /
    Khristo N. Boyadzhiev, Ohio Northern University, USA.
Description: Hackensack : World Scientific Publishing Co. Pte. Ltd., [2022] |
    Includes bibliographical references and index.
Identifiers: LCCN 2021049418 | ISBN 9789811235757 (hardcover) |
    ISBN 9789811236259 (paperback) | ISBN 9789811235764 (ebook)
Subjects: LCSH: Integrals. | Calculus.
Classification: LCC QA308 .B69 2022 | DDC 515/.43--dc23/eng/20211110
LC record available at https://lccn.loc.gov/2021049418

British Library Cataloguing-in-Publication Data
A catalogue record for this book is available from the British Library.

For any available supplementary material, please visit
https://www.worldscientific.com/worldscibooks/10.1142/12244#t=suppl

Printed in Singapore

This book is dedicated to the memory of my teacher

Yaroslav A. Tagamlitski

Preface

This book is intended for students and professionals who need to solve integrals or like to solve integrals and want to learn more about the various methods on how to do that. Readers will find here techniques of integration which are not found in standard calculus and advanced calculus books.

Undergraduate and graduate students whose studies include mathematical analysis or mathematical physics will strongly benefit from this material. Moreover, many items, examples and problems discussed here can be used in student projects and student research.

Students training for mathematical competitions (like the MIT integration bee) will find here many useful techniques and examples.

The book will be quite helpful to mathematicians involved in research and teaching in areas related to calculus, advanced calculus and real analysis. Examples from the book can be used in classwork or for home assignments. This way the book can be a helpful supplement to calculus and advanced calculus courses.

The reader will see how Parseval's theorem for Fourier series and for Fourier and Laplace transforms efficiently work for solving integrals.

Among other things the book contains solutions to about twenty problems from the *American Mathematical Monthly*, the *College Mathematics Journal*, the *Mathematics Magazine,* and the *Fibonacci Quarterly.* All these solutions were found by the author and submitted to the journals. They may differ from the published solutions. Throughout the book "Monthly Problem" means a problem from the *American Mathematical Monthly*.

The content is organized in five chapters. In the first chapter the reader will see the classical substitutions of Abel and Euler which are missing from standard calculus books. The important theorem of Chebyshev on the differential binomial is also found there. At the end of Chapter 1 there is a special bonus — the Gauss formula for the Arithmetic-Geometric Mean proved with the help of a special trigonometric substitution.

The second and the third chapters provide two efficient techniques for solving definite integrals. The second chapter is focused on differentiation with respect to a suitably introduced parameter in the integral. This method is illustrated by numerous examples. In some cases this method is combined with differential equations.

At the end of Chapter 2 there is another bonus for the reader – examples of integrals which help to solve the famous Basel problem, the evaluation of the series $\zeta(2) = 1 + 1/2^2 + 1/3^2 + \dots$.

In the third chapter we present various important integrals evaluated by using Fourier series in combination with the logarithmic series. The examples include log-gamma, log-sine, and log-cosine integrals. At the end of this chapter we prove a special Binet type formula for the log-gamma function $\ln \Gamma(z)$.

The fourth chapter deals with an unusual but powerful technique — evaluation of integrals by using Laplace and Fourier transforms. Here we demonstrate the efficiency of Parseval's theorem applied to these transforms. Some important hard integrals involving the Gamma function can be evaluated with the help of Parseval's formula and the exponential polynomials.

The fourth chapter also presents a special bonus for the reader. We give a new proof of the functional equation of the Riemann zeta function $\zeta(s)$. The proof is based on two special integrals. This is followed by a proof of the famous Euler formula relating the values $\zeta(2n)$ to the Bernoulli numbers

$$\zeta(2n) = \frac{(-1)^{n-1}(2\pi)^{2n}}{2(2n)!} B_{2n} \ (n = 1,\ 2, \dots).$$

In the fifth chapter the reader will see several special formulas (Frullani's formula, Poisson's integral formula and a special formula discovered by the author) which produce immediate results even with some very tough integrals. In that chapter we also show how some challenging series can be evaluated by using integrals.

The fifth chapter has a section dedicated to power series with harmonic and skew-harmonic numbers where special integration is also needed.

Appendix A represents a list of about 330 integrals solved in the book.

The reader will find proofs for many integrals from the popular reference tables of Gradshteyn and Ryzhik [25] and Prudnikov, Brychkov, Marichev [43] (see also [35]). There is a minimal overlap if any with the books of Boros and Moll [7] and Nahin [37].

Together with many classical formulas the book presents also some new results not published before.

Here at the end I want to express my gratitude to my wife Irina Boyadzhiev for her indispensable help in putting the book together.

About the Author

Khristo N. Boyadzhiev is Professor of Mathematics at Ohio Northern University. He has about 90 research publications in the areas of classical analysis, number theory, and operator theory. He received his Ph.D. in mathematics from Sofia University "St. Kliment Ohridski" and worked at the Bulgarian Academy of Sciences before joining ONU.

Contents

Chapter 1

Special Substitutions

1.1 Introduction

A very efficient method for solving integrals is using a substitution. That is, replacing the original variable by a new variable which helps simplify the integral. Most calculus textbooks, however, use only the "u-substitution" and the standard trigonometric substitutions. For example, the integral

$$\int (5+x)^{-3} dx$$

can be solved immediately with the u-substitution $u = 5 + x$. The integral

$$\int \sqrt{3-x^2}\, dx$$

can be solved quickly with the trigonometric substitution $x = \sqrt{3}\sin\theta$. In many important cases, however, the u-substitution and the trigonometric substitutions do not help. For example, the integral

$$\int \frac{dx}{(x+3)\sqrt{3x-x^2-2}}$$

is difficult to solve with only these two substitutions. The natural way to solve this integral is by the second Euler substitution (see below Example 1.2.2). Unfortunately, many tools of classical analysis are not present in standard calculus books. The purpose of this chapter is to fill part of this gap by reviving several classical substitutions.

1

We also demonstrate that in certain cases the hyperbolic substitutions are more efficient than the trigonometric substitutions.

Further, in Section 1.7 we present a very special substitution which helps to prove the Gauss formula for the Arithmetic-Geometric Mean of two positive numbers.

The u-substitution and the simple trigonometric substitutions are prerequisites for this chapter (and the entire book) as well as the integration by parts formula

$$\int f \, dg = fg - \int g \, df$$

and for definite integrals

$$\int_a^b f \, dg = fg\Big|_a^b - \int_a^b g \, df$$

where a or b, or both can be $\pm\infty$.

1.2 Euler Substitutions

Euler's substitutions are used mostly for solving indefinite integrals of the form

(1.1) $$\int R\left(x, \sqrt{ax^2 + bx + c}\right) dx$$

by "removing" the radical. There are three specific substitutions suggested by Euler. In each one of them the idea is to eliminate the term with x^2. We assume that $a \neq 0$ and that the polynomial $ax^2 + bx + c$ is not negative, i.e. the graph of the parabola $y = ax^2 + bx + c$ is not entirely under the x-axis (so that the radical $\sqrt{y}$ is defined at least on some interval). Note that when $b = 0$ the radical can be removed by a trigonometric or hyperbolic substitution (Section 1.5).

1.2.1 *First Euler substitution*

When $a > 0$ we introduce a new variable t by setting

(1.2) $$x\sqrt{a} + t = \sqrt{ax^2 + bx + c}.$$

Squaring both sides in this equation we get

$$ax^2 + 2xt\sqrt{a} + t^2 = ax^2 + bx + c$$

so that the term ax^2 cancels out. Solving for x we write

$$x = \frac{t^2 - c}{b - 2t\sqrt{a}}.$$

This way the integral takes the form

(1.3) $$\int Q(t)dt$$

where $Q(t)$ is a rational function. In general, this integral can be solved by partial fractions. At the end we return to the original variable by setting

$$t = \sqrt{ax^2 + bx + c} - x\sqrt{a}.$$

1.2.2 *Second Euler substitution*

Suppose the polynomial $ax^2 + bx + c$ has two different real roots $r_1 \neq r_2$ so that

$$ax^2 + bx + c = a(x - r_1)(x - r_2).$$

We set

(1.4) $$\sqrt{ax^2 + bx + c} = (x - r_1)t$$

(either one of the roots can be used). This yields

$$a(x - r_1)(x - r_2) = (x - r_1)^2 t^2$$

$$a(x - r_2) = (x - r_1)t^2$$

$$x = \frac{r_1 t^2 - r_1 r_2}{t^2 - a}$$

and again the integral (1.1) is reduced to the form (1.3).

1.2.3   *Third Euler substitution*

The third Euler substitution can be used when $c \geq 0$. In this case we set

(1.5) $$\sqrt{ax^2 + bx + c} = xt + \sqrt{c}.$$

This yields

$$ax^2 + bx + c = x^2 t^2 + 2xt\sqrt{c} + c$$

$$ax + b = xt^2 + 2t\sqrt{c}$$

$$x = \frac{b - 2t\sqrt{c}}{t^2 - a}$$

leading again to (1.3).

Remark. The first two Euler substitutions are sufficient to cover all possible cases. If $a < 0$ then the polynomial $ax^2 + bx + c$ has two different real roots (the parabola $y = ax^2 + bx + c$ opens down and is not under the x-axis). Nevertheless, the third Euler substitution is useful because it sometimes requires less computations.

Example 1.2.1

We will evaluate the integral

(1.6)
$$\int \frac{dx}{\sqrt{x^2 + x + 1}}$$

by the first Euler substitution. Therefore, we set

$$\sqrt{x^2 + x + 1} = x + t.$$

Then

$$x^2 + x + 1 = x^2 + 2xt + t^2$$

$$x + 1 = 2xt + t^2$$

$$x = \frac{t^2 - 1}{1 - 2t}$$

$$dx = \frac{2(t - t^2 - 1)}{(1 - 2t)^2} dt$$

$$\sqrt{x^2 + x + 1} = \frac{t - t^2 - 1}{1 - 2t}$$

$$\int \frac{dx}{\sqrt{x^2 + x + 1}} = 2\int \frac{dt}{1 - 2t} = -\ln|1 - 2t| + C$$

$$= -\ln|1 + 2x - 2\sqrt{x^2 + x + 1}| + C$$

$$= -\ln(2\sqrt{x^2 + x + 1} - 1 - 2x) + C.$$

Example 1.2.2

Now consider the integral

(1.7)
$$F(x) = \int \frac{dx}{(x + 3)\sqrt{3x - x^2 - 2}}$$

for $1 < x < 2$. We shall use the second Euler substitution.
Since $3x - x^2 - 2 = (2 - x)(x - 1)$ we set

$$\sqrt{3x - x^2 - 2} = t(x-1).$$

This yields

$$(2-x)(x-1) = t^2(x-1)^2$$

$$2 - x = t^2(x-1)$$

$$x = \frac{t^2 + 2}{t^2 + 1}, \quad dx = \frac{-2t}{(t^2+1)^2}dt, \quad t = \sqrt{\frac{2-x}{x-1}}$$

$$\int \frac{dx}{(x+3)\sqrt{3x - x^2 - 2}} = -2\int \frac{dt}{4t^2 + 5} = \frac{-1}{\sqrt{5}}\arctan\left(\frac{2t}{\sqrt{5}}\right) + C$$

$$= \frac{-1}{\sqrt{5}}\arctan\left(\frac{2}{\sqrt{5}}\sqrt{\frac{2-x}{x-1}}\right) + C.$$

Of course, this integral can be solved with the second Euler substitution in the form

$$\sqrt{3x - x^2 - 2} = t(2-x).$$

Example 1.2.3

Using again the second Euler substitution we will solve the integral

(1.8) $$\int \frac{dx}{x\sqrt{x^2 + 6x + 8}}.$$

Here we write

$$\sqrt{x^2 + 6x + 8} = \sqrt{(x+4)(x+2)} = t(x+2)$$

$$(x+4)(x+2) = t^2(t+2)^2$$

$$x + 4 = t^2(x+2), \quad x = \frac{4 - 2t^2}{t^2 - 1}, \quad dx = \frac{-4t}{(t^2-1)^2} dt.$$

And since

$$\sqrt{x^2 + 6x + 8} = \frac{2t}{t^2 - 1}$$

we have

$$\int \frac{dx}{x\sqrt{x^2 + 6x + 8}} = \int \frac{dt}{t^2 - 2} = \frac{1}{2\sqrt{2}} \ln\left|\frac{t - \sqrt{2}}{t + \sqrt{2}}\right| + C$$

$$= \frac{1}{2\sqrt{2}} \ln\left|\frac{\sqrt{x^2 + 6x + 8} - \sqrt{2}(x+2)}{\sqrt{x^2 + 6x + 8} + \sqrt{2}(x+2)}\right| + C.$$

Example 1.2.4

In this example we give our solution to Problem 11457 of the *American Mathematical Monthly* (October 2009). In this solution we will use the second Euler substitution.
Let $0 \le a \le b$. The problem is to evaluate the integral

$$J = \int_a^b \arccos\frac{x}{\sqrt{ax + bx - ab}} dx.$$

First we assume that $0 < a$ and factor out a from the radical to get

$$J = \int_a^b \arccos\frac{x/a}{\sqrt{x/a + bx/a^2 - b/a}} dx.$$

Next we make the substitution $x \to ax$ and set $\lambda = b/a$. The integral takes the form

$$J = a \int_0^\lambda \arccos \frac{x}{\sqrt{x + \lambda x - \lambda}}\, dx \ .$$

This integral is easier to manipulate. Integration by parts gives

$$J = ax \arccos \frac{x}{\sqrt{x + \lambda x - \lambda}} \Bigg|_1^\lambda - a \int_0^\lambda x \frac{d}{dx}\left(\arccos \frac{x}{\sqrt{x + \lambda x - \lambda}} \right) dx \ .$$

The first term is zero, as $\arccos(1) = 0$. Differentiating and simplifying inside the integral we find

$$J = \frac{a}{2} \int_1^\lambda \frac{(\lambda x + x - 2\lambda)x}{(\lambda x + x - \lambda)\sqrt{(\lambda - x)(x - 1)}}\, dx \ .$$

It is convenient now to split this into two integrals writing first

$$\lambda x + x - 2\lambda = (\lambda x + x - \lambda) - \lambda$$

$$J = \frac{a}{2} \int_1^\lambda \frac{x}{\sqrt{(\lambda - x)(x - 1)}}\, dx - \frac{a\lambda}{2} \int_1^\lambda \frac{x}{(\lambda x + x - \lambda)\sqrt{(\lambda - x)(x - 1)}}\, dx \ .$$

We will solve both integrals with the second Euler substitution

$$\sqrt{(\lambda - x)(x - 1)} = (x - 1)t \ .$$

This way

$$x = \frac{\lambda + t^2}{1 + t^2}, \quad dx = \frac{2t(1 - \lambda)}{(1 + t^2)^2}, \text{ etc.}$$

$$J = a \int_0^\infty \frac{\lambda + t^2}{(1 + t^2)^2}\, dt - a\lambda \int_0^\infty \frac{\lambda + t^2}{(1 + t^2)(\lambda^2 + t^2)}\, dt \ .$$

Simple evaluation by using partial fractions provides

$$J = \frac{\pi a}{4}(\lambda + 1) - \frac{\pi a \lambda}{\lambda + 1} = \frac{\pi a (\lambda - 1)^2}{4(\lambda + 1)} \ .$$

Replacing $\lambda = b/a$ we finally write the solution

(1.9) $$\int_a^b \arccos\frac{x}{\sqrt{ax+bx-ab}}\,dx = \frac{\pi(b-a)^2}{4(b+a)}.$$

In particular, setting $a \to 0$ we find

$$\int_0^b \arccos\sqrt{\frac{x}{b}}\,dx = \frac{\pi b}{4}.$$

Problem for the reader: Solve this last integral independently by first using the substitution $x = bt^2$ and then integrating by parts.

Example 1.2.5

Here the third Euler substitution will be used for the integral

$$\int \frac{dx}{x\sqrt{1-2x-x^2}}.$$

So we set

$$\sqrt{1-2x-x^2} = xt + 1$$

$$1 - 2x - x^2 = x^2 t^2 + 2xt + 1$$

$$x = \frac{-2(t+1)}{t^2+1}, \quad dx = \frac{2t^2+4t-2}{(t^2+1)^2}\,dt$$

$$\sqrt{1-2x-x^2} = \frac{1-2t-t^2}{t^2+1}$$

$$\int \frac{dx}{x\sqrt{1-2x-x^2}} = \int \frac{dt}{t+1} = \ln|t+1| + C$$

$$= \ln \left| \frac{\sqrt{1-2x-x^2}-1}{x} + 1 \right| + C = \ln \left| \frac{x-1+\sqrt{1-2x-x^2}}{x} \right| + C \, .$$

Problems for the reader: Solve by appropriate Euler substitutions

(a)
$$\int \frac{1}{\sqrt{x^2+5x+3}} \, dx$$

(b)
$$\int \frac{1}{x\sqrt{x^2+4x-4}} \, dx$$

(c)
$$\int \frac{1}{x\sqrt{x^2+2x-1}} \, dx$$

(d)
$$\int \frac{1}{1+\sqrt{1-2x-x^2}} \, dx$$

(e)
$$\int \frac{1}{x-\sqrt{x^2-x+1}} \, dx \, .$$

1.3 Abel's Substitution

The interesting substitution suggested by Niels Henrik Abel (1802–1829) helps to evaluate integrals of the form

$$J = \int \frac{dx}{(ax^2+bx+c)^p \sqrt{ax^2+bx+c}} = \int (ax^2+bx+c)^{-p-\frac{1}{2}} \, dx$$

where $b^2-4ac \neq 0$ and $p \geq 1$ is an integer. With $y = ax^2+bx+c$ this becomes

$$J = \int \frac{dx}{y^p \sqrt{y}}$$

We set

(1.10) $$t = \frac{d}{dx}\sqrt{y} = \frac{y'}{2\sqrt{y}} = \frac{2ax+b}{2\sqrt{y}}.$$

Squaring both sides we write

$$4t^2 y = 4a^2 x^2 + 4abx + b^2.$$

Multiplying y by $4a$ we also have

$$4ay = 4a^2 x^2 + 4abx + 4ac.$$

Subtracting these two equations we find $4y(a-t^2) = 4ac - b^2$, or

$$\frac{1}{y} = \frac{4(a-t^2)}{4ac-b^2}.$$

Next we differentiate the equation $2t\sqrt{y} = 2ax + b$ with respect to x

$$2\frac{dt}{dx}\sqrt{y} + 2t^2 = 2a$$

so that

$$\frac{dx}{\sqrt{y}} = \frac{dt}{a-t^2}.$$

This way the original integral is transformed into

$$J = \left(\frac{4}{4ac-b^2}\right)^p \int (a-t^2)^{p-1} dt$$

$$= \left(\frac{4}{4ac-b^2}\right)^p \sum_{k=0}^{p-1} \binom{p-1}{k} (-1)^k a^{p-k-1} \frac{t^{2k+1}}{2k+1} + C.$$

We return to the original variable by the substitution (1.10)

$$t = \frac{2ax+b}{2\sqrt{y}}$$

and the final result is

$$J = \left(\frac{4}{4ac - b^2}\right)^p \sum_{k=0}^{p-1} \binom{p-1}{k} \frac{(-1)^k a^{p-k-1}}{2k+1} \left(\frac{2ax+b}{2\sqrt{y}}\right)^{2k+1} + C .$$

In particular, with $p = 1$ and $p = 2$ we obtain

$$\int \frac{dx}{(ax^2 + bx + c)^{3/2}} = \frac{4ax + 2b}{(4ac - b)\sqrt{ax^2 + bx + c}} + C$$

$$\int \frac{dx}{(ax^2 + bx + c)^{5/2}}$$

$$= \left(\frac{4}{4ac - b^2}\right)^2 \left[\frac{a(2ax+b)}{2\sqrt{ax^2 + bx + c}} - \left(\frac{2ax+b}{2\sqrt{ax^2 + bx + c}}\right)^3\right] + C .$$

For example,

$$\int \frac{dx}{(x^2 + x + 1)\sqrt{x^2 + x + 1}} = \frac{4x + 2}{3\sqrt{x^2 + x + 1}} + C .$$

1.4   The Differential Binomial and Chebyshev's Theorem

Expressions of the form

$$x^m (a + bx^n)^p dx$$

where a, b are arbitrary coefficients and m, n, p are rational numbers are called differential binomials. We want to evaluate the integral

(1.11) $$F(x) = \int x^m (a + bx^n)^p dx .$$

Until 1852 it had been known for a long time that if at least one of the numbers

$$p, \quad \frac{m+1}{n}, \quad \frac{m+1}{n} + p$$

is an integer, then the integral $F(x)$ can be evaluated explicitly in terms of elementary function. In 1852 Pafnuty Chebyshev (1821-1894) proved that in all other cases the integral cannot be evaluated explicitly.

Case 1

When p is an integer, the substitution $x = t^r$, where r is an appropriate positive integer turns $x^m(a + bx^n)^p$ into a rational function of t.
For example, in the integral

$$J = \int \frac{dx}{\sqrt{x}(1 + \sqrt[3]{x})^2}$$

we make the substitution $x = t^6$ to get (integrating by parts in the next step)

$$J = 6\int \frac{t^2 dt}{(1+t^2)^2} = 3\int \frac{t\,d(1+t^2)}{(1+t^2)^2} = -3\int t\,d\frac{1}{1+t^2}$$

$$= \frac{-3t}{1+t^2} + 3\arctan t + C = \frac{-3\sqrt[6]{x}}{1+\sqrt[3]{x}} + 3\arctan \sqrt[6]{x} + C.$$

Case 2

Let now $\dfrac{m+1}{n}$ be an integer (positive or negative). We set $t = a + bx^n$ to obtain

$$x = \left(\frac{t-a}{b}\right)^{\frac{1}{n}}, \quad dx = \frac{1}{n}\left(\frac{t-a}{b}\right)^{\frac{1}{n}}\frac{dt}{t-a}$$

$$F(x) = \frac{1}{n} b^{-\frac{m+1}{n}} \int t^p (t-a)^{\frac{m+1}{n}-1} dt$$

which belongs now to Case 1.

Here are two examples. The first example is the integral

$$J = \int \frac{dx}{x\sqrt{1+x^5}}$$

where $m = -1, n = 5, p = \dfrac{-1}{2}$ and $\dfrac{m+1}{n} = 0$. We set $t = 1 + x^5$ and compute

$$J = \frac{1}{5} \int \frac{dt}{(1-t)\sqrt{t}}.$$

Now we continue with $t = u^2, u = \sqrt{t}$ to get

$$J = \frac{2}{5} \int \frac{du}{1-u^2} = \frac{1}{5} \ln \frac{u+1}{u-1} + C = \frac{1}{5} \ln \frac{\sqrt{1+x^5}+1}{\sqrt{1+x^5}-1} + C.$$

In the second example we evaluate the integral

$$J = \int \frac{\sqrt{1+x^3}}{x} dx.$$

Setting directly $1 + x^3 = u^2, u = \sqrt{1+x^3}$ we get

$$dx = \frac{2u}{3(u^2-1)^{2/3}} du$$

$$J = \frac{2}{3} \int \frac{u^2}{u^2-1} du = \frac{2}{3} \int \frac{u^2-1+1}{u^2-1} du = \frac{2}{3} u + \frac{2}{3} \int \frac{1}{u^2-1} du$$

$$= \frac{2}{3} u + \frac{1}{3} \ln \left| \frac{u-1}{u+1} \right| + C$$

and finally

$$J = \frac{2}{3}\sqrt{1+x^3} + \frac{1}{3}\ln\frac{\sqrt{1+x^3}-1}{\sqrt{1+x^3}+1} + C .$$

Note that the integral

$$\int \sqrt{1+x^3}\, dx$$

cannot be solved by the substitution $t = 1 + x^3$ or by any other substitution. Here $m = 0, n = 3, p = \frac{1}{2}$ and by Chebyshev's theorem the function $\sqrt{1+x^3}$ does not have an explicit antiderivative. For such cases we can use the convenient antiderivative

$$J(x) = \int_0^x \sqrt{1+t^3}\, dt .$$

Case 3

This is the case when $\frac{m+1}{n} + p$ is an integer (positive or negative). We first transform the integral by "factoring" out x^n this way

$$\int x^m (a + bx^n)^p\, dx = \int x^{m+np}(ax^{-n}+b)^p\, dx .$$

The result is a new integral of a differential binomial falling in Case 2, as the number

$$\frac{m+np+1}{-n} = -\left(\frac{m+1}{n}+p\right)$$

is an integer.

We illustrate this case by two examples. Consider first the integral

$$J = \int \frac{dx}{\sqrt{(1+x^2)^3}}$$

where we assume that $x > 0$. Here $m = 0, n = 2, p = \dfrac{-3}{2}$ and $\dfrac{m+1}{n} + p = -1$ is an integer. We write

$$\int \frac{dx}{\sqrt{(1+x^2)^3}} = \int \frac{dx}{x^3(x^{-2}+1)^{3/2}}$$

and set $x^{-2} + 1 = t$. Then $-2x^{-3}dx = dt, dx = \dfrac{-x^3}{2}dt$, so that

$$J = \frac{-1}{2}\int t^{-\frac{3}{2}}dt = \frac{1}{\sqrt{t}} + C = \frac{x}{\sqrt{1+x^2}} + C.$$

The second example is the integral

$$J = \int \frac{dx}{x^2\sqrt{x+x^4}} = \int \frac{dx}{x^{5/2}\sqrt{1+x^3}}$$

where $\dfrac{m+1}{n} + p = -1$. With $t = x^{-3} + 1$ we have $dt = -3x^{-4}dx$ and

$$J = \int \frac{dx}{x^4\sqrt{x^{-3}+1}} = \frac{-1}{3}\int \frac{dt}{\sqrt{t}} = \frac{-2}{3}\sqrt{t} + C = \frac{-2}{3}\sqrt{x^{-3}+1} + C.$$

Problems for the reader: Evaluate by appropriate substitutions

(a) $$\int \frac{dx}{\sqrt{1+\sqrt[3]{x^2}}}$$

(b) $$\int \frac{dx}{\sqrt[3]{1+x^3}}$$

(c) $$\int \frac{dx}{\sqrt[5]{1+x^5}}$$

(d) $$\int \sqrt{x^3 + x^4}\, dx$$

(e) $$\int \sqrt[3]{3x - x^3}\, dx$$

(f) $$\int \frac{1}{x^2 (x^4 + 1)^{3/4}}\, dx$$

(g) $$\int \frac{1}{\sqrt{x}\,(\sqrt[4]{x} + 1)^{10}}\, dx$$

(the last two integrals are from the 2006 MIT integration bee competition).

1.5 Hyperbolic Substitutions for Integrals

In order to evaluate integrals containing radicals of the form

$$\sqrt{a^2 \pm x^2} \quad \text{and} \quad \sqrt{x^2 - a^2}$$

($a > 0$), most calculus textbooks use the trigonometric substitutions

1   For $\sqrt{a^2 - x^2}$ set $x = a\sin\theta$

2   For $\sqrt{a^2 + x^2}$ set $x = a\tan\theta$

3   For $\sqrt{x^2 - a^2}$ set $x = a\sec\theta$.

The substitution in 1 works very well, but the other two sometimes require longer computations. We want to demonstrate that for cases 2 and 3 it is more natural to use the hyperbolic substitutions

2*   For $\sqrt{a^2 - x^2}$ set $x = a\sinh t$

3*   For $\sqrt{x^2 - a^2}$ set $x = a\cosh t$

where $-\infty < t < \infty$. These two substitutions are based on the important identity

$$\cosh^2 t - \sinh^2 t = 1$$

so that

$$\sqrt{x^2 + a^2} = \sqrt{a^2(\sinh^2 t - 1)} = a\cosh t$$

$$\sqrt{x^2 - a^2} = \sqrt{a^2(\cosh^2 t - 1)} = a|\sinh t|.$$

For returning to the original variable x it is convenient to use the equations

(1.12) $\sinh^{-1} z = \ln(z + \sqrt{z^2 + 1}),\ -\infty < z < \infty$

(1.13) $\cosh^{-1} z = \ln(z + \sqrt{z^2 - 1}),\ 1 \le z$

(1.14) $\tanh^{-1} z = \dfrac{1}{2}\left[\ln(1 + z) - \ln(1 - z)\right],\ -1 < z < 1$

and also the identities

(1.15) $1 - \tanh^2 t = \dfrac{1}{\cosh^2 t} = \dfrac{d}{dt}\tanh t$

(1.16) $\coth^2 t - 1 = \dfrac{1}{\sinh^2 t} = -\dfrac{d}{dt}\coth t$

(1.17) $\cosh^2 t = \dfrac{1}{2}(\cosh 2t + 1),\quad \sinh^2 t = \dfrac{1}{2}(\cosh 2t - 1)$

(1.18) $\sinh 2t = 2\sinh t \cosh t.$

Example 1.5.1

Evaluate

(1.19)
$$F(x) = \int \frac{\sqrt{x^2 - 3}}{x^2} \, dx \, .$$

Assuming without loss of generality that $x > 0$ we set $x = \sqrt{3} \cosh t$ with $t > 0$ to obtain

$$\sqrt{x^2 - 3} = \sqrt{3} \sinh t, \quad dx = \sqrt{3} \sinh t$$

$$F = \int \frac{\sinh^2 t}{\cosh^2 t} dt = \int \frac{\cosh^2 t - 1}{\cosh^2 t} dt = t - \tanh t + C \, .$$

Thus

$$F(x) = \cosh^{-1} \frac{x}{\sqrt{3}} - \tanh \left(\cosh^{-1} \frac{x}{\sqrt{3}} \right) + C$$

and according to (1.13) and (1.15)

$$F(x) = \ln(x + \sqrt{x^2 - 3}) - \frac{\sqrt{x^2 - 3}}{x} + C \, .$$

We can also use the equation (for $t > 0$)

$$\tanh t = \frac{\sinh t}{\cosh t} = \frac{\sqrt{\cosh^2 t - 1}}{\cosh t} \, .$$

Example 1.5.2

We want to evaluate here the popular integral

(1.20)
$$F(x) = \int \sqrt{x^2 + 1} \, dx \, .$$

With $x = \sinh t$ we find

$$F = \int \cosh^2 t \, dt = \frac{1}{2} \int (\cosh 2t + 1) dt = \frac{1}{4} \sinh 2t + \frac{t}{2} + C \, .$$

And in view of (1.19)

$$F(x) = \frac{x}{2}\sqrt{x^2+1} + \frac{1}{2}\ln(x+\sqrt{x^2+1}) + C.$$

Example 1.5.3

For $r > 1$ consider the integral

(1.21) $$J_r = \int_0^\infty \frac{dx}{(x+\sqrt{1+x^2})^r}.$$

The substitution $x = \sinh t$ transforms this integral into

$$J_r = \int_0^\infty \frac{\cosh t\, dt}{(\sinh t + \cosh t)^r}$$

$$= \frac{1}{2}\int_0^\infty \frac{(e^t+e^{-t})dt}{e^{rt}} = \frac{1}{2}\int_0^\infty \left(e^{-(r-1)t} + e^{-(r+1)t}\right)dt$$

because $\sinh t + \cosh t = e^t$. This way

$$J_r = \frac{1}{2}\left(\frac{1}{r-1} + \frac{1}{r+1}\right) = \frac{r}{r^2-1}.$$

Note that r does not need to be an integer.

Example 1.5.4

We will evaluate now the interesting integral

(1.22) $$J = \int_0^\infty \frac{\cos t \sin\sqrt{t^2+1}}{\sqrt{t^2+1}}dt.$$

This is Problem 12145 from the *American Mathematical Monthly* (Vol. 126, November 2019). We make the substitution $t = \sinh x$ to get

$$J = \int_0^\infty \cos(\sinh x)\sin(\cosh x)dx \ .$$

Here we use the trigonometric identity

$$\sin\alpha\cos\beta = \frac{1}{2}\left[\sin(\alpha+\beta) + \sin(\alpha-\beta)\right]$$

with $\alpha = \cosh x, \beta = \sinh(x)$. The result is

$$J = \frac{1}{2}\int_0^\infty [\sin(e^x) + \sin(e^{-x})]\,dx$$

$$= \frac{1}{2}\left[\int_0^\infty \sin(e^x)dx + \int_0^\infty \sin(e^{-x})dx\right] = \frac{1}{2}[A+B]$$

where A, B are the last two integrals correspondingly. We will show now that these integrals are convergent and therefore, the previous integrals are convergent. For this purpose we make the substitutions $e^x = u$ in A and $e^{-x} = u$ in B. This way we find

$$A = \int_0^\infty \sin(e^x)dx = \int_1^\infty \frac{\sin u}{u}du, \quad B = \int_0^\infty \sin(e^{-x})dx = \int_0^1 \frac{\sin u}{u}du$$

which are both convergent. Finally

$$J = \frac{1}{2}[A+B] = \frac{1}{2}\left[\int_0^\infty \frac{\sin u}{u}du\right] = \frac{1}{2}\left[\frac{\pi}{2}\right] = \frac{\pi}{4}.$$

The well-known integral

$$\int_0^\infty \frac{\sin u}{u}du = \frac{\pi}{2}$$

is evaluated in Example 2.2.1 in Chapter 2 below.

1.6   General Trigonometric Substitution

Calculus textbooks usually consider integrals of the form

$$\int \sin^n x \cos^m x\,dx$$

and mostly suggest one of the substitutions

$$\sin x = t, \ \cos x = t, \ \text{or} \ \tan x = t.$$

This technique, however, is not very helpful with integrals like

(1.23) $$\int \frac{dx}{5 + \sin x}$$

so we need a better method. Integrals of the form

$$\int R(\sin x, \cos x)\,dx$$

where $R(u,v)$ is a rational function of two variables can be reduced to integrals of a rational function of one variable by setting

$$\tan \frac{x}{2} = t, \ x = 2\arctan t$$

for $-\pi < x < \pi, \ -\infty < t < \infty$. Simple trigonometry yields

$$\sin x = \frac{2t}{1+t^2}, \ \cos x = \frac{1-t^2}{1+t^2}, \ dx = \frac{2dt}{1+t^2}$$

by using the formulas

$$\sin x = 2\sin \frac{x}{2}\cos \frac{x}{2}, \ \cos x = \cos^2 \frac{x}{2} - \sin^2 \frac{x}{2}$$

$$\cos^2 \frac{x}{2} = \frac{1}{1 + \tan^2 \frac{x}{2}}.$$

Therefore,

$$\int R(\sin x, \cos x)\,dx = 2\int R\left(\frac{2t}{1+t^2}, \frac{1-t^2}{1+t^2}\right)\frac{dt}{1+t^2}.$$

In integrals of the form

$$\int R(\sin^2 x, \cos^2 x)\,dx$$

it is better to use the substitution

$$\tan x = t, \ x = \arctan t$$

where

$$\sin^2 x = \frac{t^2}{1+t^2}, \ \cos^2 x = \frac{1}{1+t^2}, \ dx = \frac{dt}{1+t^2}$$

and hence

$$\int R(\sin^2 x, \cos^2 x)\,dx = \int R\left(\frac{t^2}{1+t^2}, \frac{1}{1+t^2}\right)\frac{dt}{1+t^2}.$$

Example 1.6.1

We shall evaluate the integral in (1.23) with the substitution $t = \tan\dfrac{x}{2}$.

$$\int \frac{dx}{5+\sin x} = \frac{2}{5}\int \frac{dt}{t^2 + \frac{2}{5}t + 1} = \frac{2}{5}\int \frac{dt}{\left(t+\frac{1}{5}\right)^2 + \frac{24}{25}}$$

$$= \frac{1}{\sqrt{6}}\arctan\frac{5t+1}{2\sqrt{6}} + C = \frac{1}{\sqrt{6}}\arctan\left[\frac{1}{2\sqrt{6}}\left(5\tan\frac{x}{2}+1\right)\right] + C.$$

In general, we can solve in the same way the integral

$$\int \frac{dx}{a+b\sin x}$$

where $a > |b|$. Namely, $t = \tan\dfrac{x}{2}$ gives

$$\int \frac{dx}{a+b\sin x} = \frac{2}{\sqrt{a^2-b^2}}\arctan\left[\frac{1}{\sqrt{a^2-b^2}}\left(a\tan\frac{x}{2}+b\right)\right]+C.$$

A similar computations for $a > |b|$ provides also

$$(1.24) \qquad \int \frac{dx}{a+b\cos x} = \frac{1}{\sqrt{a^2-b^2}}\arctan\left(\sqrt{\frac{a-b}{a+b}}\tan\frac{x}{2}\right)+C.$$

The case $b \geq |a|$ is left to the reader.

Even more general, we can solve the integral

$$J = \int \frac{dx}{a+b\cos x+c\sin x}$$

by transforming it to the form in (1.25) this way

$$J = \int \frac{dx}{a+\sqrt{b^2+c^2}\,\cos(x-\alpha)}$$

because of the representation

$$b\cos x+c\sin x = \sqrt{b^2+c^2}\,\cos(x-\alpha)$$

where

$$\cos\alpha = \frac{b}{\sqrt{b^2+c^2}}, \quad \sin\alpha = \frac{c}{\sqrt{b^2+c^2}}.$$

To prove this we write

$$\cos(x-\alpha) = \cos\alpha\cos x+\sin\alpha\sin x$$

and compare coefficients in front of $\cos x, \sin x$ in the equation

$$b\cos x + c\sin x = \sqrt{b^2 + c^2}\,(\cos\alpha\cos x + \sin\alpha\sin x)\,.$$

Example 1.6.2

Consider now the integral

(1.25)
$$J = \int \frac{dx}{2 - \cos^2 x}\,.$$

It is appropriate to use the substitution $t = \tan x$. This way

$$dx = \frac{dt}{1 + t^2}, \quad \frac{1}{2 - \cos^2 x} = \frac{1 + t^2}{1 + 2t^2}$$

$$J = \int \frac{dt}{1 + 2t^2} = \frac{1}{\sqrt{2}}\arctan(\sqrt{2}\,t) + C = \frac{1}{\sqrt{2}}\arctan(\sqrt{2}\,\tan x) + C\,.$$

1.6.1 *Restrictions and extensions*

We want to point out that the substitutions $t = \tan\dfrac{x}{2}$ and $t = \tan x$ bring to natural restrictions on the variable x in order to have one-to-one correspondence between x and t. For the first one, as mentioned before, we need $-\pi < x < \pi$ and for the second one we need

$$-\frac{\pi}{2} < x < \frac{\pi}{2}\,.$$

For example, we can use the antiderivative

$$J(x) = \int_0^x \frac{dt}{2 - \cos^2 t}$$

for (1.25) and this will be a continuous and differentiable function defined everywhere. At the same time the function

$$J = \frac{1}{\sqrt{2}}\arctan(\sqrt{2}\tan x) + C$$

is well-defined on the interval $-\frac{\pi}{2} < x < \frac{\pi}{2}$, but not defined at the endpoints $\pm\frac{\pi}{2}$. It can be extended as a continuous function on $-\frac{\pi}{2} \le x \le \frac{\pi}{2}$ by using limits. It can also be extended beyond $-\frac{\pi}{2} \le x \le \frac{\pi}{2}$ by choosing appropriate constants C in the neighboring intervals.

1.7 Arithmetic-Geometric Mean and the Gauss Formula

In this section we present one very interesting substitution and prove a classical formula of Carl Friedrich Gauss (1777-1855). First we need to define the arithmetic-geometric mean of two positive numbers.

1.7.1 *The arithmetic-geometric mean*

Let $a > b > 0$ and define $a_1 = \frac{a+b}{2}$ to be the arithmetic mean of these two numbers and $b_1 = \sqrt{ab}$ to be their geometric mean. Then $a_1 > b_1$ as

$$a_1 - b_1 = \frac{1}{2}\left(\sqrt{a} - \sqrt{b}\right)^2 > 0.$$

This is the well-known arithmetic-geometric inequality. Further, we set

(1.26) $$a_{n+1} = \frac{a_n + b_n}{2}, b_{n+1} = \sqrt{a_n b_n}$$

where $n = 1, 2, \ldots$ and $a_n > b_n$. The sequence $\{a_n\}$ is decreasing, while the sequence $\{b_n\}$ is increasing, since

$$a_{n+1} - a_n = \frac{1}{2}(b_n - a_n) < 0, \; b_{n+1} - b_n = \sqrt{b_n}\left(\sqrt{a_n} - \sqrt{b_n}\right) > 0 \, .$$

Both sequences are bounded and therefore, convergent. They have the same limit because for every n

$$a_{n+1} - b_{n+1} < a_{n+1} - b_n < a_{n+1} - a_n < \frac{1}{2}(a_n - b_n)$$

and by iteration

(1.27) $$0 < a_n - b_n < \frac{1}{2^n}(a - b) \, .$$

This common limit $M(a,b) = \lim a_n = \lim b_n$ is called the *arithmetic-geometric mean* of a and b. The initial restriction $a > b$ is unimportant, as we can define the two sequences starting from a_1 and b_1. We see that also $M(a,b) = \inf\{a_n\} = \sup\{b_n\}$. The inequality (1.27) shows that

(1.28) $$a_n - M(a,b) < \frac{1}{2^n}(a - b)$$

so that a_n is a decent approximation to $M(a,b)$.

1.7.2 The Gauss formula

For $a > b > 0$ consider the integral

(1.29) $$G(a,b) = \int_0^{\pi/2} \frac{dx}{\sqrt{a^2 \cos^2 x + b^2 \sin^2 x}} \, .$$

In this integral we make the substitution

(1.30) $$\sin x = \frac{2a\sin\theta}{(a+b) + (a-b)\sin^2\theta} \, .$$

Differentiating with respect to θ we simplify to get

(1.31) $\qquad \cos x\,dx = 2a\dfrac{(a+b)-(a-b)\sin^2\theta}{[(a+b)+(a-b)\sin^2\theta]^2}\cos\theta\,d\theta$.

Also, using the identity $\cos^2 x = 1 - \sin^2 x$ we compute from (1.30)

(1.32) $\qquad \cos x = \dfrac{\sqrt{(a+b)^2 - (a-b)^2\sin^2\theta}}{(a+b)+(a-b)\sin^2\theta}\cos\theta$.

Solving for dx in (1.31) and using (1.32) we find

(1.33) $\quad dx = 2a\dfrac{(a+b)-(a-b)\sin^2\theta}{(a+b)+(a-b)\sin^2\theta}\dfrac{d\theta}{\sqrt{(a+b)^2 - (a-b)^2\sin^2\theta}}$.

At the same it easy to see that

$$\sqrt{a^2\cos^2 x + b^2\sin^2 x} = a\frac{(a+b)-(a-b)\sin^2\theta}{(a+b)+(a-b)\sin^2\theta}\,.$$

In view of (1.33) this gives after some elementary computations

$$\frac{dx}{\sqrt{a^2\cos^2 x + b^2\sin^2 x}} = \frac{2d\theta}{\sqrt{(a+b)^2\cos^2 x + 4ab\sin^2 x}}\,.$$

Dividing the numerator and the denominator by 2 in the right-hand side we find the interesting equation

(1.34) $\quad \dfrac{dx}{\sqrt{a^2\cos^2 x + b^2\sin^2 x}} = \dfrac{d\theta}{\sqrt{\left(\dfrac{a+b}{2}\right)^2\cos^2 x + \left(\sqrt{ab}\right)^2\sin^2 x}}$.

At this point we note that the substitution (1.30) is one-to-one on the interval $[0, \pi/2]$ having nonzero derivative (1.31). It maps this interval onto itself and preserves the endpoints. Applying the substitution to (1.29) we find

$$\int\limits_0^{\pi/2} \frac{dx}{\sqrt{a^2\cos^2 x + b^2\sin^2 x}} = \int\limits_0^{\pi/2} \frac{d\theta}{\sqrt{\left(\dfrac{a+b}{2}\right)^2\cos^2 x + \left(\sqrt{ab}\right)^2\sin^2 x}}$$

which is the remarkable Gauss formula. Iterating this equation we obtain for every $n \geq 1$

$$G(a,b) = G(a_1,b_1) = G(a_2,b_2) = ...G(a_n,b_n) = ...$$

Passing to limits inside the integral is easily justifiable and the results is

$$\lim G(a_n,b_n) = G(M(a,b), M(a,b)) = \frac{\pi}{2M(a,b)}.$$

That is,

(1.35) $$G(a,b) = \frac{\pi}{2M(a,b)}.$$

According to (1.28) the sequence $\{a_n\}$ can be used for approximating $G(a,b)$.

The integral $G(a,b)$ can be put in a different form. With the substitution $\tan x = t$ we find

$$G(a,b) = \int\limits_0^\infty \frac{dt}{\sqrt{a^2 + b^2 t^2}\sqrt{1 + t^2}}$$

and further, with $t = \dfrac{u}{b}$ this becomes

(1.36) $$G(a,b) = \int\limits_0^\infty \frac{du}{\sqrt{a^2 + u^2}\sqrt{b^2 + t^2}}.$$

Integrals of this form are called elliptic integrals. More on elliptic integrals can be found, for example, in Akhiezer's nice little book [1].

1.8   Some Interesting Examples

Example 1.8.1

Let us try to evaluate the integral

$$J = \int_0^\pi \frac{x \sin x}{1 + \cos^2 x} \, dx \, .$$

This integral looks very difficult! The usual trigonometric substitutions do not look very promising. However, the simple substitution $x = \pi - t$ works very well

$$\int_0^\pi \frac{x \sin x}{1 + \cos^2 x} \, dx = -\int_\pi^0 \frac{(\pi - t) \sin t}{1 + \cos^2 x} \, dt$$

$$= \pi \int_0^\pi \frac{\sin x}{1 + \cos^2 x} \, dx - \int_0^\pi \frac{t \sin t}{1 + \cos^2 t} \, dt$$

$$= -\pi \arctan(\cos x) \big|_0^\pi - \int_0^\pi \frac{t \sin t}{1 + \cos^2 t} \, dt$$

and we come to the equation

$$J = \frac{\pi^2}{2} - J \, .$$

This way

(1.37) $$\int_0^\pi \frac{x \sin x}{1 + \cos^2 x} \, dx = \frac{\pi^2}{4} \, .$$

More generally, if we have a continuous function $f(x)$, the substitution $x \to b - x$ shows that

$$\int_0^b f(x) \, dx = \int_0^b f(b - x) \, dx$$

which helps in many cases with trigonometric integrals. For instance,

$$\int_0^\pi \frac{\cos x}{1 + \cos^2 x} dx = 0$$

because the substitution $x \to \pi - x$ gives immediately

$$\int_0^\pi \frac{\cos x}{1 + \cos^2 x} dx = -\int_0^\pi \frac{\cos x}{1 + \cos^2 x} dx .$$

Problem for the reader: Prove that

$$\int_0^\pi \frac{2x \sin x}{3 + \cos 2x} dx = \frac{\pi^2}{4}$$

by using the substitution $x = \pi - t$. This is calculus problem 9 from the Stanford Mathematics Tournament 2018.

Example 1.8.2

A similar trick helps to solve the integral

$$J = \int_0^{\pi/2} \frac{(\cos x)^p}{(\cos x)^p + (\sin x)^p} dx$$

where p is some number. The integral looks impossible! However, the substitution $x = \frac{\pi}{2} - t$ gives

$$J = -\int_{\pi/2}^0 \frac{(\sin t)^p}{(\cos t)^p + (\sin t)^p} dt = \int_0^{\pi/2} \frac{(\sin t)^p}{(\cos t)^p + (\sin t)^p} dx$$

and then

$$J + J = \int_0^{\pi/2} \frac{(\sin t)^p + (\cos t)^p}{(\cos t)^p + (\sin t)^p} dx = \int_0^{\pi/2} 1 dx = \frac{\pi}{2}$$

$$J = \frac{\pi}{4}$$

independent of p!

Example 1.8.3

The integral

$$\int_0^\infty \frac{\ln x}{x^2 + 1} dx$$

seems quite difficult. We can try integration by parts

$$\int_0^\infty \frac{\ln x}{x^2 + 1} dx = \int_0^\infty \ln x \, d \arctan x = \ln x \arctan x \big|_0^\infty - \int_0^\infty \frac{\arctan x}{x} dx$$

but on the right-hand side we find divergent expressions.

Fortunately, a simple substitution works! We set $x = \frac{1}{t}$ and compute

$$\int_0^\infty \frac{\ln x}{x^2 + 1} dx = \int_\infty^0 \frac{-\ln t}{(1/t^2) + 1} \left(\frac{dt}{-t^2} \right) = \int_\infty^0 \frac{\ln t}{1 + t^2} dt = -\int_0^\infty \frac{\ln t}{1 + t^2} dt .$$

Thus

(1.38) $$\int_0^\infty \frac{\ln x}{x^2 + 1} dx = 0 .$$

Example 1.8.4

Let $\beta > 0$. The integral

$$\int_0^\infty (1 + x^{2\beta})^{-\frac{1}{\beta}} dx = \int_0^\infty \frac{1}{\sqrt[\beta]{1 + x^{2\beta}}} dx$$

is invariant with respect to the same substitution, $x = \dfrac{1}{t}$. Namely,

$$\int_0^\infty (1 + x^{2\beta})^{-\frac{1}{\beta}} dx = -\int_\infty^0 (1 + t^{-2\beta})^{-\frac{1}{\beta}} \left(\frac{-1}{t^2} \right) dt$$

$$= \int_0^\infty (t^{2\beta} + 1)^{-\frac{1}{\beta}} (t^{-2\beta})^{-\frac{1}{\beta}} \left(\frac{1}{t^2} \right) dt = \int_0^\infty (t^{2\beta} + 1)^{-\frac{1}{\beta}} dt .$$

The integral can be evaluated by using Euler's beta function

$$B(u,v) = \int_0^\infty \frac{t^{u-1}}{(1+t)^{u+v}} dt = \frac{\Gamma(u)\Gamma(v)}{\Gamma(u+v)} \quad (u, v > 0).$$

First, with the substitution $x^{2\beta} = t$ we compute

$$\int_0^\infty (1 + x^{2\beta})^{-\frac{1}{\beta}} dx = \frac{1}{2\beta} \int_0^\infty \frac{t^{\frac{1}{2\beta}-1}}{(1+t)^{\frac{1}{\beta}}} dt$$

and then taking $u = v = \dfrac{1}{2\beta}$ we find

$$\int_0^\infty (1 + x^{2\beta})^{-\frac{1}{\beta}} dx = \frac{1}{2\beta} B\left(\frac{1}{2\beta}, \frac{1}{2\beta} \right) = \frac{1}{2\beta} \Gamma^2\left(\frac{1}{2\beta} \right) \Gamma\left(\frac{1}{\beta} \right)^{-1} .$$

This result can be used to evaluate some other interesting integrals. For example, consider the integral

$$J_\beta = \int_0^\infty \frac{\arctan x}{\sqrt[\beta]{1+x^{2\beta}}}\,dx.$$

With the substitution $x = \dfrac{1}{t}$ we find

$$J_\beta = \int_0^\infty \frac{\arctan(1/t)}{\sqrt[\beta]{1+t^{2\beta}}}\,dt = \int_0^\infty \frac{\pi/2 - \arctan t}{\sqrt[\beta]{1+t^{2\beta}}}\,dt = \frac{\pi}{2}\int_0^\infty \frac{1}{\sqrt[\beta]{1+t^{2\beta}}}\,dt - J_\beta$$

(using the identity $\arctan t + \arctan\dfrac{1}{t} = \dfrac{\pi}{2}$ for $t > 0$). From here

$$J_\beta = \frac{\pi}{4}\int_0^\infty \frac{1}{\sqrt[\beta]{1+t^{2\beta}}}\,dt.$$

That is,

$$J_\beta = \frac{\pi}{8\beta}\Gamma^2\left(\frac{1}{2\beta}\right)\Gamma\left(\frac{1}{\beta}\right)^{-1}.$$

In particular, for $\beta = 1, 2, 3$ we have correspondingly

$$J_1 = \int_0^\infty \frac{\arctan x}{1+x^2}\,dx = \frac{\pi^2}{8}$$

as $\Gamma(1/2) = \sqrt{\pi}$. (This integral can be solved directly with the substitution $u = \arctan x$.)

$$J_2 = \int_0^\infty \frac{\arctan x}{\sqrt{1+x^4}}\,dx = \frac{\sqrt{\pi}}{16}\Gamma^2(1/4)$$

$$J_3 = \int_0^\infty \frac{\arctan x}{\sqrt[3]{1+t^6}}\,dt = \frac{\pi}{24}\Gamma^2(1/6)\Gamma^{-1}(1/3)$$

etc.

Example 1.8.5

Let $a < b$ and $a \le x \le b$. To evaluate the integral

$$\int \frac{dx}{\sqrt{(x-a)(b-x)}}$$

we use the substitution

$$x = a\cos^2\theta + b\sin^2\theta, \quad 0 \le \theta \le \frac{\pi}{2}$$

which can be written also as

$$x = a + (b-a)\sin^2\theta$$

because of the equation $\cos^2 x = 1 - \sin^2 x$. From here

$$\sin\theta = \sqrt{\frac{x-a}{b-a}}, \quad \theta = \arcsin\sqrt{\frac{x-a}{b-a}}.$$

We compute

$$x - a = (b-a)\sin^2\theta$$

$$b - x = (b-a)\cos^2\theta$$

$$(x-a)(b-x) = (b-a)^2\cos^2\theta\sin^2\theta$$

$$\frac{dx}{d\theta} = 2(b-a)\cos\theta\sin\theta$$

and then

$$\int \frac{dx}{\sqrt{(x-a)(b-x)}} = \int \frac{2(b-a)\cos\theta\sin\theta}{(b-a)\cos\theta\sin\theta}d\theta = \int 2d\theta = 2\theta + C.$$

Therefore,

(1.39) $$\int \frac{dx}{\sqrt{(x-a)(b-x)}} = 2\arcsin\sqrt{\frac{x-a}{b-a}} + C .$$

In particular,

(1.40) $$\int_{a}^{b} \frac{dx}{\sqrt{(x-a)(b-x)}} = 2\arcsin\sqrt{\frac{x-a}{b-a}}\Bigg|_{a}^{b} = \pi .$$

Notice that this last integral does not depend on a and b!

With $a = 0$, $b = 1$ we find

$$\int_{0}^{1} \frac{dx}{\sqrt{x(1-x)}} = \pi .$$

Example 1.8.6

Here is a very amusing integral! Solve

$$F(x) = \int \sqrt{x + \sqrt{x + \sqrt{x + \ldots}}}\, dx$$

for $x > 0$. Inside the integral there are infinitely many nested radicals. The integral seems impossible!

It turns out that the integral can be solve by a simple substitution. We just need to write the integrand in a different form. Let

$$y = \sqrt{x + \sqrt{x + \sqrt{x + \ldots}}} .$$

Then $y^2 = x + y$ and we can find y from the quadratic equation

$$y^2 - y - x = 0 .$$

We have two solutions for this equation

$$y = \frac{1}{2}(1 + \sqrt{1 + 4x}), \quad y = \frac{1}{2}(1 - \sqrt{1 + 4x}).$$

The second solution we drop, because $y > 0$ by its definition. Thus

$$F(x) = \frac{1}{2}\int (1 + \sqrt{1 + 4x})\, dx$$

and the simple substitution $u = 1 + 4x$ gives

(1.41) $$\int \sqrt{x + \sqrt{x + \sqrt{x + \ldots}}}\, dx = \frac{x}{2} + \frac{1}{12}(1 + 4x)^{\frac{3}{2}} + C.$$

Unbelievable!

A legitimate question is the convergence of the infinite radical. It converges by Pólya's criterion (George Pólya, a Hungarian mathematician (1887-1985)): Given a sequence of positive numbers $a_n > 0$ ($n = 1, 2, \ldots$), the sequence

$$u_n = \sqrt{a_1 + \sqrt{a_2 + \ldots + \sqrt{a_n}}}$$

converges if $\ln a_n \le 2^n M$ for some constant $M > 0$ and every $n = 1, 2, \ldots$.

A problem for the reader: Solve the integral

$$\int \sqrt{x\sqrt{x\sqrt{x \ldots}}}\, dx$$

also with an infinite radical inside.

Example 1.8.7

In this example we evaluate the integral

$$J(a) = \int_0^\infty e^{-a^2 x^2 - x^{-2}} \, dx \quad (a > 0)$$

We can write

$$a^2 x^2 + \frac{1}{x^2} = \left(ax - \frac{1}{x} \right)^2 + 2a$$

so that

$$J(a) = e^{-2a} \int_0^\infty e^{-(ax - x^{-1})^2} \, dx \, .$$

Now we introduce the new variable $u = ax - x^{-1}$. The function $u(x)$ is a one-to-one mapping of the interval $(0, \infty)$ onto $(-\infty, \infty)$ as

$$\frac{du}{dx} = a + x^{-2} > 0 \, .$$

To change the variable fist we solve for x from the equation $ax^2 - ux - 1 = 0$ and take the positive root

$$x = \frac{u + \sqrt{u^2 + 4a}}{2a} \, .$$

Then

$$dx = \frac{1}{2a} \left(1 + \frac{u}{\sqrt{u^2 + 4a}} \right) du$$

so changing the variables we get

$$J(a) = \frac{e^{-2a}}{2a} \left(\int_{-\infty}^\infty e^{-u^2} \, du + \int_{-\infty}^\infty \frac{u e^{-u^2}}{\sqrt{u^2 + 4a}} \, du \right)$$

and here we discover a very nice thing – the second integral is zero! (Odd integrand.) Therefore,

$$J(a) = \frac{e^{-2a}}{2a} \int_{-\infty}^{\infty} e^{-u^2} du = \frac{e^{-2a}}{a} \int_{0}^{\infty} e^{-u^2} du .$$

The Gaussian integral is well-known

$$\int_{0}^{\infty} e^{-u^2} du = \frac{\sqrt{\pi}}{2}$$

and we obtain

(1.42) $$\int_{0}^{\infty} e^{-a^2 x^2 - x^{-2}} dx = \frac{\sqrt{\pi}}{2a} e^{-2a} .$$

This integral appears in [25, entry 3.325] as

$$\int_{0}^{\infty} \exp\left(-ax^2 - \frac{b}{x^2} \right) dx = \frac{1}{2} \sqrt{\frac{\pi}{a}} \exp(-2\sqrt{ab}) \quad (a,b > 0)$$

and a simple rescaling $x \to x\sqrt{b}$ brings it to the form (1.42).

(The Gaussian integral will be computed in the next chapter.)

M. Laurence Glasser has found a general formula for this interesting substitution. For details see [24].

Chapter 2

Solving Integrals
by Differentiation with Respect
to a Parameter

2.1 Introduction

In this chapter we present an efficient technique for evaluating challenging definite integrals – differentiation with respect to a parameter inside the integral (or appearing in the limits of integration). The integrals can be proper or improper. We provide numerous examples, many of which are listed in the popular handbooks of Gradshteyn and Ryzhik [25] and Prudnikov, Brychkov, Marichev [43]. Many more integrals in these tables can be evaluated by the same method. In our examples we focus on the formal manipulations. Several theorems justifying these manipulations are provided in the last section. Applying the theorems in every particular case is left to the reader.

For many of the integrals here differentiation with respect to parameter is possibly the best way to solve them.

There are several books and articles presenting this technique. First of all we want to mention the excellent book of Fikhtengolts [23]. The reader can see also [21, 28, 29, 49, 50, 51].

Below we evaluate three very different integrals in order to demonstrate the wide scope of the method.

Example 2.1.1

We start with a very simple example. It is easy to show that the popular integral

$$\int_0^\infty \frac{\sin x}{x} dx$$

is convergent (integrating by parts we have

$$\int_1^\infty \frac{\sin x}{x} dx = -\frac{\cos x}{x}\Big|_1^\infty - \int_1^\infty \frac{\cos x}{x^2} dx = \cos 1 - \int_1^\infty \frac{\cos x}{x^2} dx$$

which shows convergence of the original integral).

For its evaluation we consider the function

$$F(\lambda) = \int_0^\infty e^{-\lambda x} \frac{\sin x}{x} dx, \ \lambda > 0$$

with derivative

$$F'(\lambda) = -\int_0^\infty e^{-\lambda x} \sin x\, dx = \frac{-1}{1+\lambda^2}$$

(Laplace transform of the sine function). Integrating back we find

$$F(\lambda) = -\arctan \lambda + C$$

Setting $\lambda \to \infty$ leads to

$$0 = -\frac{\pi}{2} + C, \text{ i.e. } C = \frac{\pi}{2}.$$

Therefore,

(2.1) $$\int_0^\infty e^{-\lambda x} \frac{\sin x}{x} dx = \frac{\pi}{2} - \arctan \lambda$$

and with $\lambda \to 0$ we find

$$\int_0^\infty \frac{\sin x}{x} dx = \frac{\pi}{2}.$$

Note that for any $a \neq 0$ by rescaling $x \to ax$ we have

$$\int_0^\infty \frac{\sin ax}{x} dx = \begin{cases} \dfrac{\pi}{2} \; (a > 0) \\ -\dfrac{\pi}{2} \; (a < 0) \end{cases}.$$

From (2.1) we obtain also

$$\int_0^\infty \frac{e^{-\lambda x} - e^{-\mu x}}{x} \sin x \, dx = \arctan \mu - \arctan \lambda$$

for any $\lambda, \mu > 0$.

In the same way we can evaluate the integral

$$G(\lambda) = \int_0^\infty e^{-\lambda x} \frac{\sinh x}{x} dx, \quad \lambda > 1.$$

$$G'(\lambda) = -\int_0^\infty e^{-\lambda x} \sinh x \, dx = \frac{-1}{\lambda^2 - 1} = \frac{1}{2}\left(\frac{1}{\lambda+1} - \frac{1}{\lambda-1}\right)$$

$$G(\lambda) = \frac{1}{2} \ln \frac{\lambda+1}{\lambda-1}.$$

Problem for the reader: Integrating by parts and using the above evaluation show that

$$\int_0^\infty \frac{\cos ax - \cos bx}{x^2} dx = \frac{\pi}{2}(b - a).$$

Example 2.1.2

The 66 Annual William Lowell Putnam Mathematics Competition (2005) included the integral (Problem (A5))

(2.2) $$\int_0^1 \frac{\ln(1+x)}{1+x^2} dx$$

with a solution published in [53]. This is entry 4.291(4) in [25]. We will give a different solution by introducing a parameter. Consider the function

$$F(\lambda) = \int_0^1 \frac{\ln(1+\lambda x)}{1+x^2} dx$$

defined for $\lambda > 0$. Differentiating this function gives

$$F'(\lambda) = \int_0^1 \frac{x}{(1+\lambda x)(1+x^2)} dx .$$

This integral is easy to evaluate by using partial fractions. The results is

$$F'(\lambda) = -\frac{\ln(1+\lambda)}{1+\lambda^2} + \frac{\ln 2}{2(1+\lambda^2)} + \frac{\pi}{4}\frac{\lambda}{1+\lambda^2} .$$

Integrating we find

$$F(\lambda) = -\int_0^\lambda \frac{\ln(1+x)}{1+x^2} dx + \frac{\ln 2}{2}\arctan\lambda + \frac{\pi}{8}\ln(1+\lambda^2)$$

and setting $\lambda = 1$ we arrive at the equation

$$2F(1) = \frac{\pi}{4}\ln 2 .$$

That is,

$$\int_0^1 \frac{\ln(1+x)}{1+x^2} dx = \frac{\pi}{8}\ln 2 .$$

The similar integral

$$\int_0^1 \frac{\ln(1+x^2)}{1+x} dx = \frac{3}{4}(\ln 2)^2 - \frac{\pi^2}{48}$$

will be evaluated later in Chapter 5, Example 5.4.6.

As shown in this chapter many integrals containing logarithms and inverse trigonometric functions can be evaluated by using a parameter.

Notice that integration by parts gives

$$\int_0^1 \frac{\ln(1+x)}{1+x^2} dx = \ln(1+x)\arctan x \Big|_0^1 - \int_0^1 \frac{\arctan x}{1+x} dx$$

so that also

(2.3) $$\int_0^1 \frac{\arctan x}{1+x} dx = \frac{\pi}{8}\ln 2 \ .$$

These integrals will be used later in section 5.5 in the evaluation of the interesting series

$$\sum_{n=1}^{\infty} (-1)^n \left(\ln 2 - \frac{1}{n+1} - \frac{1}{n+2} - \dots - \frac{1}{2n} \right)^2$$

$$= \frac{\pi^2}{48} - \frac{7}{8}(\ln 2)^2 - \frac{\pi}{8}\ln 2 + \frac{G}{2} \ .$$

where

$$G = \sum_{n=0}^{\infty} \frac{(-1)^n}{(2n+1)}$$

is Catalan's constant.

The two integrals (2.2) and (2.3) also appeared in Problem 883 of the College Mathematics Journal – see [15].

Example 2.1.3

This is Problem 1997 from the *Mathematics Magazine*, 89(3), 2016, p. 223. Evaluate

$$\int_0^\infty \left(\frac{1-e^{-x}}{x} \right)^2 dx .$$

Solution. We show that for every $\lambda > 0$

$$F(\lambda) = \int_0^\infty \left(\frac{1-e^{-\lambda x}}{x} \right)^2 dx = \lambda \ln 4 .$$

Remarkably, this integral is a linear function! Indeed, differentiation with respect to λ gives

$$F'(\lambda) = 2\int_0^\infty \left(\frac{1-e^{-\lambda x}}{x} \right) e^{-\lambda x} dx = 2\int_0^\infty \frac{e^{-\lambda x} - e^{-2\lambda x}}{x} dx = 2\ln 2$$

by using Frullani's formula for the last integral. We conclude that $F(\lambda)$ is a linear function and since $F(0) = 0$ we can write $F(\lambda) = \lambda \ln 4$. With $\lambda = 1$ we find $F(1) = \ln 4$.

Note that the above differentiation is legitimate because the integral $F(\lambda)$ is uniformly convergent on every interval $0 < a < \lambda < b$.

Frullani's formula says that for appropriate functions $f(x)$ we have

$$\int_0^\infty \frac{f(ax) - f(bx)}{x} dx = [f(0) - f(\infty)] \ln \frac{b}{a}$$

(this formula will be discussed later in Chapter 5).

2.2 General Examples

Example 2.2.1

We start this section with a popular example. Consider the integral

$$J(\alpha) = \int_0^1 \frac{x^\alpha - 1}{\ln x} dx, \ \alpha \geq 0.$$

Note that the integrand is a continuous function on $[0,1]$ when defined as zero for $x = 0$ and as α for $x = 1$ $\left(\alpha = \lim_{x \to 1} \frac{x^\alpha - 1}{\ln x} \right)$. Since

$$\frac{d}{d\alpha} x^\alpha = x^\alpha \ln x$$

we find

$$J'(\alpha) = \int_0^1 x^\alpha dx = \frac{1}{1 + \alpha}$$

and from here $J(\alpha) = \ln(1 + \alpha) + C$. Now $C = 0$ because $J(0) = 0$.

Finally,

(2.4)
$$\int_0^1 \frac{x^\alpha - 1}{\ln x} dx = \ln(1 + \alpha).$$

The similar integral

$$\int_0^1 \frac{x^\alpha - x^\beta}{\ln x} dx$$

where $\alpha, \beta \geq 0$ can be reduced to (2.4) by writing $x^\alpha - x^\beta = x^\alpha - 1 - (x^\beta - 1)$. Thus

$$\int_0^1 \frac{x^\alpha - x^\beta}{\ln x} dx = \ln \frac{1 + \alpha}{1 + \beta}.$$

Example 2.2.2

We evaluate here the improper integral

$$J(\lambda) = \int_0^1 \frac{\arctan \lambda x}{x\sqrt{1-x^2}} dx .$$

Differentiation gives

$$J'(\lambda) = \int_0^1 \frac{dx}{(1+\lambda^2 x^2)\sqrt{1-x^2}}$$

and with the substitution $x = \cos\theta$ this transforms into

$$J'(\lambda) = \int_0^{\pi/2} \frac{d\theta}{1+\lambda^2 \cos^2\theta} = \int_0^{\pi/2} \frac{d\theta}{(1+\tan^2\theta+\lambda^2)\cos^2\theta}$$

$$= \int_0^{\pi/2} \frac{d\tan\theta}{1+\lambda^2+\tan^2\theta} = \frac{1}{\sqrt{1+\lambda^2}}\arctan\frac{\tan\theta}{\sqrt{1+\lambda^2}}\bigg|_0^{\frac{\pi}{2}} = \frac{\pi}{2\sqrt{1+\lambda^2}} .$$

Therefore, since $J(0) = 0$,

(2.5) $\qquad J(\lambda) = \int_0^1 \frac{\arctan \lambda x}{x\sqrt{1-x^2}} dx = \frac{\pi}{2}\ln\left(\lambda+\sqrt{1+\lambda^2}\right).$

In particular, with $\lambda = 1$

$$\int_0^1 \frac{\arctan x}{x\sqrt{1-x^2}} dx = \frac{\pi}{2}\ln\left(1+\sqrt{2}\right)$$

which is entry 4.531(12) in [25].

Note that the similar integral

(2.6) $\qquad J(\lambda) = \int_0^1 \frac{\arctan \lambda x}{\sqrt{1-x^2}} dx$

cannot be evaluated in the same manner. The derivative here is

$$J'(\lambda) = \int_0^1 \frac{xdx}{(1+\lambda^2 x^2)\sqrt{1-x^2}} = \int_0^{\pi/2} \frac{\cos\theta d\theta}{1+\lambda^2 \cos^2\theta}$$

$$= \frac{\pi}{2\lambda\sqrt{1+\lambda^2}} \ln \frac{\sqrt{1+\lambda^2}+\lambda}{\sqrt{1+\lambda^2}-\lambda}$$

which is not easy to integrate. The integral (2.6) will be evaluated later in Section 2.4 by a more sophisticated method.

Example 2.2.3

For all $\lambda \geq 0$ consider the function

$$G(\lambda) = \int_0^\infty \frac{\arctan \lambda x}{x(1+x^2)} dx.$$

We compute (under the temporary restriction $\lambda \neq 1$)

$$G'(\lambda) = \int_0^\infty \frac{dx}{(1+\lambda^2 x^2)(1+x^2)} = \frac{1}{1-\lambda^2} \int_0^\infty \left(\frac{1}{1+x^2} - \frac{\lambda^2}{1+\lambda^2 x^2} \right) dx$$

$$= \frac{1}{1-\lambda^2} \left(\frac{\pi}{2} - \frac{\pi\lambda}{2} \right) = \frac{\pi}{2(1+\lambda)}$$

and since $G(0) = 0$

(2.7) $$G(\lambda) = \int_0^\infty \frac{\arctan \lambda x}{x(1+x^2)} dx = \frac{\pi}{2} \ln(1+\lambda).$$

At this point we can drop the restriction $\lambda \neq 1$ and for $\lambda = 1$ we find

$$\int_0^\infty \frac{\arctan x}{x(1+x^2)} dx = \frac{\pi}{2} \ln 2.$$

Example 2.2.4

Related to (2.7) is the following integral ($\lambda, \mu > 0$)

$$G(\lambda, \mu) = \int_0^\infty \frac{\arctan(\lambda x)\arctan(\mu x)}{x^2} dx \, .$$

Using (2.7) we compute and evaluate the partial derivative

$$G_\mu(\lambda, \mu) = \int_0^\infty \frac{\arctan \lambda x}{x(1 + \mu^2 x^2)} dx = \int_0^\infty \frac{\arctan(\lambda / \mu)x}{x(1 + x^2)} dx = \frac{\pi}{2} \ln\left(1 + \frac{\lambda}{\mu}\right).$$

Integrating this logarithm with respect to μ (integration by parts) we find

$$G(\lambda, \mu) = \frac{\pi}{2} \left[(\lambda + \mu)\ln(\lambda + \mu) - \mu \ln \mu \right] + C(\lambda) \, .$$

Here $\mu \to 0$ yields $C(\lambda) = -\frac{\pi}{2} \lambda \ln \lambda$ and finally we have

$$\int_0^\infty \frac{\arctan(\lambda x)\arctan(\mu x)}{x^2} dx$$

$$= \frac{\pi}{2} \left[(\lambda + \mu)\ln(\lambda + \mu) - \lambda \ln \lambda - \mu \ln \mu \right].$$

In particular, for $\lambda = \mu = 1$

$$\int_0^\infty \frac{(\arctan x)^2}{x^2} dx = \pi \ln 2 \, .$$

Example 2.2.5

Now consider for $\lambda \ge 0$ the function

$$J(\lambda) = \int_0^\infty \frac{\ln(1 + \lambda^2 x^2)}{1 + x^2} dx$$

where

$$J'(\lambda) = \int_0^\infty \frac{2\lambda x^2}{(1+\lambda^2 x^2)(1+x^2)} dx$$

$$= \frac{2\lambda}{1-\lambda^2} \int_0^\infty \left(\frac{1}{1+\lambda^2 x^2} - \frac{1}{1+x^2} \right) dx = \frac{2\lambda}{1-\lambda^2} \left(\frac{\pi}{2\lambda} - \frac{\pi}{2} \right) = \frac{\pi}{1+\lambda}$$

under the temporary restriction $\lambda \neq 1$. This way, for $\lambda \geq 0$

(2.8) $$\int_0^\infty \frac{\ln(1+\lambda^2 x^2)}{1+x^2} dx = \pi \ln(1+\lambda).$$

In particular, for $\lambda = 1$

$$\int_0^\infty \frac{\ln(1+x^2)}{1+x^2} dx = \pi \ln 2.$$

Comparing (2.8) to (2.7) we see that for any $\lambda \geq 0$

$$\int_0^\infty \frac{\ln(1+\lambda^2 x^2)}{1+x^2} dx = 2 \int_0^\infty \frac{\arctan \lambda x}{x(1+x^2)} dx.$$

Example 2.2.6

Consider the integral 3.943 from [25]

$$F(\lambda) = \int_0^\infty e^{-\beta x} \frac{1-\cos \lambda x}{x} dx$$

where $\beta > 0$ is fixed. Differentiating we find

$$F'(\lambda) = \int_0^\infty e^{-\beta x} \sin \lambda x \, dx = \frac{\lambda}{\lambda^2 + \beta^2}$$

and integrating back

$$F(\lambda) = \frac{1}{2}\ln(\lambda^2 + \beta^2) + C(\beta)$$

To compute $C(\beta)$ we set $\lambda = 0$ and this gives $C(\beta) = \frac{-1}{2}\ln(\beta^2)$.

Therefore,

(2.9) $$\int_0^\infty e^{-\beta x} \frac{1-\cos\lambda x}{x} dx = \frac{1}{2}\ln\left(1 + \frac{\lambda^2}{\beta^2}\right).$$

From here we conclude that also

$$\int_0^\infty e^{-\beta x} \frac{\cos\lambda x - \cos\mu x}{x} dx = \frac{1}{2}\ln\left(\frac{\beta^2 + \mu^2}{\beta^2 + \lambda^2}\right)$$

as a difference of two integrals like (2.9). When $\lambda > 0$ and $\mu > 0$ in this integral we can set $\beta \to 0$ to obtain

$$\int_0^\infty \frac{\cos\lambda x - \cos\mu x}{x} dx = \ln\frac{\mu}{\lambda}$$

(this is entry 3.784(1) in [25] and entry 2.5.29(16) in [43]).

A problem for the reader: Prove entry 2.4.22(1) in [43]

$$\int_0^\infty e^{-\beta x} \frac{1-\cosh\lambda x}{x} dx = \frac{1}{2}\ln\left(1 - \frac{\lambda^2}{\beta^2}\right) \quad (\text{Re}\,\beta > |\,\text{Re}\,\lambda\,|).$$

Example 2.2.7

A symmetric analogue to (2.9) is the integral

(2.10) $$F(\lambda) = \int_0^\infty \frac{1-e^{-\lambda x}}{x} \cos\beta x\, dx$$

defined for $\lambda > 0$ and $\beta \neq 0$. The integral is divergent at infinity when $\beta = 0$. We have

$$F'(\lambda) = \int_0^\infty e^{-\lambda x} \cos \beta x \, dx = \frac{\lambda}{\lambda^2 + \beta^2}$$

and integrating like in the previous example

$$F(\lambda) = \int_0^\infty \frac{1 - e^{-\lambda x}}{x} \cos \beta x \, dx = \frac{1}{2} \ln \left(1 + \frac{\lambda^2}{\beta^2} \right).$$

We see that for every $\lambda \geq 0$, $\beta > 0$

$$\int_0^\infty \frac{1 - e^{-\lambda x}}{x} \cos \beta x \, dx = \int_0^\infty e^{-\beta x} \frac{1 - \cos \lambda x}{x} \, dx.$$

The integral 3.951(3) from [25]

$$\int_0^\infty \frac{e^{-\lambda x} - e^{-\mu x}}{x} \cos \beta x \, dx$$

($\lambda > 0, \mu > 0$) can be reduced to (2.10) by writing $e^{-\lambda x} - e^{-\mu x} = (e^{-\lambda x} - 1) + (1 - e^{-\mu x})$. Thus

$$\int_0^\infty \frac{e^{-\lambda x} - e^{-\mu x}}{x} \cos \beta x \, dx = \frac{1}{2} \ln \frac{\mu^2 + \beta^2}{\lambda^2 + \beta^2}.$$

Example 2.2.8

Here we will use the well-known Gaussian integral

(2.11) $$\int_0^\infty e^{-x^2} \, dx = \frac{\sqrt{\pi}}{2}$$

to evaluate for every $\lambda \geq 0$ the integral

$$F(\lambda) = \int_0^\infty \frac{1 - e^{-\lambda x^2}}{x^2} dx \,.$$

For convenience, the Gaussian integral will be solved at the end of the example.

We have for $\lambda > 0$

$$F'(\lambda) = \int_0^\infty e^{-\lambda x^2} dx = \frac{1}{\sqrt{\lambda}} \int_0^\infty \exp\left(-(x\sqrt{\lambda})^2\right) d(x\sqrt{\lambda}) = \frac{\sqrt{\pi}}{2\sqrt{\lambda}}$$

so that

$$\int_0^\infty \frac{1 - e^{-\lambda x^2}}{x^2} dx = \sqrt{\lambda \pi} \,.$$

Now we will evaluate (2.11). We present a very short and eloquent proof by using double integrals.

The Gaussian integral

Let

$$G = \int_0^\infty e^{-x^2} dx \,.$$

We write

$$G^2 = \left(\int_0^\infty e^{-x^2} dx \right)^2 = \left(\int_0^\infty e^{-x^2} dx \right)\left(\int_0^\infty e^{-y^2} dy \right) = \int_0^\infty \int_0^\infty e^{-x^2 - y^2} dx dy.$$

In this integral we introduce polar coordinates

$$x = r\cos\theta, \ y = r\sin\theta, \ r^2 = x^2 + y^2, \ dx dy = r dr d\theta$$

and since the integration is over the first quadrant we have

$$0 \le \theta \le \frac{\pi}{2}, \ 0 \le r < \infty.$$

Now we compute

$$G^2 = \int\limits_0^{\pi/2} \int\limits_0^\infty e^{-r^2} r \, dr \, d\theta = \frac{\pi}{2} \left\{ \frac{-1}{2} \int\limits_0^\infty e^{-r^2} d(-r^2) \right\} = -\frac{\pi}{4} e^{-r^2} \Big|_0^\infty = \frac{\pi}{4}.$$

Finally

$$\int\limits_0^\infty e^{-x^2} dx = \frac{\sqrt{\pi}}{2}.$$

The Gaussian integral can be written in terms of the gamma function

$$\Gamma(z) = \int\limits_0^\infty t^{z-1} e^{-t} dt \quad (\mathrm{Re}\, z > 0).$$

Namely, with the substitution $u = x^2$ we have

$$\int\limits_0^\infty e^{-x^2} dx = \frac{1}{2} \Gamma\left(\frac{1}{2}\right)$$

and more generally, with the substitution $u = x^p$, $p > 0$, we have

$$\int\limits_0^\infty e^{-x^p} dx = \frac{1}{p} \Gamma\left(\frac{1}{p}\right)$$

([25, Entry 3.325]).

Example 2.2.9

Sometimes we can use partial derivatives. Consider the function

$$F(\lambda, \mu) = \int\limits_0^\infty \frac{e^{-px} \cos qx - e^{-\lambda x} \cos \mu x}{x} dx$$

with four parameters. Here $\lambda > 0, \mu$ will be variables and $p > 0, q$ will be fixed. The partial derivatives are

$$F_\lambda(\lambda,\mu) = \int_0^\infty e^{-\lambda x} \cos \mu x \, dx = \frac{\lambda}{\lambda^2 + \mu^2}$$

$$F_\mu(\lambda,\mu) = \int_0^\infty e^{-\lambda x} \sin \mu x \, dx = \frac{\mu}{\lambda^2 + \mu^2}.$$

It is easy to restore the function from these derivatives

$$F(\lambda,\mu) = \frac{1}{2} \ln(\lambda^2 + \mu^2) + C(p,q)$$

where $C(p,q)$ is unknown. We notice that $F(p,q) = 0$ (from the definition of $F(\lambda,\mu)$). From the last equation $C(p,q) = \frac{-1}{2} \ln(p^2 + q^2)$ and finally

$$\int_0^\infty \frac{e^{-px} \cos qx - e^{-\lambda x} \cos \mu x}{x} dx = \frac{1}{2} \ln \frac{\lambda^2 + \mu^2}{p^2 + q^2}.$$

In all following examples containing $e^{-\lambda x}$ we assume that $\lambda > 0$.

Example 2.2.10

Now for $a > b > 0$ we evaluate

$$J(\lambda) = \int_0^\infty e^{-\lambda x} \frac{\sin(ax)\sin(bx)}{x} dx.$$

Clearly,

$$J'(\lambda) = -\int_0^\infty e^{-\lambda x} \sin(ax)\sin(bx) \, dx$$

$$= \frac{1}{2} \left\{ \int_0^\infty e^{-\lambda x} \cos(a+b)x \, dx - \int_0^\infty e^{-\lambda x} \cos(a-b)x \, dx \right\}$$

$$= \frac{1}{2} \left\{ \frac{\lambda}{\lambda^2 + (a+b)^2} - \frac{\lambda}{\lambda^2 + (a-b)^2} \right\}.$$

Integrating this with respect to λ and computing the constant of integration by setting $\lambda \to \infty$ we find

(2.12) $$\int_0^\infty e^{-\lambda x} \frac{\sin(ax)\sin(bx)}{x} \, dx = \frac{1}{4} \ln \frac{\lambda^2 + (a+b)^2}{\lambda^2 + (a-b)^2}.$$

This is entry 3.947(1) in [25].

Example 2.2.11

Using the previous example we shall evaluate now entry 3.947(2) from [25].

$$G(\lambda) = \int_0^\infty e^{-\lambda x} \frac{\sin(ax)\sin(bx)}{x^2} \, dx$$

where again $a > b > 0$. We have from (2.12)

$$G'(\lambda) = -J(\lambda) = \frac{1}{4} \ln \frac{\lambda^2 + (a+b)^2}{\lambda^2 + (a-b)^2}$$

and integrating by parts

$$G(\lambda) = \frac{\lambda}{4} \ln \frac{\lambda^2 + (a+b)^2}{\lambda^2 + (a-b)^2} - \frac{1}{4} \int \left(\frac{\lambda^2}{\lambda^2 + (a-b)^2} - \frac{\lambda^2}{\lambda^2 + (a+b)^2} \right) d\lambda.$$

With simple algebra we find

$$\frac{\lambda^2}{\lambda^2+(a-b)^2}-\frac{\lambda^2}{\lambda^2+(a+b)^2}=\frac{(a+b)^2}{\lambda^2+(a+b)^2}-\frac{(a-b)^2}{\lambda^2+(a-b)^2}$$

and the integration becomes easy. The result is

$$G(\lambda)=\frac{\lambda}{4}\ln\frac{\lambda^2+(a+b)^2}{\lambda^2+(a-b)^2}+\frac{a-b}{2}\arctan\frac{\lambda}{a-b}-\frac{a+b}{2}\arctan\frac{\lambda}{a+b}+\frac{\pi b}{2}$$

(the constant of integration is computed by setting $\lambda\to\infty$).

This answer is simpler than the one given in [25]. Setting $\lambda\to 0$ we prove also 3.741(3) in [25], namely,

$$\int_0^\infty\frac{\sin(ax)\sin(bx)}{x^2}dx=\frac{\pi b}{2}\ \ (a>b>0).$$

Example 2.2.12

Let again $a>b>0$. The integral

$$G(\lambda)=\int_0^\infty e^{-\lambda x}\frac{\sin(ax)\cos(bx)}{x}dx$$

is similar to the one in (2.12) (entry 3.947(3) in [25]). Here

$$G'(\lambda)=-\int_0^\infty e^{-\lambda x}\sin(ax)\cos(bx)\,dx$$

$$=\frac{-1}{2}\left\{\int_0^\infty e^{-\lambda x}\sin(a+b)x\,dx+\int_0^\infty e^{-\lambda x}\sin(a-b)x\,dx\right\}$$

$$=\frac{-1}{2}\left\{\frac{a+b}{\lambda^2+(a+b)^2}+\frac{a-b}{\lambda^2+(a-b)^2}\right\}$$

and integration gives

$$G(\lambda) = \frac{\pi}{2} - \frac{1}{2}\left(\arctan\frac{\lambda}{a+b} + \arctan\frac{\lambda}{a-b} \right)$$

(again evaluating the constant of integration with $\lambda \to \infty$).

Setting $b \to a$ we find also

$$\int_0^\infty e^{-\lambda x}\frac{\sin(ax)\cos(ax)}{x}dx = \frac{\pi}{4} - \frac{1}{2}\arctan\frac{\lambda}{2a}.$$

Example 2.2.13

$$F(\lambda) = \int_0^\infty e^{-\lambda x}\frac{\cos(ax)-\cos(bx)}{x^2}dx$$

(entry 3.948(3) in [25]). Differentiation gives

$$F'(\lambda) = \int_0^\infty e^{-\lambda x}\frac{\cos(bx)-\cos(ax)}{x}dx = \frac{1}{2}\ln\frac{\lambda^2+a^2}{\lambda^2+b^2}$$

as $\cos(bx) - \cos(ax) = [\cos(bx)-1] + [1-\cos(ax)]$ and we can use (2.9). Integration by parts yields

(2.13) $$F(\lambda) = \frac{\lambda}{2}\ln\frac{\lambda^2+a^2}{\lambda^2+b^2} + b\arctan\frac{b}{\lambda} - a\arctan\frac{a}{\lambda}$$

(again the constant of integration is found by setting $\lambda \to \infty$).

Many other integrals of similar structure can be evaluated this way or can be reduced to those already evaluated here. For example, entry 3.948(4) from [25]

$$A(\lambda) = \int_0^\infty e^{-\lambda x}\frac{\sin^2(ax)-\sin^2(bx)}{x^2}dx$$

can be reduced to (2.13) by using the identity $2\sin^2\alpha = 1-\cos 2\alpha$. Thus

$$A(\lambda) = \frac{1}{2}\int_0^\infty e^{-\lambda x}\frac{\cos(2bx)-\cos(2ax)}{x^2}dx$$

$$= \frac{\lambda}{4}\ln\frac{\lambda^2+4b^2}{\lambda^2+4a^2} + a\arctan\frac{2a}{\lambda} - b\arctan\frac{2b}{\lambda}.$$

The next several examples show some interesting logarithmic integrals involving trigonometric functions as well.

Example 2.2.14

Consider the integral

$$(2.14) \qquad J(\alpha) = \int_0^{\pi/2} \ln(\alpha^2 - \cos^2\theta)d\theta$$

for $\alpha > 1$. Differentiation with respect to α gives

$$J'(\alpha) = 2\alpha\int_0^{\pi/2}\frac{d\theta}{\alpha^2-\cos^2\theta}$$

and now the substitution $x = \tan\theta$ turns this into

$$J'(\alpha) = 2\alpha\int_0^\infty \frac{dx}{\alpha^2-1+\alpha^2 x^2} = \frac{2}{\sqrt{\alpha^2-1}}\arctan\frac{\alpha x}{\sqrt{\alpha^2-1}}\Bigg|_0^\infty = \frac{\pi}{\sqrt{\alpha^2-1}}.$$

Therefore,

$$J(\alpha) = \int_0^{\pi/2}\ln(\alpha^2-\cos^2\theta)d\theta = \pi\ln\left(\alpha+\sqrt{\alpha^2-1}\right) + C.$$

Now the question is how to evaluate the constant C. For this purpose we factor out α^2 inside the logarithm in the integral and likewise we factor out α on the right-hand side. Using the properties of the logarithm we write

$$\pi \ln \alpha + \int_0^{\pi/2} \ln\left(1 - \frac{\cos^2 \theta}{\alpha^2}\right) d\theta = \pi \ln \alpha + \pi \ln\left(1 + \sqrt{1 - \frac{1}{\alpha^2}}\right) + C .$$

Removing $\pi \ln \alpha$ from both sides and setting $\alpha \to \infty$ we compute $C = -\pi \ln 2$. As a result, two integrals are evaluated (see below). For the second one we set $\beta = 1 / \alpha$

$$(2.15) \qquad \int_0^{\pi/2} \ln(\alpha^2 - \cos^2 \theta) d\theta = \pi \ln \frac{\alpha + \sqrt{\alpha^2 - 1}}{2} \quad (\alpha > 1)$$

$$(2.16) \qquad \int_0^{\pi/2} \ln(1 - \beta^2 \cos^2 \theta) d\theta = \pi \ln \frac{1 + \sqrt{1 - \beta^2}}{2} \quad (0 \le \beta \le 1) .$$

In particular, with $\beta = 1$ in (2.16) we come to the important log-sine integral

$$(2.17) \qquad \int_0^{\pi/2} \ln(\sin \theta) d\theta = -\frac{\pi}{2} \ln 2 .$$

Example 2.2.15

By the same method (as in the previous example) we prove that

$$(2.18) \qquad \int_0^{\pi/2} \ln(1 + \alpha \sin^2 \theta) d\theta = \pi \ln \frac{1 + \sqrt{1 + \alpha}}{2} \quad (-1 \le \alpha) .$$

Calling the left-hand side $F(\alpha)$ we differentiate to find

$$F'(\alpha) = \int_0^{\pi/2} \frac{\sin^2 \theta}{1 + \alpha \sin^2 \theta} d\theta .$$

To solve this integral we divide top and bottom of the integrand by $\cos^2 \theta$ and then make the substitution $x = \tan \theta$. The result is

$$F'(\alpha) = \int_0^\infty \frac{x^2}{(1+x^2)(1+(1+\alpha)x^2)}\, dx \, .$$

Assuming for the moment that $\alpha \neq 0$ and using partial fractions we compute

$$F'(\alpha) = \frac{1}{\alpha}\int_0^\infty \left(\frac{1}{1+x^2} - \frac{1}{1+(1+\alpha)x^2} \right) dx$$

$$= \frac{1}{\alpha}\left(\arctan x - \frac{1}{\sqrt{1+\alpha}}\arctan(x\sqrt{1+\alpha}) \right)\Bigg|_0^\infty = \frac{\pi}{2\alpha}\left(1 - \frac{1}{\sqrt{1+\alpha}} \right).$$

Simple algebra shows that

$$F'(\alpha) = \frac{\pi}{2}\frac{\sqrt{1+\alpha}-1}{\alpha\sqrt{1+\alpha}} = \frac{\pi}{2(1+\sqrt{1+\alpha})\sqrt{1+\alpha}} \, .$$

Which is exactly the derivative of $\pi \ln\left(1 + \sqrt{1+\alpha} \right)$. This way

$$F(\alpha) = \pi \ln\left(1 + \sqrt{1+\alpha} \right) + C \, .$$

At this point we can drop the restriction $\alpha \neq 0$ and set $\alpha = 0$. We find $C = -\pi \ln 2$. The evaluation (2.18) is proved.

Example 2.2.16

Let $|\alpha| < 1$. Now we prove entry 4.397(3) in [25]

(2.19) $$F(\alpha) = \int_0^\pi \frac{\ln(1+\alpha\cos\theta)}{\cos\theta}\, d\theta = \pi \arcsin \alpha \, .$$

Assuming that the value of the integrand at $\theta = \pi / 2$ is α, the integrand becomes a continuous function on $[0, \pi]$. Then

(2.20) $$F'(\alpha) = \int_0^\pi \frac{1}{1 + \alpha \cos\theta} d\theta$$

which is easily solved by the substitution $t = \tan\dfrac{\theta}{2}$ (see Section 1.6)

$$F'(\alpha) = 2\int_0^\infty \frac{dt}{1 + \alpha + (1-\alpha)t^2} = \frac{2}{1+\alpha}\int_0^\infty \frac{dt}{1 + \dfrac{1-\alpha}{1+\alpha}t^2}$$

$$= \frac{2}{\sqrt{1-\alpha^2}} \arctan\left(t\sqrt{\frac{1-\alpha}{1+\alpha}}\right)\Bigg|_0^\infty = \frac{\pi}{\sqrt{1-\alpha^2}}$$

and (2.19) follows by integration.

Equation (2.19) holds also for $\alpha = \pm 1$, the improper integral being convergent. Thus

$$\int_0^\pi \frac{\ln(1 \pm \cos\theta)}{\cos\theta} d\theta = \pm\frac{\pi^2}{2}.$$

Example 2.2.17

Let again $|\alpha| < 1$. With the result from the previous example we can prove the evaluation

(2.21) $$G(\alpha) = \int_0^\pi \ln(1 + \alpha\cos\theta)d\theta = \pi \ln\frac{1 + \sqrt{1-\alpha^2}}{2}.$$

Indeed, assuming for the moment that $\alpha \neq 0$ we write

$$G'(\alpha) = \int_0^\pi \frac{\cos\theta}{1 + \alpha\cos\theta} d\theta = \frac{1}{\alpha}\int_0^\pi \frac{1 + \alpha\cos\theta - 1}{1 + \alpha\cos\theta} d\theta.$$

Splitting this integral in two parts and borrowing (2.20) from the previous example we have

$$\frac{\pi}{\alpha} - \frac{1}{\alpha}\int_0^\pi \frac{1}{1+\alpha\cos\theta}d\theta = \frac{\pi}{\alpha} - \frac{F'(\alpha)}{\alpha} = \frac{\pi}{\alpha} - \frac{\pi}{\alpha\sqrt{1-\alpha^2}}.$$

Integration is easy

$$G(\alpha) = \pi\left(\ln\alpha + \int\frac{d\alpha^{-1}}{\sqrt{\alpha^{-2}-1}}\right) = \pi\left(\ln\alpha + \ln(\alpha^{-1}+\sqrt{\alpha^{-2}-1})\right) + C$$

or

$$G(\alpha) = \pi\ln\left(1+\sqrt{1-\alpha^2}\right) + C.$$

Dropping the restriction $\alpha \neq 0$ and setting $\alpha = 0$ we find $C = -\pi\ln 2$. With this (2.21) is proved. With $\alpha \to \pm 1$ we obtain the well-known

$$\int_0^\pi \ln(1\pm\cos\theta)d\theta = -\pi\ln 2.$$

Example 2.2.18

We work here with a very interesting integral

$$(2.22) \qquad F(\alpha) = \int_0^\pi \ln(1-2\alpha\cos x + \alpha^2)dx.$$

It can be found, for example, in [23] and [50]. At first we assume $\alpha \neq 0$ and $\alpha \neq 1$. Then

$$F'(\alpha) = \int_0^\pi \frac{-2\cos x + 2\alpha}{1-2\alpha\cos x + \alpha^2}dx = \frac{1}{\alpha}\int_0^\pi\left(1 - \frac{1-\alpha^2}{1-2\alpha\cos x + \alpha^2}\right)dx$$

$$= \frac{\pi}{2} - \frac{1-\alpha^2}{\alpha}\int_0^\pi \frac{1}{1-2\alpha\cos x + \alpha^2}dx.$$

The last integral can be evaluated by the substitution $t = \tan\dfrac{x}{2}$. Simple work gives

$$F'(\alpha) = \frac{\pi}{\alpha} - \frac{2}{\alpha}\arctan\frac{1+\alpha}{1-\alpha}t\,\Bigg|_0^\infty$$

and we find from here that $F'(\alpha) = 0$ when $|\alpha| < 1$. When $|\alpha| > 1$ we have $F'(\alpha) = \dfrac{2\pi}{\alpha}$. Thus $F(\alpha) = C_1$, a constant, when $|\alpha| < 1$ and $F(\alpha) = C_2 + \pi\ln(\alpha^2)$ when $|\alpha| > 1$. Since $F(0) = 0$ (as (2.22) shows) we have

$$F(\alpha) = 0 \quad \text{for all } |\alpha| < 1.$$

Next, let $|\alpha| > 1$. In order to evaluate the constant C_2 we factor out α^2 inside the logarithm in (2.22) and write

$$F(\alpha) = \int_0^\pi \ln\left(\alpha^2\left(\frac{1}{\alpha^2} - \frac{2\cos x}{\alpha} + 1\right)\right) dx$$

$$= \pi\ln(\alpha^2) + F\left(\frac{1}{\alpha}\right) = \pi\ln(\alpha^2)$$

as $F(1/\alpha) = 0$ (here $|1/\alpha| < 1$). This also extends to $\alpha = \pm 1$. To summarize,

$$F(\alpha) = \pi\ln(\alpha^2) \quad \text{for all } |\alpha| \geq 1.$$

This integral will be considered also in Chapter 3 by using a series representation.

The integral can be written in a symmetrical form (cf. entry 2.6.36(14) in [43]). With $|\beta| \leq |\alpha|$

$$\int_0^\pi \ln(\beta^2 - 2\alpha\beta\cos x + \alpha^2)\,dx = 2\pi\ln|\alpha|.$$

Example 2.2.19

One more integral with a logarithm. We will evaluate

$$f(\alpha) = \int_0^1 \frac{\ln(1-\alpha^2 x^2)}{x^2\sqrt{1-x^2}}\,dx$$

where $|\alpha| < 1$. After differentiation

$$f'(\alpha) = -2\alpha\int_0^1 \frac{1}{(1-\alpha^2 x^2)\sqrt{1-x^2}}\,dx.$$

In this integral we make the substitution $x = \sin\theta$ to get

$$f'(\alpha) = -2\alpha\int_0^{\pi/2} \frac{1}{(1-\alpha^2\sin^2\theta)}\,d\theta.$$

Dividing top and bottom of the integrand by $\cos^2\theta$ and using the equations

$$\frac{1}{\cos^2\theta}\,d\theta = d\tan\theta, \quad \frac{1}{\cos^2\theta} = 1+\tan^2\theta$$

we bring the integral to the form

$$f'(\alpha) = -2\alpha\int_0^{\pi/2} \frac{1}{(1+(1-\alpha^2)\tan^2\theta)}\,d\tan\theta$$

$$= -\frac{2\alpha}{\sqrt{1-\alpha^2}}\int_0^{\pi/2} \frac{1}{(1+(1-\alpha^2)\tan^2\theta)}\,d(\sqrt{1-\alpha^2}\tan\theta).$$

$$= -\frac{2\alpha}{\sqrt{1-\alpha^2}} \arctan(\sqrt{1-\alpha^2} \tan\theta)\Big|_0^{\frac{\pi}{2}} = -\frac{\pi\alpha}{\sqrt{1-\alpha^2}} \cdot$$

Integrating we find

$$f(\alpha) = \pi\sqrt{1-\alpha^2} + C$$

and with $\alpha = 0$ we compute $C = -\pi$. Finally,

$$\int_0^1 \frac{\ln(1-\alpha^2 x^2)}{x^2\sqrt{1-x^2}} dx = \pi\left(\sqrt{1-\alpha^2} - 1\right).$$

At this point we see that both sides are defined also for $\alpha = \pm 1$, so the initial restriction on α can be relaxed and the equation holds for all $|\alpha| \le 1$.

A problem for the reader: in a similar way evaluate

$$\int_0^1 \frac{\ln(1-\alpha^2 x^2)}{\sqrt{1-x^2}} dx.$$

Example 2.2.20

Here we present our solution to Problem 11101 from the *American Mathematical Monthly* (2006, p. 270). Let $a,b > 0$. The problem is to prove the identity

$$\int_0^\infty \frac{a}{\sqrt{a^2+x^2}} \arctan\frac{b}{\sqrt{a^2+x^2}} dx = \frac{\pi a}{2}\left(\ln(b+\sqrt{b^2+a^2}) - \ln a\right).$$

Solution: With the substitution $x \to ax$ we can write this in the form

$$\int_0^\infty \frac{1}{\sqrt{1+x^2}} \arctan\frac{b/a}{\sqrt{1+x^2}} dx = \frac{\pi}{2}\ln\left(\frac{b}{a} + \sqrt{\frac{b^2}{a^2}+1}\right).$$

Now for $\alpha \ge 0$ define the function

$$F(\alpha) = \int_0^\infty \frac{1}{\sqrt{1+x^2}} \arctan \frac{\alpha}{\sqrt{1+x^2}} \, dx$$

where

$$F'(\alpha) = \int_0^\infty \frac{1}{\sqrt{1+x^2}} \left(\frac{1}{\sqrt{1+x^2}} \right) \left(1 + \frac{\alpha^2}{1+x^2} \right)^{-1} dx = \int_0^\infty \frac{1}{1+\alpha^2+x^2} \, dx$$

$$= \frac{1}{\sqrt{1+\alpha^2}} \arctan \frac{x}{\sqrt{1+\alpha^2}} \bigg|_0^\infty = \frac{\pi}{2\sqrt{1+\alpha^2}} \, .$$

Therefore,

$$F(\alpha) = \frac{\pi}{2} \ln(\alpha + \sqrt{1+\alpha^2}) + C \, .$$

Setting $\alpha = 0$ we find $C = 0$, so that

$$F(\alpha) = \frac{\pi}{2} \ln(\alpha + \sqrt{1+\alpha^2}) \, .$$

Replacing $\alpha = b / a$ we come to the desired result.

Example 2.2.21

Entry 2.7.5 (20) in [43] is a very strange integral

$$\int_0^\infty \frac{1}{x^2} \left(1 - \frac{1}{x} \arctan x \right) dx = \frac{\pi}{4} \, .$$

It looks like the substitution $1 / x = t$ will help to solve it. With this substitution, however, the integral becomes

$$\int_0^\infty \left(1 - t \arctan \frac{1}{t} \right) dx$$

which does not look easier. Using the identity $\arctan\dfrac{1}{t} = \dfrac{\pi}{2} - \arctan t$ here will not help.

We will try something different. The integral can be written in the form

$$\int_0^\infty \frac{x - \arctan x}{x^3}\,dx$$

and here we introduce a parameter. Consider the function

$$F(\lambda) = \int_0^\infty \frac{\lambda x - \arctan \lambda x}{x^3}\,dx$$

for $\lambda \geq 0$. The parameter should be put in both places. The reader can easily check that if we put the parameter only in the arctangent it will not be very helpful. Now we differentiate with respect to λ, compute the resulting integral, and then integrate with respect to λ

$$F'(\lambda) = \int_0^\infty \frac{1}{x^3}\left(x - \frac{x}{1+\lambda^2 x^2}\right)dx = \int_0^\infty \frac{1}{x^3}\left(\frac{\lambda^2 x^3}{1+\lambda^2 x^2}\right)dx$$

$$= \int_0^\infty \frac{\lambda^2}{1+\lambda^2 x^2}\,dx = \lambda \arctan(\lambda x)\Big|_0^\infty = \frac{\pi \lambda}{2}.$$

Integration gives

$$F(\lambda) = \int_0^\infty \frac{\lambda x - \arctan \lambda x}{x^3}\,dx = \frac{\pi \lambda^2}{4}$$

(the constant of integration is zero, because $F(0) = 0$). With $\lambda = 1$ we prove the original integral.

2.3 Using Differential Equations

Example 2.3.1

For every x consider the integral

$$y(x) = \int_0^\infty e^{-t^2} \cos(2xt)\, dt \,.$$

Here

$$y'(x) = -\int_0^\infty 2t e^{-t^2} \sin(2xt)\, dt$$

and integration by parts leads to the separable differential equation $y' = -2xy$, or

$$\frac{dy}{dx} = -2xy, \quad \frac{dy}{y} = -2x\, dx$$

with general solution $y(x) = Ce^{-x^2}$. According to (2.11)

$$y(0) = \int_0^\infty e^{-t^2}\, dt = \frac{\sqrt{\pi}}{2}$$

and so $C = \sqrt{\pi}/2$. Finally,

$$\int_0^\infty e^{-t^2} \cos(2xt)\, dt = \frac{\sqrt{\pi}}{2} e^{-x^2}\,.$$

This important integral shows the invariance of the function e^{-x^2} under Fourier's cosine transform. It appears in rescaled form as entry 3.896(4) in [25].

The integrand is an even function and with a simple rescaling we can write the result in the form

$$\int_{-\infty}^\infty e^{-at^2} \cos(xt)\, dt = \sqrt{\frac{\pi}{a}}\, e^{-x^2/4a} \quad (\operatorname{Re} a > 0)$$

This integral was used by the author to solve Problem 1896 in the

Mathematics Magazine (vol. 86, June 2013, 228-230). The problem is to evaluate the integral

$$A = \int_0^\infty \frac{\cos\sqrt{t}}{\sqrt{t}} \cos t \, dt \, .$$

First, the substitution $t \to t^2$ transforms this integral to

$$A = 2\int_0^\infty \cos(t^2)\cos t \, dt = \int_{-\infty}^\infty \cos(t^2)\cos t \, dt \, .$$

Next we extend by continuity the identity

$$\int_{-\infty}^\infty e^{-at^2}\cos(xt)\,dt = \sqrt{\frac{\pi}{a}}\, e^{-x^2/4a}$$

to complex numbers $a \neq 0$ even when $\mathrm{Re}\,a = 0$. Setting $x = 1$ and $a = i$ we find

$$\int_{-\infty}^\infty \left(\cos(t^2) - i\sin(t^2)\right)\cos t \, dt = \int_{-\infty}^\infty e^{-it^2}\cos(t)\,dt = \sqrt{\frac{\pi}{i}}\, e^{-1/4i}$$

$$= \sqrt{\pi}\left(\frac{\sqrt{2}}{2} - i\frac{\sqrt{2}}{2}\right)\left(\cos\frac{1}{4} + i\sin\frac{1}{4}\right).$$

Comparing real parts gives

$$\int_{-\infty}^\infty \cos(t^2)\cos t \, dt = \sqrt{\frac{\pi}{2}}\left(\cos\frac{1}{4} + \sin\frac{1}{4}\right).$$

This way

$$\int_0^\infty \frac{\cos\sqrt{t}}{\sqrt{t}}\cos t \, dt = \sqrt{\frac{\pi}{2}}\left(\cos\frac{1}{4} + \sin\frac{1}{4}\right).$$

Example 2.3.2

In this example we evaluate two interesting integrals (3.723(2) and 3.723(3) from [25])

$$F(\lambda) = \int_0^\infty \frac{\cos \lambda x}{a^2 + x^2} dx, \quad G(\lambda) = \int_0^\infty \frac{x \sin \lambda x}{a^2 + x^2} dx$$

where $a > 0, \lambda \geq 0$. The two integrals can be viewed as Fourier cosine and sine transforms. We will find a second order differential equation for $F(\lambda)$. First we notice that

(2.23) $$F'(\lambda) = -G(\lambda)$$

Further direct differentiation brings to a divergent integral. To avoid this obstacle we shall use a special trick, adding to both sides of (2.23) the number

$$\frac{\pi}{2} = \int_0^\infty \frac{\sin x}{x} dx$$

(see Example 2.1.1), After a simple calculation we find

$$F'(\lambda) + \frac{\pi}{2} = a^2 \int_0^\infty \frac{\sin \lambda x}{x(a^2 + x^2)} dx.$$

Differentiating again we come to the second order differential equation

$$F'' = a^2 F$$

with general solution (A, B - arbitrary parameters)

$$F(\lambda) = Ae^{a\lambda} + Be^{-a\lambda}.$$

Then $A = 0$ because $F(\lambda)$ is a bounded function when $\lambda \to \infty$. To find B we set $\lambda = 0$ and compute

$$B = F(0) = \int_0^\infty \frac{dx}{a^2 + x^2} = \frac{1}{a}\arctan\frac{x}{a}\Big|_0^\infty = \frac{\pi}{2a}.$$

Finally,

(2.24)
$$F(\lambda) = \int_0^\infty \frac{\cos \lambda x}{a^2 + x^2} dx = \frac{\pi}{2a} e^{-a\lambda}.$$

And from (2.23) also

(2.25)
$$G(\lambda) = \int_0^\infty \frac{x \sin \lambda x}{a^2 + x^2} dx = \frac{\pi}{2} e^{-a\lambda}.$$

Notice the interesting equation $G(\lambda) = aF(\lambda)$!

This result can be used to evaluate other similar integrals. Integrating (2.24) with respect to λ and adjusting the constant of integration we prove entry 3.725(1) in [25], namely,

$$\int_0^\infty \frac{\sin \lambda x}{x(a^2 + x^2)} dx = \frac{\pi}{2a^2}(1 - e^{-a\lambda}).$$

Differentiating this integral with respect to a we prove also entry 3.735

$$\int_0^\infty \frac{\sin \lambda x}{x(a^2 + x^2)^2} dx = \frac{\pi}{2a^4}(1 - e^{-a\lambda}) - \frac{\pi\lambda}{4a^3} e^{-a\lambda}.$$

Problems for the reader: Prove that

$$\int_0^\infty \frac{\cos \lambda x}{(a^2 + x^2)^2} dx = \frac{\pi}{4a^3}(1 + a\lambda)e^{-a\lambda}$$

$$\int_0^\infty \frac{\cos^2 x}{(1 + x^2)^2} dx = \frac{\pi}{8}(1 + 3e^{-2}).$$

Example 2.3.3

We shall evaluate here two Laplace integrals. For $s > 0$ and $a > 0$ consider

$$F(s) = \int_0^\infty \frac{e^{-st}}{a^2 + t^2} dt \quad \text{and} \quad G(s) = \int_0^\infty \frac{te^{-st}}{a^2 + t^2} dt .$$

Differentiating twice the first integral we find

(2.26) $F'(s) = -G(s), \ F''(s) = -G'(s) .$

At the same time

$$-G'(s) = \int_0^\infty \frac{t^2 e^{-st}}{a^2 + t^2} dt = \int_0^\infty \frac{(-a^2 + a^2 + t^2)e^{-st}}{a^2 + t^2} dt = -a^2 F(s) + \frac{1}{s}$$

which leads to the second order differential equation

$$F'' + a^2 F = \frac{1}{s} .$$

This equation can be solved by variation of parameters. The solution is

$$F(s) = \frac{1}{a}\left[\text{ci}(as)\sin(as) - \text{si}(as)\cos(as) \right]$$

involving the special sine and cosine integrals

$$\text{si}(x) = -\int_x^\infty \frac{\sin t}{t} dt = -\frac{\pi}{2} + \int_0^x \frac{\sin t}{t} dt$$

$$\text{ci}(x) = -\int_x^\infty \frac{\cos t}{t} dt .$$

The choice of integral limits here is dictated by the initial conditions $F(\infty) = G(\infty) = 0$.

From (2.26) we find also

$$G(s) = -\text{ci}(as)\cos(as) - \text{si}(as)\sin(as).$$

The integral $F(s)$ can be used to construct an interesting extension of the integral

$$\int_0^\infty \frac{\sin x}{x} dt = \frac{\pi}{2}.$$

Namely, for $b > 0$

$$\int_0^\infty \frac{\sin ax}{x+b} dx = \int_0^\infty \sin ax \left\{ \int_0^\infty e^{-t(x+b)} dt \right\} dx$$

$$= \int_0^\infty \left\{ \int_0^\infty e^{-xt} \sin ax\, dx \right\} e^{-bt} dt$$

$$= a \int_0^\infty \frac{e^{-bt} dt}{a^2 + t^2} = aF(b) = \text{ci}(ab)\sin(ab) - \text{si}(ab)\cos(ab).$$

That is,

$$\int_0^\infty \frac{\sin ax}{x+b} dx = \text{ci}(ab)\sin(ab) - \text{si}(ab)\cos(ab).$$

This is entry 3.772(1) in [25]. In the same way we prove 3.722(3)

$$\int_0^\infty \frac{\cos ax}{x+b} dx = -\text{ci}(ab)\cos(ab) - \text{si}(ab)\sin(ab).$$

Example 2.3.4

Here we evaluate the integral

$$H(\alpha) = \int_0^\infty \exp\left(-x - \frac{\alpha}{x}\right) \frac{dx}{\sqrt{x}}, \quad \alpha > 0$$

by using a differential equation (cf. [29]). Differentiation yields

$$H'(\alpha) = -\int_0^\infty \exp\left(-x - \frac{\alpha}{x}\right) \frac{dx}{x\sqrt{x}}.$$

The substitution $x = \alpha / t$ transforms this into

$$H'(\alpha) = -\frac{1}{\sqrt{\alpha}} \int_0^\infty \exp\left(-t - \frac{\alpha}{t}\right) \frac{dt}{\sqrt{t}}$$

so we have the separable differential equation

$$H'(\alpha) = \frac{-1}{\sqrt{\alpha}} H(\alpha), \quad \frac{dH}{H} = \frac{-d\alpha}{\sqrt{\alpha}}$$

with solution

$$H(\alpha) = M \exp(-2\sqrt{\alpha})$$

where $M > 0$ is a constant. Setting $\alpha \to 0$ and using the fact that

$$H(0) = \Gamma\left(\frac{1}{2}\right) = \sqrt{\pi}$$

we find

$$H(\alpha) = \sqrt{\pi} \exp(-2\sqrt{\alpha}).$$

Remark. With the substitution $x = \alpha t^2$ our integral becomes

$$H(\alpha) = \int_0^\infty \exp\left(-\alpha t^2 - \frac{1}{t^2}\right) dt = \frac{\sqrt{\pi}}{2\sqrt{\alpha}} \exp(-2\sqrt{\alpha})$$

and this is exactly integral (1.42) solved in Example 1.8.7 in the previous chapter by a special substitution.

This work gives also the value of another important integral, namely

$$\int_0^\infty \exp\left(-x - \frac{\alpha}{x}\right)\frac{dx}{x\sqrt{x}} = -H'(\alpha) = \frac{1}{\sqrt{\alpha}}H(\alpha)$$

$$\int_0^\infty \exp\left(-x - \frac{\alpha}{x}\right)\frac{dx}{x\sqrt{x}} = \sqrt{\frac{\pi}{\alpha}}\exp(-2\sqrt{\alpha})\,.$$

Problem for the reader: Show that

$$J(\alpha) = \int_0^\infty \exp\left(-x - \frac{\alpha}{x}\right)\sqrt{x}\,dx = \sqrt{\pi}\left(\sqrt{\alpha} + \frac{1}{2}\right)\exp(-2\sqrt{\alpha})\,.$$

Hint: $J'(\alpha) = -H(\alpha)$.

Example 2.3.5

Here we work with the two integrals

$$U(\alpha) = \int_0^\infty \exp(-x^2)\cos\left(\frac{\alpha^2}{x^2}\right)dx$$

$$V(\alpha) = \int_0^\infty \exp(-x^2)\sin\left(\frac{\alpha^2}{x^2}\right)dx$$

where $\alpha > 0$. We differentiate $U(\alpha)$ and then set $y = \alpha/x$ to find

$$U'(\alpha) = -2\int_0^\infty \exp(-x^2)\cos\left(\frac{\alpha^2}{x^2}\right)\frac{\alpha}{x^2}dx = 2\int_\infty^0 \exp\left(\frac{-\alpha^2}{y^2}\right)\sin(y^2)dy$$

$$= -2\int_0^\infty \exp\left(\frac{-\alpha^2}{y^2}\right)\sin(y^2)dy\,.$$

A second differentiation gives (with $x = \alpha/y$)

$$U''(\alpha) = -2\int_0^\infty \exp\left(\frac{-\alpha^2}{y^2}\right)\sin(y^2)\frac{-2\alpha}{y^2}\,dy$$

$$= 4\int_0^\infty \exp(-x^2)\sin\left(\frac{\alpha^2}{x^2}\right)\,dx$$

that is,

$$U''(\alpha) = 4V(\alpha).$$

In the same way we obtain $V''(\alpha) = -4U(\alpha)$. We define now the complex function $W(\alpha) = U(\alpha) + iV(\alpha)$. This function satisfies the second order differential equation

$$W''(\alpha) = -4iW(\alpha)$$

with characteristic equation $r^2 + 4i = 0$ and roots $r_1 = -\sqrt{2} + i\sqrt{2}$ and $r_2 = \sqrt{2} - i\sqrt{2}$. With these roots we construct the general solution

$$W(\alpha) = A\exp(r_1\alpha) + B\exp(r_2\alpha)$$

with parameters A, B. Explicitly,

$$W(\alpha) = Ae^{-\sqrt{2}\alpha}(\cos\sqrt{2}\alpha + i\sin\sqrt{2}\alpha) + Be^{\sqrt{2}\alpha}(\cos\sqrt{2}\alpha - i\sin\sqrt{2}\alpha).$$

At this point we conclude that $B = 0$ since $W(\alpha)$ is a bounded function. This way

$$W(\alpha) = Ae^{-\sqrt{2}\alpha}(\cos\sqrt{2}\alpha + i\sin\sqrt{2}\alpha).$$

Setting $\alpha = 0$ we find $W(0) = A$. At the same time from the definition of $U(\alpha)$ and $V(\alpha)$ we have

$$W(0) = U(0) = \int_0^\infty \exp(-x^2)\,dx = \frac{\sqrt{\pi}}{2} = A.$$

Thus

$$W(\alpha) = \frac{\sqrt{\pi}}{2} e^{-\sqrt{2}\alpha} (\cos\sqrt{2}\alpha + i\sin\sqrt{2}\alpha).$$

Taking real and complex parts we conclude that

$$U(\alpha) = \int_0^\infty \exp(-x^2)\cos\left(\frac{\alpha^2}{x^2}\right) dx = \frac{\sqrt{\pi}}{2} e^{-\sqrt{2}\alpha} \cos\sqrt{2}\alpha$$

$$V(\alpha) = \int_0^\infty \exp(-x^2)\sin\left(\frac{\alpha^2}{x^2}\right) dx = \frac{\sqrt{\pi}}{2} e^{-\sqrt{2}\alpha} \sin\sqrt{2}\alpha.$$

2.4 Advanced Techniques

In certain cases we can use the Leibniz Integral Rule

$$\frac{d}{d\alpha} \int_{\varphi(\alpha)}^{\psi(\alpha)} f(\alpha,x)\,dx$$

$$= \int_{\varphi(\alpha)}^{\psi(\alpha)} \frac{d}{d\alpha} f(\alpha,x)\,dx + f(\alpha,\psi(\alpha))\psi'(\alpha) - f(\alpha,\varphi(\alpha))\varphi'(\alpha)$$

where $f(\alpha,x)$, $\varphi(\alpha)$, $\psi(\alpha)$ are appropriate functions (see [23]).

Example 2.4.1

We will evaluate the challenging integral

$$\int_0^1 \frac{\arctan x}{\sqrt{1-x^2}} dx$$

by using the function

$$J(\alpha) = \int_{\varphi(\alpha)}^1 \frac{\arctan(\alpha x)}{\sqrt{1-x^2}} dx$$

where $\alpha > 1$ and

$$\varphi(\alpha) = \sqrt{1 - \frac{1}{\alpha^2}} = \frac{\sqrt{\alpha^2 - 1}}{\alpha}$$

with derivative

$$\varphi'(\alpha) = \frac{1}{\alpha^2 \sqrt{\alpha^2 - 1}}.$$

Applying the Leibniz rule we find

$$J'(\alpha) = \int_{\varphi(\alpha)}^{1} \frac{x}{(1 + \alpha^2 x^2)\sqrt{1 - x^2}} dx - \frac{\arctan\sqrt{\alpha^2 - 1}}{\alpha\sqrt{\alpha^2 - 1}}.$$

Let us call the integral here $A(\alpha)$. We will evaluate it by the substitution $1 - x^2 = u^2$, $u \geq 0$

$$A(\alpha) = \int_{\varphi(\alpha)}^{1} \frac{x}{(1 + \alpha^2 x^2)\sqrt{1 - x^2}} dx = \int_{0}^{1/\alpha} \frac{du}{\alpha^2 + 1 - \alpha^2 u^2}$$

$$= \frac{1}{2\alpha\sqrt{\alpha^2 + 1}} \ln \frac{\sqrt{\alpha^2 + 1} + \alpha u}{\sqrt{\alpha^2 + 1} - \alpha u} \Bigg|_{0}^{1/\alpha} = \frac{1}{2\alpha\sqrt{\alpha^2 + 1}} \ln \frac{\sqrt{\alpha^2 + 1} + 1}{\sqrt{\alpha^2 + 1} - 1}.$$

This function is easy to integrate by noticing that

$$\frac{d}{d\alpha} \ln \frac{\sqrt{\alpha^2 + 1} + 1}{\sqrt{\alpha^2 + 1} - 1} = \frac{-2}{\alpha\sqrt{\alpha^2 + 1}}$$

and therefore, one antiderivative is

$$\int A(\alpha) d\alpha = \frac{-1}{8} \left(\ln \frac{\sqrt{\alpha^2 + 1} + 1}{\sqrt{\alpha^2 + 1} - 1} \right)^2.$$

We also compute

$$\frac{d}{d\alpha}\arctan\sqrt{\alpha^2-1} = \frac{1}{\alpha\sqrt{\alpha^2-1}}$$

and so we have the antiderivative

$$\int \frac{\arctan\sqrt{\alpha^2-1}}{\alpha\sqrt{\alpha^2-1}}\,d\alpha = \frac{1}{2}\left(\arctan\sqrt{\alpha^2-1}\right)^2$$

Now we can integrate $J'(\alpha)$ to get

$$J(\alpha) = \frac{-1}{8}\left(\ln\frac{\sqrt{\alpha^2+1}+1}{\sqrt{\alpha^2+1}-1}\right)^2 - \frac{1}{2}\left(\arctan\sqrt{\alpha^2-1}\right)^2 + C.$$

Using the limit $\lim\limits_{\alpha\to\infty} J(\alpha) = 0$ we compute $C = \dfrac{\pi^2}{8}$. Finally,

$$(2.27)\quad J(\alpha) = \frac{-1}{8}\left(\ln\frac{\sqrt{\alpha^2+1}+1}{\sqrt{\alpha^2+1}-1}\right)^2 - \frac{1}{2}\left(\arctan\sqrt{\alpha^2-1}\right)^2 + \frac{\pi^2}{8}.$$

Setting here $\alpha \to 1$ we find after a simple computation

$$(2.28)\qquad \int_0^1 \frac{\arctan x}{\sqrt{1-x^2}}\,dx = \frac{\pi^2}{8} - \frac{1}{2}\left(\ln(1+\sqrt{2})\right)^2.$$

With $\alpha = 2$ in (2.27) we find also

$$\int_{\sqrt{3}/2}^1 \frac{\arctan(2x)}{\sqrt{1-x^2}}\,dx = \frac{\pi^2}{8} - \frac{1}{8}\left(\ln\frac{\sqrt{5}+1}{\sqrt{5}-1}\right)^2 - \frac{1}{2}\left(\arctan\sqrt{3}\right)^2.$$

Remark. Integration by parts in (2.28) yields

$$\int_0^1 \frac{\arctan x}{\sqrt{1-x^2}}\,dx = \frac{\pi^2}{8} - \int_0^1 \frac{\arcsin x}{1+x^2}\,dx$$

and so

$$\int_0^1 \frac{\arcsin x}{1+x^2}\,dx = \frac{1}{2}\left(\ln(1+\sqrt{2})\right)^2.$$

Using the identity

$$\arctan x = \frac{\pi}{2} - \arctan\frac{1}{x} \quad (x>0)$$

we can write

$$\int_0^1 \frac{\arctan x}{\sqrt{1-x^2}}\,dx = \frac{\pi^2}{4} - \int_0^1 \frac{\arctan(1/x)}{\sqrt{1-x^2}}\,dx.$$

In the second integral we make the substitution $x = 1/t$ to find also

(2.29) $$\int_1^\infty \frac{\arctan t}{t\sqrt{t^2-1}}\,dt = \frac{\pi^2}{8} + \frac{1}{2}\left(\ln(1+\sqrt{2})\right)^2.$$

This integral was evaluated in [18] by using a parameter, much like (2.28) was evaluated.

Example 2.4.2

We evaluate here in explicit form the function

$$F(\alpha) = \int_{1/\alpha}^\infty \frac{\ln(\alpha x + \sqrt{\alpha^2 x^2 - 1})}{x(1+x^2)}\,dx$$

for $\alpha > 0$. Differentiating by the Leibniz rule we compute

$$F'(\alpha) = \int_{1/\alpha}^{\infty} \frac{dx}{(1+x^2)\sqrt{\alpha^2 x^2 - 1}}$$

(notice that $\ln(\alpha x + \sqrt{\alpha^2 x^2 - 1})$ becomes zero for $x = 1/\alpha$).

To solve this integral we first write it in the form

$$F'(\alpha) = \frac{-1}{2} \int_{1/\alpha}^{\infty} \frac{dx^{-2}}{(x^{-2}+1)\sqrt{\alpha^2 - x^{-2}}}$$

and then make the substitution $\alpha^2 - x^{-2} = t^2$, $t \geq 0$. This gives

$$F'(\alpha) = \int_0^{\alpha} \frac{dt}{1+\alpha^2 - t^2} = \frac{1}{2\sqrt{1+\alpha^2}} \ln \frac{\sqrt{1+\alpha^2} + \alpha}{\sqrt{1+\alpha^2} - \alpha}.$$

This function is easy to integrate, as

$$\frac{d}{d\alpha} \ln \frac{\sqrt{1+\alpha^2} + \alpha}{\sqrt{1+\alpha^2} - \alpha} = \frac{2}{\sqrt{1+\alpha^2}}.$$

Thus we find

$$F(\alpha) = \frac{1}{8} \left(\ln \frac{\sqrt{1+\alpha^2} + \alpha}{\sqrt{1+\alpha^2} - \alpha} \right)^2 + C.$$

To evaluate the constant of integration we compute the limit

$$\lim_{\alpha \to 0} F(\alpha) = 0.$$

Finding this limit is a good exercise for the reader. The evaluation of the limit becomes easier, if we make the substitution $x = \dfrac{1}{\alpha t}$ in the integral $F(\alpha)$ and transform it into an integral with finite limits. Thus we come to

$$F(\alpha) = \frac{1}{8}\left(\ln\frac{\sqrt{1+\alpha^2}+\alpha}{\sqrt{1+\alpha^2}-\alpha}\right)^2.$$

The expression in the logarithm can be simplified by writing

$$\frac{\sqrt{1+\alpha^2}+\alpha}{\sqrt{1+\alpha^2}-\alpha} = \frac{(\sqrt{1+\alpha^2}+\alpha)^2}{(\sqrt{1+\alpha^2}-\alpha)(\sqrt{1+\alpha^2}+\alpha)} = (\sqrt{1+\alpha^2}+\alpha)^2$$

so finally we have

$$F(\alpha) = \frac{1}{2}\ln^2(\sqrt{1+\alpha^2}+\alpha)$$

that is,

(2.30) $$\int_{1/\alpha}^{\infty} \frac{\ln(\alpha x + \sqrt{\alpha^2 x^2 -1})}{x(1+x^2)}dx = \frac{1}{2}\ln^2(\sqrt{1+\alpha^2}+\alpha).$$

In particular, for $\alpha = 1$

(2.31) $$\int_{1}^{\infty} \frac{\ln(x+\sqrt{x^2-1})}{x(1+x^2)}dx = \frac{1}{2}\ln^2(\sqrt{2}+1).$$

For $\alpha = 1/2$ in (2.30) we find also

(2.32) $$\int_{2}^{\infty} \frac{\ln(x+\sqrt{x^2-4})}{x(1+x^2)}dx = \frac{1}{2}\ln^2\frac{\sqrt{5}+1}{2} + \ln 2\ln\frac{\sqrt{5}}{2}.$$

Remark. The similar integral

(2.33) $$\int_{1}^{\infty} \frac{\ln(x+\sqrt{x^2-1})}{x\sqrt{x^2-1}}dx$$

cannot be evaluated this way. The value of this integral is $2G$, where

$$G = \sum_{n=0}^{\infty} \frac{(-1)^n}{(2n+1)^2} = 1 - \frac{1}{3^2} + \frac{1}{5^2} + \ldots$$

is Catalan's constant. The substitution $x = \cosh t$ with $\ln(x + \sqrt{x^2 - 1}) = t$ turns (2.33) into

$$\int_0^{\infty} \frac{t}{\cosh t} dt = 2G$$

which is a well-known result. This last integral will be evaluated in Chapter 5, Section 5.5.

2.5 The Basel Problem and Related Integrals

2.5.1 *Introduction*

The famous Basel problem posed by Pietro Mengoli in 1644 and solved by Leonhard Euler in 1735 asked for a closed form evaluation of the series

$$1 + \frac{1}{2^2} + \frac{1}{3^2} + \ldots$$

- see [31]. Euler proved that

(2.34)
$$1 + \frac{1}{2^2} + \frac{1}{3^2} + \ldots = \frac{\pi^2}{6}.$$

In the meantime, trying to evaluate the series Leibniz discovered the representation

$$1 + \frac{1}{2^2} + \frac{1}{3^2} + \ldots = -\int_0^1 \frac{\ln(1-t)}{t} dt$$

but was unable to find the exact numerical value of this integral. It was shown in [31] that using complex logarithms this integral can be evaluated as $\pi^2/6$.

There exist, however other integrals which give a quick solution to the Basel problem without using complex numbers. A very good example is the integral

$$(2.35) \qquad \int_0^1 \frac{\arcsin \alpha x}{\sqrt{1-x^2}}\, dx = \frac{1}{2}\left[\mathrm{Li}_2(\alpha) - \mathrm{Li}_2(-\alpha)\right]$$

for $|\alpha| \leq 1$. Here

$$\mathrm{Li}_2(x) = \sum_{n=1}^\infty \frac{x^n}{n^2}$$

is the dilogarithm ([33], [52]). Setting $\alpha = 1$ in (2.35) we find

$$\int_0^1 \frac{\arcsin x}{\sqrt{1-x^2}}\, dx = \frac{1}{2}(\arcsin x)^2 \Big|_0^1 = \frac{\pi^2}{8} = \frac{1}{2}\left[\mathrm{Li}_2(1) - \mathrm{Li}_2(-1)\right]$$

and (2.34) follows immediately from here, as

$$\mathrm{Li}_2(1) - \mathrm{Li}_2(-1) = \frac{3}{2}\mathrm{Li}_2(1) = \frac{3}{2}\left(1 + \frac{1}{2^2} + \frac{1}{3^2} + \dots\right).$$

This clever solution to the Basel problem was published by Habib Bin Muzaffar in [36].

The series in (2.34) is usually denoted by $\zeta(2)$ as a particular value of the Riemann zeta function

$$\zeta(s) = \sum_{n=1}^\infty \frac{1}{n^s} \qquad (\mathrm{Re}\, s > 1).$$

Here is a proof of (2.35). Let the integral on the left-hand side be $F(\alpha)$. Then

$$F'(\alpha) = \int_0^1 \frac{x\,dx}{\sqrt{1-\alpha^2 x^2}\sqrt{1-x^2}} = \int_0^{\pi/2} \frac{\sin\theta\,d\theta}{\sqrt{1-\alpha^2 \sin^2\theta}}$$

with the substitution $x = \sin\theta$. Next we write

$$\int_0^{\pi/2} \frac{\sin\theta\,d\theta}{\sqrt{1-\alpha^2 \sin^2\theta}} = -\int_0^{\pi/2} \frac{d\cos\theta}{\sqrt{1-\alpha^2 + \alpha^2 \cos^2\theta}}$$

$$= -\frac{1}{\alpha}\int_0^{\pi/2} \frac{d(\alpha\cos\theta)}{\sqrt{1-\alpha^2 + (\alpha\cos\theta)^2}}$$

and setting $t = \alpha\cos\theta$

$$= \frac{1}{\alpha}\int_0^\alpha \frac{dt}{\sqrt{1-\alpha^2+t^2}} = \frac{1}{\alpha}\ln(t+\sqrt{1-\alpha^2+t^2})\Big|_0^\alpha$$

$$= \frac{1}{\alpha}\left(\ln(1+\alpha) - \frac{1}{2}\ln(1-\alpha^2)\right) = \frac{1}{2}\left(\frac{\ln(1+\alpha)}{\alpha} - \frac{\ln(1-\alpha)}{\alpha}\right)$$

$$= \frac{1}{2}\left(\sum_{n=1}^\infty \frac{\alpha^{n-1}}{n} + \sum_{n=1}^\infty \frac{(-1)^{n-1}\alpha^{n-1}}{n}\right)$$

and (2.35) follows by simple integration.

A problem for the reader: Show that for every $|\alpha| < 1$

$$2\alpha \int_0^1 \frac{\arccos x}{\sqrt{1-\alpha^2 x^2}}\,dx = \mathrm{Li}_2(\alpha) - \mathrm{Li}_2(-\alpha)$$

(hint: start with the substitution $t = \alpha x$). This is entry 4.1.6 (58) in [20].

Example 2.5.1

For $\alpha \geq 1$ consider the integral

$$\int_0^1 \frac{x \arccos x}{\sqrt{\alpha^2 - x^2}} dx$$

which is entry 4.1.6 (57) in [20]. This integral resembles the previous two, but has a different solution. Not differentiation with respect to the parameter α, but integration by parts.

$$\int_0^1 \frac{x \arccos x}{\sqrt{\alpha^2 - x^2}} dx = -\frac{1}{2} \int_0^1 \frac{\arccos x}{\sqrt{\alpha^2 - x^2}} d(\alpha^2 - x^2)$$

$$= -\int_0^1 \arccos x \, d\sqrt{\alpha^2 - x^2}$$

$$= -\arccos x \sqrt{\alpha^2 - x^2} \Big|_0^1 - \int_0^1 \frac{\sqrt{\alpha^2 - x^2}}{\sqrt{1 - x^2}} dx$$

(continuing here with the substitution $x = \sin t$)

$$= \frac{\pi \alpha}{2} - \int_0^{\pi/2} \sqrt{\alpha^2 - \sin^2 t} \, dt = \frac{\pi \alpha}{2} - \alpha \int_0^{\pi/2} \sqrt{1 - \alpha^{-2} \sin^2 t} \, dt$$

$$= \frac{\pi \alpha}{2} - \alpha \, \mathrm{E}\left(\frac{\pi}{2}, \frac{1}{\alpha}\right)$$

where

$$\mathrm{E}(\mu, k) = \int_0^\mu \sqrt{1 - k^2 \sin^2 t} \, dt$$

is the (incomplete) elliptic integral of the second kind. That is,

$$\int_0^1 \frac{x \arccos x}{\sqrt{\alpha^2 - x^2}} dx = \frac{\pi \alpha}{2} - \alpha \, \mathrm{E}\left(\frac{\pi}{2}, \frac{1}{\alpha}\right).$$

A problem for the reader: Check that for $\alpha \geq 1$

$$\int_0^1 \frac{x \arcsin x}{\sqrt{\alpha^2 - x^2}} dx = -\frac{\pi \sqrt{\alpha^2 - 1}}{2} + \alpha \, \mathrm{E}\!\left(\frac{\pi}{2}, \frac{1}{\alpha}\right).$$

2.5.2 Special integrals with arctangents

We will consider now some integrals similar to the one in (2.35) which can be associated with the Basel problem, either solving it or leading to similar results.

Example 2.5.2

We will prove here that

$$(2.36) \qquad 2\int_0^\infty \frac{\arctan \alpha x}{1 + x^2} dx = \ln \alpha \ln \frac{1 - \alpha}{1 + \alpha} + \mathrm{Li}_2(\alpha) - \mathrm{Li}_2(-\alpha)$$

for any $0 \le \alpha \le 1$. Indeed, it is easy to see (by using limits) that the right-hand side extends to $\alpha = 0$ and $\alpha = 1$. The function $\ln \alpha \ln \dfrac{1 - \alpha}{1 + \alpha}$ becomes zero for $\alpha \to 0$ and for $\alpha \to 1$. With $\alpha \to 1$ we compute immediately from (2.36)

$$\left.(\arctan x)^2\right|_0^1 = \frac{\pi^2}{4} = \frac{3}{4}\mathrm{Li}_2(1) = \frac{3}{4}\zeta(2)$$

and (2.34) follows. This integral also solves the Basel problem.

Proof of (2.36). Let $J(\alpha)$ be the left-hand side in (2.36). Differentiation yields

$$J'(\alpha) = \int_0^\infty \frac{2x\,dx}{(1 + \alpha^2 x^2)(1 + x^2)} = \frac{1}{1 - \alpha^2}\int_0^\infty \left(\frac{2x}{1 + x^2} - \frac{2\alpha^2 x}{1 + \alpha^2 x^2}\right) dx$$

$$= \frac{1}{1 - \alpha^2}\left[\ln \frac{1 + x^2}{1 + \alpha^2 x^2}\right]\Bigg|_0^\infty = \frac{-2\ln \alpha}{1 - \alpha^2}.$$

Thus, since $J(\alpha)$ is defined for $\alpha = 0$ and $J(0) = 0$

$$J(\alpha) = \int_0^\alpha \frac{-2\ln t}{1-t^2}\, dt\;.$$

Integrating by parts we find

$$J(\alpha) = \ln\alpha \ln\frac{1-\alpha}{1+\alpha} + \int_0^\alpha \left(\frac{\ln(1+t)}{t} - \frac{\ln(1-t)}{t} \right) dt$$

$$= \ln\alpha \ln\frac{1-\alpha}{1+\alpha} + \mathrm{Li}_2(\alpha) - \mathrm{Li}_2(-\alpha)$$

as desired.

This integral appears in the table [43] as entry 2.7.4(12) in a different form

$$2\int_0^\infty \frac{\arctan\alpha x}{1+x^2}\, dx = \frac{\pi^2}{3} - \frac{1}{2}\ln^2(1+\alpha) - \mathrm{Li}_2\left(\frac{1}{1+\alpha} \right) - \mathrm{Li}_2(1-\alpha)$$

which is less helpful for solving the Basel problem.

Example 2.5.3

For every $|\alpha| \le 1$ we have

(2.37) $\qquad 2\int_0^1 \frac{\arctan\alpha x}{1+x^2}\, dx = \sum_{n=0}^\infty (\ln 2 - H_n^-)\frac{\alpha^{2n+1}}{2n+1}$

where

$$H_n^- = 1 - \frac{1}{2} + \frac{1}{3} + \ldots + \frac{(-1)^{n-1}}{n} \;\; (n \ge 1),\; H_0^- = 0$$

are the skew-harmonic numbers. In particular, when $\alpha = 1$

(2.38)
$$\frac{\pi^2}{16} = \sum_{n=0}^{\infty} (\ln 2 - H_n^-) \frac{1}{2n+1}.$$

The numbers H_n^- are the partial sums in the series

$$\ln 2 = \sum_{k=1}^{\infty} \frac{(-1)^{k-1}}{k}$$

and we can also write

$$\ln 2 - H_n^- = \sum_{k=n+1}^{\infty} \frac{(-1)^{k-1}}{k}.$$

This integral does not solve the Basel problem, but the representation (2.38) deserves to be mentioned.

Proof of (2.38). As before, we first assume that $|\alpha| < 1$ and at the end we drop this restriction. Calling the left-hand side $G(\alpha)$ we have

$$G'(\alpha) = \int_0^1 \frac{2x\,dx}{(1+\alpha^2 x^2)(1+x^2)} = \frac{2}{1-\alpha^2} \int_0^1 \left\{ \frac{x}{1+x^2} - \frac{\alpha^2 x}{1+\alpha^2 x^2} \right\} dx$$

$$= \frac{\ln 2}{1-\alpha^2} - \frac{\ln(1+\alpha^2)}{1-\alpha^2} = \sum_{n=0}^{\infty} (\ln 2 - H_n^-)\alpha^{2n}$$

since the generating function of the skew-harmonic numbers is

$$\frac{\ln(1+t)}{1-t} = \sum_{n=0}^{\infty} H_n^- t^n \quad (|t| < 1)$$

(see Section 5.7.2). Now integration brings to (2.38). The proof is completed.

2.5.3 Several integrals with logarithms

We present here several integrals which are associated with the integrals in the previous examples.

Example 2.5.4

For every $0 \le \alpha \le 1$

$$(2.39) \qquad h(\alpha) = \int_0^\infty \frac{\ln(1+\alpha x)}{x(1+x)} dx = \ln \alpha \ln(1-\alpha) + \mathrm{Li}_2(\alpha)$$

where the function $\ln \alpha \ln(1-\alpha)$ is defined for $\alpha = 0$ and $\alpha = 1$ in terms of limits and these limits are zeros. For $\alpha = 1$ we have

$$\int_0^\infty \frac{\ln(1+x)}{x(1+x)} dx = \mathrm{Li}_2(1) = \zeta(2) = \frac{\pi^2}{6}$$

but we cannot use this result for directly solving the Basel problem, because we cannot obtain a multiple of π^2 on the left-hand side. Such things were possible in the equations (2.35) and (2.36) because of the trigonometric nature of those integrals.

The last integral is entry 4.291(13) in [25] and is a particular case of 2.6.10 (52) in [43].

For the proof of (2.39) we compute

$$h'(\alpha) = \int_0^\infty \frac{dx}{(1+\alpha x)(1+x)} = \frac{1}{1-\alpha} \ln \frac{1+x}{1+\alpha x} \Big|_0^\infty$$

$$= \frac{1}{1-\alpha} \ln \frac{1}{\alpha} = \frac{-\ln \alpha}{1-\alpha}.$$

Integrating by parts

$$h(\alpha) = \int_0^\alpha \frac{-\ln t}{1-t} dt = \ln \alpha \ln(1-\alpha) - \int_0^\alpha \frac{\ln(1-t)}{t} dt$$

$$= \ln \alpha \ln(1-\alpha) + \text{Li}_2(\alpha).$$

A problem for the reader: Prove that

$$\int_0^\infty \frac{\ln(1+x)}{x(1+x)}\, dx = \frac{\pi^2}{6}$$

by using the substitution $x = \dfrac{t}{1-t}$.

Example 2.5.5

We feel obliged to investigate a close cousin to the above integral. Here we prove that for every $-1 \le \alpha \le 1$

$$(2.40) \qquad g(\alpha) = \int_0^1 \frac{\ln(1+\alpha x)}{x(1+x)}\, dx = \text{Li}_2\left(\frac{1}{2}\right) - \text{Li}_2\left(\frac{1-\alpha}{2}\right).$$

When $\alpha = 1$ we get

$$\int_0^1 \frac{\ln(1+x)}{x(1+x)}\, dx = \text{Li}_2\left(\frac{1}{2}\right)$$

and we will use our integral to evaluate this value of the dilogaithm.

Proof of (2.40). Like in the previous example we have

$$g'(\alpha) = \int_0^1 \frac{dx}{(1+\alpha x)(1+x)} = \frac{1}{1-\alpha} \ln \frac{1+x}{1+\alpha x}\bigg|_0^1 = \frac{1}{1-\alpha} \ln \frac{2}{1+\alpha}.$$

We notice that

$$\frac{d}{d\alpha} \text{Li}_2\left(\frac{1-\alpha}{2}\right) = \frac{-1}{1-\alpha} \sum_{n=1}^\infty \frac{1}{n}\left(\frac{1-\alpha}{2}\right)^n = \frac{1}{1-\alpha} \ln\left(\frac{1+\alpha}{2}\right)$$

and the conclusion is

$$g'(\alpha) = -\frac{d}{d\alpha} \text{Li}_2\left(\frac{1-\alpha}{2}\right)$$

$$g(\alpha) = -\text{Li}_2\left(\frac{1-\alpha}{2}\right) + C.$$

With $\alpha = 0$ we find $C = \text{Li}_2\left(\frac{1}{2}\right)$. Done!

Now we write

$$\text{Li}_2\left(\frac{1}{2}\right) = \int_0^1 \frac{\ln(1+x)}{x(1+x)} dx = \int_0^1 \frac{\ln(1+x)}{x} dx - \int_0^1 \frac{\ln(1+x)}{1+x} dx$$

$$= \int_0^1 \frac{1}{x}\left\{\sum_{n=1}^\infty \frac{(-1)^{n-1}x^n}{n}\right\} dx - \frac{1}{2}\ln^2(1+x)\Big|_0^1$$

$$= \sum_{n=1}^\infty \frac{(-1)^{n-1}}{n}\int_0^1 x^{n-1}dx - \frac{1}{2}\ln^2 2 = \sum_{n=1}^\infty \frac{(-1)^{n-1}}{n^2} - \frac{1}{2}\ln^2 2.$$

This way

$$\text{Li}_2\left(\frac{1}{2}\right) = \frac{1}{2}\zeta(2) - \frac{1}{2}\ln^2 2$$

or, in terms of series

$$\sum_{n=1}^\infty \frac{1}{2^n n^2} = \frac{1}{2}\left(\sum_{n=1}^\infty \frac{1}{n^2} - \ln^2 2\right).$$

A nice and useful result by itself!

We can evaluate the integral in (2.40) also in a different way. Using the Taylor series for $\ln(1+\alpha x)$ we write

$$\int_0^1 \frac{\ln(1+\alpha x)}{x(1+x)} dx = \int_0^1 \left\{\sum_{n=0}^\infty \frac{(-1)^n \alpha^{n+1} x^n}{n+1}\right\} \frac{dx}{1+x}$$

$$= \sum_{n=0}^{\infty} \frac{(-1)^n \alpha^{n+1}}{n+1} \left\{ \int_0^1 \frac{x^n dx}{1+x} \right\} = \sum_{n=0}^{\infty} \frac{\alpha^{n+1}}{n+1} \left(\ln 2 - H_n^- \right)$$

since

$$\int_0^1 \frac{x^n dx}{1+x} = \int_0^1 \left\{ \sum_{k=0}^{\infty} (-1)^k x^{n+k} \right\} dx$$

$$= \sum_{k=0}^{\infty} \frac{(-1)^k}{n+k+1} = (-1)^n \sum_{j=n+1}^{\infty} \frac{(-1)^{j-1}}{j}.$$

Comparing this to (2.40) we conclude that for $-1 \le \alpha \le 1$

(2.41) $$\mathrm{Li}_2\left(\frac{1}{2}\right) - \mathrm{Li}_2\left(\frac{1-\alpha}{2}\right) = \sum_{n=0}^{\infty} \frac{\left(\ln 2 - H_n^- \right) \alpha^{n+1}}{n+1}.$$

In particular, for $\alpha = 1$ we obtain an interesting series which is a good company to (2.38)

(2.42) $$\sum_{n=0}^{\infty} \frac{\ln 2 - H_n^-}{n+1} = \frac{\pi^2}{12} - \frac{1}{2} \ln^2 2.$$

With $\alpha = -1$ in (2.41) we find also

(2.43) $$\sum_{n=0}^{\infty} \frac{(-1)^n \left(\ln 2 - H_n^- \right)}{n+1} = \frac{\pi^2}{6}.$$

More series with skew-harmonic numbers like (2.38), (2.42), and (2.43) can be found in Section 5.7.2.

A problem for the reader: Prove that

$$\int_0^{\infty} \frac{\sqrt{x} \ln x}{(1+x)^2} dx = \pi.$$

Example 2.5.6.

In contrast to the evaluation of $\zeta(2)$ in (2.34), the exact value of

$$\zeta(3) = 1 + \frac{1}{2^3} + \frac{1}{3^3} + \dots$$

is still unknown. Any information about this series is quite valuable. The number $\zeta(3)$ is known as Apéry's constant. The French mathematician Roger Apéry (1916-1994) proved in 1979 that this series represents an irrational number.

Euler found the exact values of $\zeta(2), \zeta(4), \dots$ (see Euler's formula in Section 4.9). The exact values at the odd integers $\zeta(2n+1)$, $n = 1, 2, \dots$ are unknown.

In this example we will present two logarithmic integrals which lead to four interesting representations of Apéry's constant $\zeta(3)$.

Lemma 2.1. *For every* $|\beta| \le 1$ *and every* $p \ge 0$

(2.44) $$\int_0^1 \frac{1}{x} (\log x)^p \log \frac{1 - \beta x}{1 + \beta x} dx$$

$$= (-1)^{p+1} \Gamma(p+1)\left\{\mathrm{Li}_{p+2}(\beta) - \mathrm{Li}_{p+2}(-\beta)\right\},$$

$$\int_0^1 \frac{1}{x} (\log x)^p \log(1 - \beta x)\, dx = (-1)^{p+1} \Gamma(p+1) \mathrm{Li}_{p+2}(\beta).$$

(The second integral is entry 2.6.19 (6) in [43].) Here

$$\mathrm{Li}_p(x) = \sum_{n=1}^{\infty} \frac{x^n}{n^p}$$

is the polylogarithm.

Proof. We start with the first integral. The substitution $x = e^{-t}$ transforms it in the following manner

$$\int_0^1 \frac{1}{x} (\log x)^p \, \log \frac{1-\beta x}{1+\beta x} \, dx$$

$$= (-1)^{p+1} \int_0^\infty t^p \left\{ -\log(1-\beta e^{-t}) + \log(1+\beta e^{-t}) \right\} dt$$

$$= (-1)^{p+1} \sum_{n=1}^\infty \frac{\beta^n}{n} \left\{ \int_0^\infty t^p \, e^{-nt} dt \right\} + (-1)^{p+1} \sum_{n=1}^\infty \frac{(-1)^{n-1} \beta^n}{n} \left\{ \int_0^\infty t^p \, e^{-nt} dt \right\}$$

$$= (-1)^{p+1} \Gamma(p+1) \left(\sum_{n=1}^\infty \frac{\beta^n}{n} \left\{ \frac{1}{n^{p+1}} \right\} + \sum_{n=1}^\infty \frac{(-1)^{n-1} \beta^n}{n} \left\{ \frac{1}{n^{p+1}} \right\} \right)$$

$$= (-1)^{p+1} \Gamma(p+1) \left(\mathrm{Li}_{p+2}(\beta) - \mathrm{Li}_{p+2}(-\beta) \right).$$

The same substitution in the second integral provides

$$\int_0^1 \frac{1}{x} (\log x)^p \, \log(1-\beta x) \, dx = (-1)^{p+1} \int_0^\infty t^p \left\{ -\log(1-\beta e^{-t}) \right\} dt$$

$$= (-1)^{p+1} \sum_{n=1}^\infty \frac{\beta^n}{n} \left\{ \int_0^\infty t^p e^{-nt} dt \right\} = (-1)^{p+1} \Gamma(p+1) \sum_{n=1}^\infty \frac{\beta^n}{n} \left\{ \frac{1}{n^{p+1}} \right\}$$

$$= (-1)^{p+1} \Gamma(p+1) \mathrm{Li}_{p+2}(\beta)$$

and the lemma is proved.

Now we present four representations of Apéry's constant $\zeta(3)$.

Proposition 2.2. *We have the representations*

$$(2.45) \qquad \zeta(3) = \frac{4}{7} \int_0^\infty \frac{1}{t} \arctan t \, \arctan \frac{1}{t} \, dt$$

$$(2.46) \qquad \zeta(3) = \frac{8}{7} \int_0^1 \frac{1}{t} \arctan t \, \arctan \frac{1}{t} \, dt$$

(2.47) $$\zeta(3) = \frac{1}{2}\int_0^\infty \frac{1}{t}\log(1+t)\log\left(1+\frac{1}{t}\right)dt$$

(2.48) $$\zeta(3) = \int_0^1 \frac{1}{t}\log(1+t)\log\left(1+\frac{1}{t}\right)dt .$$

The starting point in the proof is equation (2.36). With the substitution $t = \alpha x$ it takes the form

$$2\alpha\int_0^\infty \frac{\arctan t}{\alpha^2 + t^2}\,dt = \log\alpha\,\log\frac{1-\alpha}{1+\alpha} + \mathrm{Li}_2(\alpha) - \mathrm{Li}_2(-\alpha).$$

We divide both sides by α and integrate with respect to α from 0 to 1

$$2\int_0^\infty \arctan t\left\{\int_0^1 \frac{d\alpha}{\alpha^2 + t^2}\right\}dt$$

$$= \int_0^1 \frac{1}{\alpha}\log\alpha\,\log\frac{1-\alpha}{1+\alpha}\,d\alpha + \int_0^1 \frac{1}{\alpha}\{\mathrm{Li}_2(\alpha) - \mathrm{Li}_2(-\alpha)\}\,d\alpha .$$

Evaluating these integrals (using the lemma for the second one) we come to the equation

$$\int_0^\infty \arctan t\,\arctan\frac{1}{t}\frac{dt}{t} = \mathrm{Li}_3(1) - \mathrm{Li}_3(-1) = \frac{7}{4}\zeta(3)$$

which is (2.45). Note that

$$\int_0^1 \frac{\mathrm{Li}_2(t)}{t}\,dt = \mathrm{Li}_3(t)\big|_0^1 = \mathrm{Li}_3(1) = \zeta(3),\ \ \mathrm{Li}_3(-1) = -\frac{3}{4}\zeta(3).$$

We will transform the integral in (2.45) now. First we split it this way

$$\int_0^\infty = \int_0^1 + \int_1^\infty$$

and then in the last integral we make the substitution $x = \dfrac{1}{t}$ to get

$$\int_0^\infty \frac{1}{t} \arctan t \, \arctan \frac{1}{t} \, dt = 2\int_0^1 \frac{1}{t} \arctan t \, \arctan \frac{1}{t} \, dt$$

and now (2.46) comes from (2.45).

For equation (2.47) we use the integral

$$\alpha \int_0^\infty \frac{\log(1+t)}{t(\alpha+t)} dt = \log \alpha \log(1-\alpha) + \mathrm{Li}_2(\alpha)$$

which is equation (2.39) from Example 2.5.4 after the substitution $t = \alpha x$. Dividing by α and integrating with respect to α from 0 to 1 we write

$$\int_0^\infty \frac{\log(1+t)}{t} \left\{ \int_0^1 \frac{d\alpha}{\alpha+t} \right\} dt$$

$$= \int_0^1 \frac{1}{\alpha} \log \alpha \log(1-\alpha) d\alpha + \int_0^1 \frac{\mathrm{Li}_2(\alpha)}{\alpha} d\alpha.$$

This way we have from (2.44)

$$\int_0^\infty \frac{\log(1+t)}{t} \log\left(1+\frac{1}{t}\right) dt = \mathrm{Li}_3(1) + \mathrm{Li}_3(1) = 2\zeta(3).$$

In the same way as above we show that

$$\int_0^\infty \frac{\log(1+t)}{t} \log\left(1+\frac{1}{t}\right) dt = 2\int_0^1 \frac{\log(1+t)}{t} \log\left(1+\frac{1}{t}\right) dt$$

and the proposition is proved.

(See also [10].)

Problems for the reader: Prove the following representations of Apéry's constant

$$\zeta(3) = \frac{1}{2}\int_0^\infty \frac{x^2}{e^x - 1}\,dx$$

(hint: expand $(1 - e^{-x})^{-1}$ as geometric series and integrate term by term)

$$\zeta(3) = \frac{2}{3}\int_0^\infty \frac{x^2}{e^x + 1}\,dx$$

(expand $(1 + e^{-x})^{-1}$ as geometric series and integrate term by term)

$$\zeta(3) = \frac{2}{7}\int_0^\infty \frac{x^2}{\sinh x}\,dx$$

(write $\sinh x$ in terms of exponentials)

$$\zeta(3) = 2\int_0^\infty x^2(\coth x - 1)\,dx$$

$$\zeta(3) = \frac{8}{7}\int_0^\infty x(\ln 4 - x\cot x)\,dx$$

(hint: Example 3.2.1 from the next chapter could be helpful).

Example 2.5.7

This is in fact a problem for the reader. We introduce one symmetrical and nice integral whose proof will be left to the reader.

In Example 2.5.5. we showed that

$$\frac{d}{dt}\text{Li}_2\left(\frac{1-t}{2}\right) = \frac{1}{1-t}\ln\left(\frac{1+t}{2}\right)$$

which gives

(2.49) $$\int_0^x \frac{1}{1-t} \ln\left(\frac{1+t}{2}\right) dt = \mathrm{Li}_2\left(\frac{1-x}{2}\right) - \mathrm{Li}_2\left(\frac{1}{2}\right).$$

The logarithm on the left-hand side can be written as $\ln(1+t) - \ln 2$, so the equation can be put in the form

(2.50) $$\int_0^x \frac{\ln(1+t)}{1-t} dt = \mathrm{Li}_2\left(\frac{1-x}{2}\right) - \mathrm{Li}_2\left(\frac{1}{2}\right) - \ln 2 \ln(1-x).$$

Here is the problem for the reader. Prove that for every $|x| < 1$

(2.51) $$\int_0^x \ln(1+t) \ln(1-t) dt$$

$$= \mathrm{Li}_2\left(\frac{1-x}{2}\right) - \mathrm{Li}_2\left(\frac{1+x}{2}\right) + (1 - \ln 2) \ln \frac{1-x}{1+x}$$

$$+ x \ln(1+x) \ln(1-x) - x \ln(1-x^2) + 2x.$$

Evaluate the limit of the right-hand side for $x \to 1$ to show that

$$\int_0^1 \ln(1+t) \ln(1-t) dt = 2 - \frac{\pi^2}{6} + (\ln 2)^2 - 2\ln 2.$$

Hint: Use integration by parts and equation (2.50).

Note that (2.51) can also be proved by differentiating both sides.

2.6   Some Theorems

Here we present some theorems that can be used to justify differentiation with respect to a parameter inside the integral.

Theorem A. *Suppose the function $f(\alpha, x)$ is defined and continuous on the rectangle $[a,b] \times [c,d]$ together with its partial derivative $f_\alpha(\alpha, x)$. Then*

$$\frac{d}{d\alpha}\int_c^d f(\alpha,x)\,dx = \int_c^d f_\alpha'(\alpha,x)\,dx.$$

In order to apply this theorem in the case of improper integrals we have to require uniform convergence of the integral with respect to the variable α. A simple sufficient condition for uniform convergence is given by the next theorem.

Theorem B. *Suppose the function $f(\alpha,x)$ is continuous on $[a,b]\times[0,\infty)$ and the function $g(x)\geq 0$ is integrable on $[0,\infty)$. If*

$$|f(\alpha,x)| \leq g(x)$$

for all $\alpha \in [a,b]$ and all $x \geq 0$, then the integral

$$\int_0^\infty f(\alpha,x)\,dx$$

is uniformly convergent on $[a,b]$.

Theorem C. *Suppose the function $f(\alpha,x)$ is defined and continuous on $[a,b]\times[0,\infty)$ together with its partial derivative $f_\alpha(\alpha,x)$. In this case*

$$\frac{d}{d\alpha}\int_0^\infty f(\alpha,x)\,dx = \int_0^\infty f_\alpha(\alpha,x)\,dx$$

When the first integral is convergent and the second integral is uniformly convergent on $[a,b]$.

The case of improper integrals on finite intervals is treated in the same way. For proofs and details we refer to [23, 28, 29, 50].

Chapter 3

Solving Logarithmic Integrals by Using Fourier Series

3.1 Introduction

Many integrals can be evaluated by expanding the integrand in an appropriate series and integrating this series term by term. This is a classical method, some examples can be found in the books [7] and [37].

In this chapter we present a selection of examples using this method. We solve integrals which are interesting and challenging. Most of these integrals can be found in the popular tables [25] and [43], but some are rare, and some are possibly new. We use standard concepts from analysis, simple Fourier series, and the standard expansion of the logarithm $\ln(1 + x)$. Focusing mostly on the technical part, we leave some details to the reader.

A prerequisite for this chapter is some general knowledge about the gamma function

$$\Gamma(z) = \int_0^\infty t^{z-1}e^{-t}dt \quad (\operatorname{Re} z > 0).$$

The digamma function $\psi(z) = \Gamma'(z)/\Gamma(z)$ will also be used, as well as the Riemann zeta function $\zeta(z)$ defined by the series

$$\zeta(z) = \sum_{n=1}^\infty \frac{1}{n^z}$$

103

for $\operatorname{Re} z > 1$ and then extended as analytic on the entire complex plane with a simple pole at $z = 1$.

Euler's dilogarithm $\operatorname{Li}_2(x)$ and the polylogarithm $\operatorname{Li}_p(x)$ also appear in this chapter

$$\operatorname{Li}_2(x) = \sum_{n=1}^{\infty} \frac{x^n}{n^2}, \quad \operatorname{Li}_p(x) = \sum_{n=1}^{\infty} \frac{x^n}{n^p}.$$

We start with some well-known log-sine integrals and proceed with more advanced cases. Among other things we consider Euler's log-gamma integral

$$\int_0^1 \ln \Gamma(x)\,dx$$

and also the related integrals

$$\int_0^{1/2} \ln \Gamma(x)\,dx \quad \text{and} \quad \int_0^{1/4} \ln \Gamma(x)\,dx.$$

Section 3.3 is devoted to a special Binet type formula for the log-gamma function $\ln \Gamma(x)$ which is then used for evaluating a challenging integral. The term by term integration that appears in our examples is justified by the theorems collected in Section 3.4.

3.2 Examples

Example 3.2.1

In this introductory example we evaluate several well-known log-sine and log-cosine integrals. The evaluation is based on two Fourier series (see, for instance, [46, p. 148])

$$(3.1) \qquad \ln(\sin t) = -\ln 2 - \sum_{n=1}^{\infty} \frac{\cos(2nt)}{n}, \ 0 < t < \pi$$

$$\ln(\cos t) = -\ln 2 - \sum_{n=1}^{\infty} \frac{(-1)^{n-1} \cos(2nt)}{n}, \quad \frac{-\pi}{2} < t < \frac{\pi}{2}.$$

Integrating the first series between 0 and $\frac{\pi}{2}$ and then between 0 and π we find the two integrals

$$\int_0^{\pi/2} \ln(\sin t) dt = \frac{-\pi}{2} \ln 2, \quad \int_0^{\pi} \ln(\sin t) dt = -\pi \ln 2$$

while integration between 0 and $\frac{\pi}{4}$ yields

$$\int_0^{\pi/4} \ln(\sin t) dt = \frac{-\pi}{4} \ln 2 - \frac{1}{2} \sum_{n=1}^{\infty} \frac{1}{n^2} \sin \frac{\pi n}{2} = \frac{-\pi}{4} \ln 2 - \frac{1}{2} G.$$

Here

$$G = \sum_{n=1}^{\infty} \frac{1}{n^2} \sin \frac{\pi n}{2} = \sum_{k=0}^{\infty} \frac{(-1)^k}{(2k+1)^2}$$

is Catalan's constant.

Since the function $\ln(\sin t)$ is not defined at zero, the integration can be justified by first integrating on the interval $[\alpha, \pi/2]$, $0 < \alpha < \pi/2$, and then setting $\alpha \to 0$. This remark applies to the second and third integrals and to all similar cases later. The integrals are convergent improper integrals.

In the same way, integrating the second series in (3.1) we find

$$\int_0^{\pi/2} \ln(\cos t) dt = \frac{-\pi}{2} \ln 2$$

$$\int_0^{\pi/4} \ln(\cos t) dt = \frac{-\pi}{4} \ln 2 + \frac{1}{2} \sum_{n=1}^{\infty} \frac{1}{n^2} \sin \frac{\pi n}{2} = \frac{-\pi}{4} \ln 2 + \frac{1}{2} G.$$

All these integrals are present in [25], in the group of entries 4.224.

When we multiply the first series in (3.1) by t and integrate between 0 and $\pi/4$ we find

$$\int_0^{\pi/4} t\ln(\sin t)dt = \frac{-\pi^2}{32}\ln 2 - \frac{1}{2}\sum_{n=1}^{\infty}\frac{1}{n}\int_0^{\pi/4} t\cos(2nt)dt$$

$$= \frac{-\pi^2}{32}\ln 2 - \frac{\pi}{8}\sum_{n=1}^{\infty}\frac{1}{n^2}\sin\frac{\pi n}{2} + \frac{1}{4}\sum_{n=1}^{\infty}\frac{1}{n^3}\left(1-\cos\frac{\pi n}{2}\right)$$

$$= \frac{-\pi^2}{32}\ln 2 - \frac{\pi}{8}G + \frac{35}{128}\zeta(3).$$

That is,

$$\int_0^{\pi/4} t\ln(\sin t)dt = \frac{-\pi^2}{32}\ln 2 - \frac{\pi}{8}G + \frac{35}{128}\zeta(3)$$

where $\zeta(s)$ is Riemann's zeta function. This interesting integral connecting three important constants is entry 2.6.34(2) in [43]. Two more simple integrals (entries 2.6.34(9) and 2.6.34(28) in [43]) are obtained in the same way

$$\int_0^{\pi/2} t\ln(\sin t)dt = \frac{-\pi^2}{8}\ln 2 + \frac{7}{16}\zeta(3)$$

$$\int_0^{\pi} t\ln(\sin t)dt = \frac{-\pi^2}{2}\ln 2 .$$

We will return to log-sine integrals later in Example 3.9.

Problem for the reader: Using the first series in (3.1) prove that

$$\int_0^{\pi/2} \sin t\ln(\sin t)dt = \ln 2 - 1 .$$

Hint: The series

$$\sum_{n=1}^{\infty} \frac{1}{n(4n^2-1)} = 2\ln 2 - 1$$

might be helpful.

Another way to prove this integral is to compute the antiderivative

$$F(t) = \int \sin t \ln(\sin t) dt = -\int \ln(\sin t) d \cos t$$

$$= -\cos t \ln(\sin t) + \int \frac{\cos^2 t}{\sin t} dt$$

$$= -\cos t \ln(\sin t) + \int \frac{1}{\sin t} dt - \int \sin t dt$$

$$= -\cos t \ln(\sin t) + \ln(\tan \frac{t}{2}) + \cos t$$

and then to evaluate

$$F\left(\frac{\pi}{2}\right) - \lim_{t \to 0} F(t) = -\lim_{t \to 0} F(t).$$

Computing this limit is a decent calculus exercise.

Note that the substitution $x = \sin t$ gives also

$$\int_0^1 \frac{x \ln x}{\sqrt{1-x^2}} dx = \ln 2 - 1.$$

Another problem for the reader: Show that

$$\int_0^1 \frac{\ln x}{\sqrt{4-x^2}} dx = \frac{\pi}{6} \ln 2 + \int_0^{\pi/6} \ln(\sin t) dt$$

by using the substitution $x = 2\sin t$. Further, using the series in (3.1) show that

$$\int_0^1 \frac{\ln x}{\sqrt{4-x^2}}\,dx = -\frac{1}{2}\sum_{n=1}^\infty \frac{1}{n^2}\sin\frac{\pi n}{3} = -\frac{1}{2}\mathrm{Cl}_2\left(\frac{\pi}{3}\right)$$

where

$$\mathrm{Cl}_2(x) = \sum_{n=1}^\infty \frac{1}{n^2}\sin(nx)$$

is Clausen's function. For another evaluation of this integral see entry 4.1.5(1) in [20].

Next we evaluate several log-cotangent integrals.

Example 3.2.2

We evaluate here the integral

$$\int_0^{\pi p} \cos x \ln \cot \frac{x}{2}\,dx$$

for $p = 1$, $p = \frac{1}{2}$, and $p = \frac{1}{4}$. We will use the series representation

(3.2) $$\cos x \ln \cot \frac{x}{2} + \frac{\pi}{2}\sin x = 2\sum_{k=0}^\infty \frac{\cos(2kx)}{2k+1}$$

$$= 2 + 2\sum_{k=1}^\infty \frac{\cos(2kx)}{2k+1}$$

for $0 < x < \pi$. This series is a slight modification of the series in entry 17.2.14 in Hansen's table [26]. Integrating we find

$$\int_0^\pi \cos x \ln \cot \frac{x}{2}\,dx - \frac{\pi}{2}\cos x \Big|_0^\pi = 2\pi + \sum_{k=1}^\infty \frac{\sin(4k\pi)}{k(2k+1)} = 2\pi$$

that is,

(3.3)
$$\int_0^\pi \cos x \ln \cot \frac{x}{2} dx = \pi .$$

Integration of (3.2) between 0 and $\pi/2$ gives

$$\int_0^{\pi/2} \cos x \ln \cot \frac{x}{2} dx - \frac{\pi}{2} \cos x \Big|_0^{\pi/2} = \pi$$

and therefore,

(3.4)
$$\int_0^{\pi/2} \cos x \ln \cot \frac{x}{2} dx = \frac{\pi}{2} .$$

Integration of (3.2) between 0 and $\pi/4$ brings to

$$\int_0^{\pi/4} \cos x \ln \cot \frac{x}{2} dx - \frac{\pi}{2} \cos x \Big|_0^{\pi/4} = \frac{\pi}{2} + \sum_{k=1}^\infty \frac{1}{k(2k+1)} \sin\left(\frac{k\pi}{2}\right)$$

and from here, changing the index of summation to $k = 2n + 1$ (for even k the terms are zeros) we write

$$\int_0^{\pi/4} \cos x \ln \cot \frac{x}{2} dx = \frac{\pi\sqrt{2}}{4} + \sum_{n=0}^\infty \frac{(-1)^n}{(2n+1)(4n+3)} .$$

We can evaluate this series by using partial fractions

(3.5)
$$\sum_{n=0}^\infty \frac{(-1)^n}{(2n+1)(4n+3)} = \sum_{n=0}^\infty \frac{(-1)^n}{2n+1} - 2\sum_{n=0}^\infty \frac{(-1)^n}{4n+3}$$

$$= \frac{\pi}{4} - \frac{\sqrt{2}}{4}(\pi - 2\ln(1+\sqrt{2})) .$$

The value of the second series is taken from entry 5.4.24 in [26].

(3.6)
$$\sum_{n=0}^\infty \frac{(-1)^n}{4n+3} = \frac{\sqrt{2}}{8}[\pi - 2\ln(1+\sqrt{2})] .$$

Finally,

(3.7) $$\int_0^{\pi/4} \cos x \ln \cot \frac{x}{2} dx = \frac{\pi}{4} + \frac{\sqrt{2}}{2} \ln(1 + \sqrt{2}).$$

Comment. The values of (3.3) and (3.4) are recognized by Maple. Equation (3.3) can be derived from entry 2.6.39 in [43]. The integral in (3.7) seems to be new.

Example 3.2.3

Very nicely, it turns out that the series (3.5) can be used to evaluate the challenging integral

$$\int_0^1 \arctan(x^2) dx .$$

Starting from the Maclaurin series for this arctangent

$$\arctan(x^2) = \sum_{n=0}^{\infty} \frac{(-1)^n}{2n+1} x^{4n+2} \quad (|x| \le 1)$$

and integrating both sides between 0 and 1 we find

$$\int_0^1 \arctan(x^2) dx = \sum_{n=0}^{\infty} \frac{(-1)^n}{(2n+1)(4n+3)}$$

$$= \frac{\pi}{4} - \frac{\sqrt{2}}{4} \left(\pi - 2 \ln(1 + \sqrt{2}) \right)$$

or

(3.8) $$\int_0^1 \arctan(x^2) dx = \frac{\pi(1-\sqrt{2})}{4} + \frac{\sqrt{2}}{2} \ln(1 + \sqrt{2}).$$

The series (3.6) which is crucial for the above evaluation has an integral form

(3.9)
$$\int_0^1 \frac{x^2}{1+x^4}\,dx = \sum_{n=0}^{\infty} \frac{(-1)^n}{4n+3}\,.$$

This follows immediately from the geometric series expansion

$$\frac{x^2}{1+x^4} = x^2 \sum_{n=0}^{\infty}(-x^4)^n = \sum_{n=0}^{\infty}(-1)^n x^{4n+2}, \quad |x|<1\,.$$

Integrals like (3.9) are usually solved by partial fractions. However, we can reduce (3.9) to (3.8) using integration by parts.

$$\int_0^1 \frac{x^2}{1+x^4}\,dx = \frac{1}{2}\int_0^1 \frac{x}{1+x^4}\,dx^2 = \frac{1}{2}\int_0^1 x\,d\arctan(x^2)$$

$$= \frac{1}{2}x\arctan(x^2)\Big|_0^1 - \frac{1}{2}\int_0^1 \arctan(x^2)\,dx$$

$$= \frac{\pi}{8} - \frac{1}{2}\left(\frac{\pi(1-\sqrt{2})}{4} + \frac{\sqrt{2}}{2}\ln(1+\sqrt{2})\right)$$

and after simplification

(3.10)
$$\int_0^1 \frac{x^2}{1+x^4}\,dx = \frac{\pi\sqrt{2}}{8} - \frac{\sqrt{2}}{4}\ln(1+\sqrt{2})\,.$$

Even more! We can use this result to evaluate one seemingly unrelated tough integral

$$\int_0^{\pi/4} \sqrt{\tan\theta}\,d\theta\,.$$

With the substitution $\tan\theta = x^2$, or $\theta = \arctan(x^2)$ we compute

(3.11)
$$\int_0^{\pi/4} \sqrt{\tan\theta}\,d\theta = 2\int_0^1 \frac{x^2}{1+x^4}\,dx = \frac{\pi\sqrt{2}}{4} - \frac{\sqrt{2}}{2}\ln(1+\sqrt{2})\,.$$

A remarkable connection!

Example 3.2.4

Now we consider another important series. For all $|\alpha| < 1$ and all real x

$$(3.12) \qquad \log(1 - 2\alpha\cos x + \alpha^2) = -2\sum_{k=1}^{\infty} \frac{\alpha^k \cos kx}{k}.$$

This series is well-known. It comes from the standard expansion of the logarithm

$$\log(1 - \alpha e^{ix}) = -\sum_{k=1}^{\infty} \frac{\alpha^k e^{ikx}}{k}$$

by equating the real parts on both sides when α is real. After that (3.12) extends also to complex α in the unit disk by analytic continuation.

We will use this series to evaluate several integrals. From (3.12) we find immediately

$$(3.13) \qquad \int_0^{\pi p} \log(1 - 2\alpha\cos x + \alpha^2)\, dx = 0$$

for every integer p. Setting $\alpha \to 1$ here yields

$$\int_0^{\pi p} \ln(2 - 2\cos x)\, dx = \int_0^{\pi p} \ln(1 - \cos x)\, dx + \int_0^{\pi p} \ln 2\, dx = 0$$

that is,

$$(3.14) \qquad \int_0^{\pi p} \ln(1 - \cos x)\, dx = -\pi p \ln 2.$$

Also, from (3.12)

$$(3.15) \qquad \int_0^{\pi/2} \log(1 - 2\alpha\cos x + \alpha^2)\, dx = -2\sum_{k=1}^{\infty} \frac{\alpha^k}{k^2}\sin\left(\frac{k\pi}{2}\right)$$

$$= -2\sum_{n=0}^{\infty} \frac{(-1)^n \alpha^{2n+1}}{(2n+1)^2}.$$

This series can be expressed in terms of the dilogarithm

$$\text{Li}_2(x) = \sum_{n=1}^{\infty} \frac{x^n}{n^2}.$$

We have

(3.16) $$\sum_{n=0}^{\infty} \frac{(-1)^n \alpha^{2n+1}}{(2n+1)^2} = \frac{1}{2i}\left(\text{Li}_2(i\alpha) - \text{Li}_2(-i\alpha)\right)$$

(3.17) $$\int_0^{\pi/2} \log(1 - 2\alpha\cos x + \alpha^2)dx = i\left(\text{Li}_2(i\alpha) - \text{Li}_2(-i\alpha)\right).$$

Setting $\alpha \to 1$ in (3.15) we derive also

$$\int_0^{\pi/2} \ln(1 - \cos x)\,dx + \int_0^{\pi/2} \ln 2\,dx = -2G$$

that is,

(3.18) $$\int_0^{\pi/2} \ln(1 - \cos x)\,dx = -\frac{\pi}{2}\ln 2 - 2G.$$

Now we return to (3.13). When α with $|\alpha| > 1$ is real we write

$$\ln(1 - 2\alpha\cos x + \alpha^2) = \ln[\alpha^2(\alpha^{-2} - 2\alpha^{-1}\cos x + 1)]$$

$$= 2\ln|\alpha| + \ln(1 - 2\alpha^{-1} + \alpha^{-2})$$

so that for $|\alpha| > 1$

(3.19) $$\int_0^{\pi p} \ln(1 - 2\alpha\cos x + \alpha^2)dx = 2\pi p \ln|\alpha|.$$

This is also true for $|\alpha| = 1$ by continuity. The two integrals (3.13) and (3.19) can be united in one, replacing α by β / α

(3.20) $\displaystyle\int_0^{\pi p} \ln(\beta^2 - 2\alpha\beta\cos x + \alpha^2)\,dx = 2\pi p \ln|\alpha|$

where $|\beta| \le |\alpha|$ and p is an integer. This integral is entry 2.6.36(14) in [43] and entry 4.224(14) in [25] (the answer in [43] is correct, while the answer in [25] needs a correction). For $p = 1$ this integral was evaluated in Chapter 2 by a differentiation with respect to α (see Example 2.2.18).

Example 3.2.5

We will use the series (3.12) for some more evaluations. Multiplying both sides in (3.12) by $\cos(mx)$, where m is a positive integer, we write

(3.21) $\displaystyle\cos(mx)\log(1 - 2\alpha\cos x + \alpha^2) = -2\sum_{k=1}^{\infty}\frac{\alpha^k \cos(mx)\cos(kx)}{k}$

$$= -\sum_{k=1}^{\infty}\frac{\alpha^k}{k}\big(\cos(m+k)x + \cos(m-k)x\big).$$

Now we integrate this equation with respect to x between 0 and πp, where p is any integer. There will be only one non-zero term on the right-hand side, the term where $k = m$. Thus we find

(3.22) $\displaystyle\int_0^{\pi p} \cos mx \log(1 - 2\alpha\cos x + \alpha^2)\,dx = \frac{-\pi p}{m}\alpha^m.$

This is entry 2.6.36(15) in [43] and entry 4.397(6) in [25].

It is remarkable that (3.13) does not follow from (3.22) for $m = 0$. This phenomenon is explained by the nature of the series in (3.21). We leave it to the reader to evaluate this integral for $|\alpha| > 1$.

These results can be extended in the following way: using the representations 1.320(1,5,7) in [25]

$$\cos^{2n}(mx) = \frac{1}{4^n}\left\{\binom{2n}{n} + 2\sum_{k=0}^{n-1}\binom{2n}{n}\cos(2nm - 2km)x\right\}$$

$$\cos^{2n-1}(mx) = \frac{1}{4^{n-1}}\sum_{k=0}^{n-1}\binom{2n-1}{n}\cos(2nm - 2km - m)x$$

$$\sin^{2n}(mx) = \frac{1}{4^n}\left\{\binom{2n}{n} + 2\sum_{k=0}^{n-1}\binom{2n}{n}(-1)^{n-k}\cos(2nm - 2km)x\right\}$$

we find correspondingly

$$\int_0^{\pi p}\cos^{2n}(mx)\log(1 - 2\alpha\cos x + \alpha^2)\,dx = \frac{-\pi p}{4^n m}\sum_{k=0}^{n-1}\binom{2n}{k}\frac{\alpha^{2m(n-k)}}{n-k}$$

$$\int_0^{\pi p}\cos^{2n-1}(mx)\log(1 - 2\alpha\cos x + \alpha^2)\,dx$$

$$= \frac{-\pi p}{4^{n-1} m}\sum_{k=0}^{n-1}\binom{2n-1}{k}\frac{\alpha^{m(2n-2k-1)}}{n-k-1}$$

$$\int_0^{\pi p}\sin^{2n}(mx)\log(1 - 2\alpha\cos x + \alpha^2)\,dx$$

$$= \frac{-\pi p}{4^n m}\sum_{k=0}^{n-1}\binom{2n}{k}\frac{(-1)^{n-k}\alpha^{2m(n-k)}}{n-k}$$

for any two positive integers n, m and any $|\alpha| < 1$.

Example 3.2.6

Substituting $\alpha = \pm e^{-\beta}$, $\beta > 0$, in (3.13) and writing

$$\ln(1 \pm 2e^{-\beta}\cos x + e^{-2\beta}) = \ln\left[e^{-\beta}2(\cosh\beta \pm \cos x)\right]$$

$$= -\beta + \ln 2 + \ln(\cosh\beta \pm \cos x)$$

we find for every integer p

(3.23) $$\int_0^{\pi p} \ln(\cosh \beta \pm \cos x) dx = (\beta - \ln 2) p\pi .$$

Example 3.2.7

We continue working with the expansion (3.12). Dividing both sides by $x^2 + b^2$, where $\operatorname{Re} b > 0$, we write

(3.24) $$\frac{\log(1 - 2\alpha \cos x + \alpha^2)}{x^2 + b^2} = -2\sum_{k=1}^{\infty} \frac{\alpha^k \cos kx}{k(x^2 + b^2)} .$$

Integrating term by term and changing the order of integration and summation we arrive at

$$\int_0^{\infty} \frac{\log(1 - 2\alpha \cos x + \alpha^2)}{x^2 + b^2} dx = -2\sum_{k=1}^{\infty} \frac{\alpha^k}{k} \int_0^{\infty} \frac{\cos kx}{x^2 + b^2} dx .$$

The integrals on the right-hand side are well-known (see (2.24))

(3.25) $$\int_0^{\infty} \frac{\cos kx}{x^2 + b^2} dx = \frac{\pi}{2b} e^{-bk}$$

so that

$$\int_0^{\infty} \frac{\log(1 - 2\alpha \cos x + \alpha^2)}{x^2 + b^2} dx = -\frac{\pi}{b}\sum_{k=1}^{\infty} \frac{\alpha^k e^{-bk}}{k} = -\frac{\pi}{b}\sum_{k=1}^{\infty} \frac{(\alpha e^{-b})^k}{k}$$

and finally

(3.26) $$\int_0^{\infty} \frac{\log(1 - 2\alpha \cos x + \alpha^2)}{x^2 + b^2} dx = \frac{\pi}{b}\log(1 - \alpha e^{-b}) .$$

This is entry 2.6.36(20) in [43]. Several other integrals in [25] and [43] similar to (3.26) can be solved in the same way.

Example 3.2.8

We will evaluate here some integrals by using the standard logarithmic series. Let $\operatorname{Re}\lambda > 0$. Then for every $x > 0$, $|a| \le 1$

(3.27) $$\log(1 - ae^{-\lambda x}) = -\sum_{n=1}^{\infty} \frac{a^n e^{-\lambda x n}}{n}.$$

Multiplying both sides by x^p ($p > -1$) and integrating form 0 to ∞ we find

$$\int_0^{\infty} x^p \log(1 - ae^{-\lambda x})\, dx = -\sum_{n=1}^{\infty} \frac{a^n}{n} \int_0^{\infty} x^p e^{-\lambda x n}\, dx = -\sum_{n=1}^{\infty} \frac{a^n \, \Gamma(p+1)}{n(\lambda n)^{p+1}}$$

$$= -\frac{\Gamma(p+1)}{\lambda^{p+1}} \sum_{n=1}^{\infty} \frac{a^n}{n^{p+2}} = -\frac{\Gamma(p+1)}{\lambda^{p+1}} \operatorname{Li}_{p+2}(a).$$

That is,

$$\int_0^{\infty} x^p \log(1 - ae^{-\lambda x})\, dx = -\frac{\Gamma(p+1)}{\lambda^{p+1}} \operatorname{Li}_{p+2}(a)$$

where $\operatorname{Li}_q(x)$ is the polylogarithm.

$$\operatorname{Li}_q(a) = \sum_{n=1}^{\infty} \frac{x^n}{n^q}.$$

This integral is equivalent to entry 4.316 (1) in [25] and entry 2.6.28 (1) in [43].

Next, for any real $b \ne 0$ we have

$$\sin bx \log(1 - e^{-\lambda x}) = -\sum_{n=1}^{\infty} \frac{e^{-\lambda x n} \sin bx}{n}.$$

From here, using the Laplace transform formula for the sine we find

$$\int_0^\infty \sin bx \log(1 - e^{-\lambda x}) \, dx = -\sum_{n=1}^\infty \frac{1}{n} \int_0^\infty e^{-\lambda x n} \sin bx \, dx$$

$$= -\sum_{n=1}^\infty \frac{b}{n(\lambda^2 n^2 + b^2)}.$$

The value of this series is known (entry 6.1.67 in Hansen's table [26])

$$(3.28) \quad \sum_{n=1}^\infty \frac{b}{n(\lambda^2 n^2 + b^2)} = \frac{1}{2b} \left[\psi\left(1 + \frac{ib}{\lambda}\right) + \psi\left(1 - \frac{ib}{\lambda}\right) + 2\gamma \right]$$

where $\psi(s) = \Gamma'(s)/\Gamma(s)$ is the digamma function and $\gamma = -\psi(1)$ is Euler's constant. Thus

$$(3.29) \quad \int_0^\infty \sin bx \log(1 - e^{-\lambda x}) \, dx = \frac{-1}{2b} \left[\psi\left(1 + \frac{ib}{\lambda}\right) + \psi\left(1 - \frac{ib}{\lambda}\right) + 2\gamma \right].$$

This integral is entry 2.6.40(6) in [43]. The similar integral 2.640(5) in [43] is proved the same way

$$\int_0^\infty \sin bx \log(1 + e^{-\lambda x}) \, dx$$

$$= \frac{-1}{4b} \left[4\ln 2 - \psi\left(1 + \frac{ib}{2\lambda}\right) - \psi\left(1 - \frac{ib}{2\lambda}\right) + \psi\left(\frac{\lambda + ib}{2\lambda}\right) + \psi\left(\frac{\lambda - ib}{2\lambda}\right) \right].$$

Next we integrate the expansion

$$\cos bx \log(1 - e^{-\lambda x}) = -\sum_{n=1}^\infty \frac{e^{-\lambda x n} \cos bx}{n}$$

to get

$$\int_0^\infty \cos bx \log(1 - e^{-\lambda x}) \, dx = -\sum_{n=1}^\infty \frac{1}{n} \int_0^\infty e^{-\lambda x n} \cos bx \, dx = -\sum_{n=1}^\infty \frac{\lambda}{\lambda^2 n^2 + b^2}.$$

Using again Hansen's table for the above series (entry 6.1.32)

$$\sum_{n=1}^{\infty} \frac{\lambda}{\lambda^2 n^2 + b^2} = \frac{-\lambda}{2b^2} + \frac{\pi}{2b} \coth\left(\frac{\pi b}{\lambda}\right)$$

we find

(3.30) $\qquad \displaystyle\int_0^{\infty} \cos bx \log(1 - e^{-\lambda x})\, dx = \frac{\lambda}{2b^2} - \frac{\pi}{2b} \coth\left(\frac{\pi b}{\lambda}\right).$

The expansion of $\coth(x)$ in partial fraction series will be presented later in Section 4.5.1.

Example 3.2.9

Here we show another application of the series in (3.28). Multiplying the first series in (3.1) by e^{at}, $a \neq 0$, we write

$$e^{at} \ln(\sin t) = -e^{at} \ln 2 - \sum_{n=1}^{\infty} \frac{e^{at} \cos(2nt)}{n}, \ 0 < t < \pi.$$

We will use now the antiderivative

$$\int e^{at} \cos(2nt)\, dt = \frac{e^{at}}{a^2 + 4n^2} \left(a \cos(2nt) + 2n \sin(2nt) \right)$$

to evaluate several log-sine integrals. Integrating the above series between 0 and π we compute

$$\int_0^{\pi} e^{at} \ln(\sin t)\, dt = \frac{1 - e^{a\pi}}{a} \left(\ln 2 + a^2 \sum_{n=1}^{\infty} \frac{1}{n(4n^2 + a^2)} \right).$$

According to (3.28) this becomes

(3.31) $\qquad \displaystyle\int_0^{\pi} e^{at} \ln(\sin t)\, dt$

$$= \frac{1 - e^{a\pi}}{a} \left\{ \ln 2 + \gamma + \frac{1}{2}\left[\psi\left(1 + \frac{ia}{\lambda}\right) + \psi\left(1 - \frac{ia}{\lambda}\right) \right] \right\}.$$

Multiplying the second series in (3.1) by e^{at} and integrating between $\dfrac{-\pi}{2}$ and $\dfrac{\pi}{2}$ we find

$$\int_{-\pi/2}^{\pi/2} e^{at} \ln(\cos t)\, dt = \frac{-2}{a}\sinh\frac{\pi a}{2}\left(\ln 2 + a^2 \sum_{n=1}^{\infty} \frac{1}{n(4n^2 + a^2)} \right)$$

and correspondingly,

(3.32) $$\int_{-\pi/2}^{\pi/2} e^{at} \ln(\cos t)\, dt$$

$$= \frac{-1}{a}\sinh\frac{\pi a}{2}\left\{ 2\ln 2 + 2\gamma + \psi\left(1 + \frac{ia}{\lambda}\right) + \psi\left(1 - \frac{ia}{\lambda}\right)\right\}.$$

Integrating between 0 and $\dfrac{\pi}{2}$ we find also

$$\int_{0}^{\pi/2} e^{at} \ln(\cos t)\, dt$$

$$= \frac{1 - e^{a\pi/2}}{a}\ln 2 - \sum_{n=1}^{\infty}\frac{(-1)^n}{n}\left[\frac{(-1)^n a e^{a\pi/2}}{4n^2 + a^2} - \frac{a}{4n^2 + a^2} \right]$$

$$= \frac{1 - e^{a\pi/2}}{a}\ln 2 - a e^{a\pi/2}\sum_{n=1}^{\infty}\frac{1}{n(4n^2 + a^2)} + a\sum_{n=1}^{\infty}\frac{(-1)^n}{n(4n^2 + a^2)}.$$

From Hansen's table [26] we take now entry 6.1.77, namely,

(3.33) $$\sum_{n=1}^{\infty}\frac{(-1)^n}{n(4n^2 + a^2)} = \frac{1}{2a^2}\left[\beta\left(1 + \frac{ia}{2}\right) + \beta\left(1 - \frac{ia}{2}\right) - 2\beta(1) \right]$$

where $\beta(s)$ is Nielsen's beta function – entry 8.370 in [25] (see below Section 4.5.2 for this function)

$$\beta(s) = \sum_{n=0}^{\infty} \frac{(-1)^n}{n+s} = \frac{1}{2}\left[\psi\left(\frac{s+1}{2}\right) + \psi\left(\frac{s}{2}\right)\right].$$

Therefore, from (3.28) and (3.33) we find

$$(3.34) \qquad \int_0^{\pi/2} e^{at}\ln(\cos t)\,dt$$

$$= \frac{1-e^{a\pi/2}}{a}\ln 2 - \frac{e^{a\pi/2}}{2a}\left[\psi\left(1+\frac{ia}{\lambda}\right) + \psi\left(1-\frac{ia}{\lambda}\right) + 2\gamma\right]$$

$$+ \frac{1}{2a}\left[\beta\left(1+\frac{ia}{2}\right) + \beta\left(1-\frac{ia}{2}\right) - 2\beta(1)\right].$$

With the same method, using again the first series in (3.1) we get also

$$(3.35) \qquad \int_0^{\pi/2} e^{at}\ln(\sin t)\,dt$$

$$= \frac{1-e^{a\pi/2}}{a}\ln 2 - \frac{e^{a\pi/2}}{2a}\left[\beta\left(1+\frac{ia}{\lambda}\right) + \beta\left(1-\frac{ia}{\lambda}\right) - 2\beta(1)\right]$$

$$+ \frac{1}{2a}\left[\psi\left(1+\frac{ia}{\lambda}\right) + \psi\left(1-\frac{ia}{\lambda}\right) + 2\gamma\right].$$

The simple computations are left to the reader.

Example 3.2.10

Using the logarithmic series we shall prove now the curious result

$$(3.36) \qquad \int_0^1 \frac{\ln(1+x+x^2)}{x} = \frac{\pi^2}{9}.$$

First we write for $0 \le x \le 1$

$$\log(1 + xe^{i\pi/3}) = \sum_{n=1}^{\infty} \frac{(-1)^{n-1} x^n}{n} e^{in\pi/3}$$

and from here, dividing both sides of the equation by x and integrating

$$\int_0^1 \frac{\log(1 + xe^{i\pi/3})}{x} dx = \sum_{n=1}^{\infty} \frac{(-1)^{n-1}}{n^2} e^{in\pi/3}.$$

Taking the real parts from both sides we find

$$\int_0^1 \frac{\ln(1 + x + x^2)}{x} dx = 2\sum_{n=1}^{\infty} \frac{(-1)^{n-1}}{n^2} \cos\left(\frac{n\pi}{3}\right).$$

This series can be easily evaluated

$$2\sum_{n=1}^{\infty} \frac{(-1)^{n-1}}{n^2} \cos\left(\frac{n\pi}{3}\right) = 2\left(\frac{1}{1^2}\frac{1}{2} + \frac{1}{2^2}\frac{1}{2} - \frac{1}{3^2} + \frac{1}{4^2}\frac{1}{2} + \frac{1}{5^2}\frac{1}{2} - \frac{1}{6^2} + \ldots\right)$$

$$= \zeta(2) - 3\left(\frac{1}{3^2} + \frac{1}{6^2} + \ldots\right) = \zeta(2) - \frac{1}{3}\zeta(2) = \frac{2}{3}\zeta(2) = \frac{\pi^2}{9}.$$

Done!

Example 3.2.11

This is, in fact, a counterexample. Consider the integral

(3.37) $$\int_0^{\infty} \frac{\ln(1 - e^{-2\pi\lambda x})}{1 + x^2} dx$$

for $\operatorname{Re}\lambda > 0$. Trying to evaluate it we can start as above by using the expansion (3.27) to write

(3.38) $$\int_0^{\infty} \frac{\ln(1 - e^{-2\pi\lambda x})}{1 + x^2} dx = -\sum_{k=1}^{\infty} \frac{1}{k} \int_0^{\infty} \frac{e^{-2\pi\lambda x}}{1 + x^2} dx.$$

However, then we come to the evaluation

$$\int_0^\infty \frac{e^{-2\pi\lambda x}}{1+x^2}dx = \mathrm{ci}(2\pi\lambda n)\sin(2\pi\lambda n) - \mathrm{si}(2\pi\lambda n)\cos(2\pi\lambda n)$$

where

$$\mathrm{si}(x) = -\int_x^\infty \frac{\sin t}{t}dt = \frac{-\pi}{2} + \int_0^x \frac{\sin t}{t}dt, \quad \mathrm{ci}(x) = -\int_x^\infty \frac{\cos t}{t}dt$$

are the sine and cosine integrals. It becomes clear that the series in (3.38) will be difficult to sum. Our counterexample shows that we cannot always use the logarithmic series for integral evaluation.

A quick evaluation of the integral (3.37) is possible if we first integrate by parts

$$\int_0^\infty \frac{\ln(1-e^{-2\pi\lambda x})}{1+x^2}dx = \int_0^\infty \ln(1-e^{-2\pi\lambda x})\,d\arctan x$$

$$= \ln(1-e^{-2\pi\lambda x})\arctan x\Big|_0^\infty + 2\pi\lambda\int_0^\infty \frac{e^{-2\pi\lambda x}\arctan x}{1-e^{-2\pi\lambda x}}dx$$

$$= 2\pi\lambda\int_0^\infty \frac{\arctan x}{e^{2\pi\lambda x}-1}dx.$$

Now we can use Binet's second formula for the logarithm of the Gamma function

$$\ln\Gamma(\lambda) = \left(\lambda - \frac{1}{2}\right)\ln\lambda - \lambda + \frac{\ln 2\pi}{2} + 2\int_0^\infty \frac{\arctan(t/\lambda)}{e^{2\pi t}-1}dt$$

(entry 8.341(2) in [25]). The last term with the substitution $t = \lambda x$ becomes

$$2\lambda\int_0^\infty \frac{\arctan(x)}{e^{2\pi\lambda x}-1}dx = \ln\Gamma(\lambda) - \left(\lambda - \frac{1}{2}\right)\ln\lambda + \lambda - \frac{\ln 2\pi}{2}$$

and we come to the evaluation

(3.39) $$\int_0^\infty \frac{\ln(1-e^{-2\pi\lambda x})}{1+x^2}dx = \pi\left[\ln\Gamma(\lambda) - \left(\lambda - \frac{1}{2}\right)\ln\lambda + \lambda - \frac{\ln 2\pi}{2}\right]$$

which is entry 4.319(1) in [25].

With $\lambda = 1$ we have

$$\int_0^\infty \frac{\arctan(x)}{e^{2\pi x}-1}dx = \frac{1}{2} - \frac{\ln 2\pi}{4}.$$

In Section 3.3 below we present another version of Binet's formula and evaluate a similar integral with $e^{2\pi x}+1$ in the denominator.

Example 3.2.12

In this example we use the classical expansion (entry 6.1.33 in [26])

(3.40) $$\left(\frac{\pi}{\sinh \pi y} - \frac{1}{y}\right)\frac{1}{y} = 2\sum_{k=1}^\infty \frac{(-1)^k}{k^2+y^2}$$

for $t > 0$. With the help of this series we will evaluate two challenging integrals

(3.41) $$\int_0^\infty \left(\frac{1}{\sinh t} - \frac{1}{t}\right)\frac{\cos\mu t}{t}dt = -\ln(1+e^{-\pi\mu})$$

(3.42) $$\int_0^\infty \left(\frac{1}{\sinh t} - \frac{1}{t}\right)\sin\mu t\,dt = \frac{-\pi e^{-\pi\mu}}{1+e^{-\pi\mu}} = \frac{-\pi}{e^{\pi\mu}+1}$$

for every $\mu > 0$. Here (3.42) follows from (3.41) by differentiation with respect to μ.

Now we prove (3.41). Multiplying both sides in (3.40) by $\cos\lambda y$ and integrating term by term with respect to t we find according to (3.25)

$$\int_0^\infty \left(\frac{\pi}{\sinh \pi y} - \frac{1}{y} \right) \frac{\cos \lambda y}{y} \, dy = 2 \sum_{k=1}^\infty (-1)^k \int_0^\infty \frac{\cos \lambda y}{k^2 + y^2} \, dt$$

$$= \pi \sum_{k=1}^\infty \frac{(-1)^k e^{-\lambda k}}{k} = -\pi \ln(1 + e^{-\lambda}) .$$

With the substitutions $\pi y = t$ and $\lambda = \pi \mu$ this becomes (3.41). In particular, with $\mu \to 0$ equation (3.41) turns into

$$\int_0^\infty \left(\frac{1}{\sinh t} - \frac{1}{t} \right) \frac{dt}{t} = -\ln 2$$

Which is entry 3.529(1) in [25].

Note that while equation (3.41) is true for $\mu = 0$, equation (3.42) is not.

Example 3.2.13

In this example we evaluate some important log-gamma integrals. Our starting point is Kummer's series for $0 < x < 1$

$$\ln \Gamma(x) = \left(\frac{1}{2} - x \right)(\gamma + \ln 2) + (1 - x) \ln \pi - \frac{1}{2} \ln \sin \pi x$$

$$+ \frac{1}{\pi} \sum_{n=1}^\infty \frac{\ln n}{n} \sin 2\pi n x$$

(see section 1.9.1 in the handbook of Arthur Erdelyi and Harry Bateman, *Higher Transcendental Functions Volume I*, McGraw-Hill, 1953).

Integration between 0 and z yields

(3.43) $\displaystyle \int_0^z \ln \Gamma(x) dx = \frac{1}{2}(z - z^2)(\gamma + \ln 2) + \left(z - \frac{z^2}{2} \right) \ln \pi$

$$-\frac{1}{2\pi}\int_0^{\pi z}\ln\sin t\,dt+\frac{1}{2\pi^2}\sum_{n=1}^{\infty}\frac{\ln n}{n^2}(1-\cos 2\pi nz)$$

with the convenient replacement

$$\int_0^z\ln\sin\pi x\,dx=\frac{1}{\pi}\int_0^{\pi z}\ln\sin t\,dt.$$

From here, setting $z=1$ and using the results from Example 3.2.1

$$\int_0^1\ln\Gamma(x)dx=\frac{\ln\pi}{2}+\frac{\ln 2}{2}=\frac{1}{2}\ln 2\pi$$

as computed originally by Euler.

With $z=\dfrac{1}{2}$ in (3.43) we find

$$\int_0^{1/2}\ln\Gamma(x)\,dx=\frac{\gamma}{8}+\frac{3\ln 2}{8}+\frac{3\ln\pi}{8}+\frac{1}{2\pi^2}\sum_{n=1}^{\infty}\frac{\ln n}{n^2}\left(1-(-1)^n\right).$$

In order to compute the series in this equation we use Riemann's zeta function

$$\zeta(s)=\sum_{n=1}^{\infty}\frac{1}{n^s},\quad \operatorname{Re}s>1\ .$$

Clearly,

$$\zeta'(s)=-\sum_{n=1}^{\infty}\frac{\ln n}{n^s}\quad\text{and}\quad\sum_{n=1}^{\infty}\frac{\ln n}{n^2}=-\zeta'(2).$$

Also,

$$\sum_{n=1}^{\infty}\frac{(-1)^n}{n^s}=(2^{1-s}-1)\zeta(s)$$

and differentiation with respect to s gives

$$-\sum_{n=1}^{\infty}\frac{(-1)^n \ln n}{n^s} = \frac{d}{ds}(2^{1-s}-1)\zeta(s) =$$

$$-2^{1-s}(\ln 2)\zeta(s) + (2^{1-s}-1)\zeta'(s).$$

Therefore,

$$\sum_{n=1}^{\infty}\frac{(-1)^n \ln n}{n^2} = \frac{-\ln 2}{2}\zeta(2) - \frac{1}{2}\zeta'(2).$$

Putting together all these results we come to the evaluation

(3.44) $\qquad \int_0^{1/2} \ln\Gamma(x)\,dx = \frac{\gamma+3\ln\pi}{8} + \frac{\ln 2}{3} - \frac{3}{4\pi^2}\zeta'(2) \approx 0.8037$.

This integral can also be evaluated in terms of the Glaisher-Kinkelin constant A. The relation of this constant to $\zeta'(2)$ is given by

$$\zeta'(2) = \pi^2 \left(\frac{\gamma+\ln 2\pi}{6} - 2\ln A \right).$$

We continue working with equation (3.43). With $z = \frac{1}{4}$ there we have

$$\int_0^{1/4} \ln\Gamma(x)\,dx = \frac{3\gamma+7\ln 2\pi}{32} + \frac{G}{4\pi} - \frac{\zeta'(2)}{2\pi^2} - \frac{1}{2\pi^2}\sum_{n=1}^{\infty}\frac{\ln n}{n^2}\cos\frac{\pi n}{2}.$$

It is easy to see that

$$\sum_{n=1}^{\infty}\frac{1}{n^s}\cos\frac{\pi n}{2} = (2^{1-2s}-2^{-s})\zeta(s)$$

and therefore, differentiation gives

$$-\sum_{n=1}^{\infty}\frac{\ln n}{n^s}\cos\frac{\pi n}{2} = -(2^{2-2s}-2^{-s})(\ln 2)\zeta(s) + (2^{1-2s}-2^{-s})\zeta'(s).$$

From here

$$-\sum_{n=1}^{\infty}\frac{\ln n}{n^2}\cos\frac{\pi n}{2}=\frac{-\zeta'(2)}{8}$$

and finally,

(3.45) $$\int_0^{1/4}\ln\Gamma(x)\,dx=\frac{3\gamma+7\ln 2\pi}{32}+\frac{G}{4\pi}-\frac{9\zeta'(2)}{16\pi^2}\approx 0.58247\,.$$

Gosper evaluated the above integrals and also several others in terms of $\zeta'(2)$ by a similar method (see Ralph W. Gosper, Jr. $\int_{n/4}^{m/6}\ln\Gamma(z)\,dz$, *Fields Inst. Commun.*, 14 (1997), 71-76).

Example 3.2.14

In this example we give a solution to Problem 904 from the *College Mathematics Journal* (May 2009). The problem is to evaluate the double integral

$$\int_0^1\int_0^1\ln\Gamma(x+y)\,dx\,dy\,.$$

Solution: Our starting point is Euler's integral

$$\int_0^1\ln\Gamma(x)\,dx=\ln\sqrt{2\pi}$$

from the previous example. First we prove the formula

(3.46) $$\int_y^{y+1}\ln\Gamma(x)\,dx=y\ln y-y+\ln\sqrt{2\pi}$$

for $y\geq 0$ (the value of $y\ln y$ at $y=0$ is considered zero). Differentiating the left-hand side in in (3.46) with respect to y we find

$$\frac{d}{dy} \int_{y}^{y+1} \ln \Gamma(x)\,dx = \ln \Gamma(y+1) - \ln \Gamma(y) = \ln y \,.$$

This is true for all $y > 0$ because

$$\ln \Gamma(y+1) - \ln \Gamma(y) = \ln \frac{\Gamma(y+1)}{\Gamma(y)} = \ln \frac{y\Gamma(y)}{\Gamma(y)} = \ln y \,.$$

Also

$$\frac{d}{dy}(y \ln y - y) = \ln y \,.$$

We see that the functions

$$\int_{y}^{y+1} \ln \Gamma(x)\,dx \quad \text{and} \quad y \ln y - y$$

have the same derivative; therefore, they differ by a constant

$$\int_{y}^{y+1} \ln \Gamma(x)\,dx = y \ln y - y + C \,.$$

Setting $y \to 0$ and using Euler's integral we conclude that $C = \ln\sqrt{2\pi}$. This way the identity (3.46) is proved.

Now with the substitution $x = u - y,\ dx = du$ we write

$$\int_{0}^{1} \ln \Gamma(x+y)\,dx = \int_{y}^{y+1} \ln \Gamma(u)\,du = y \ln y - y + \ln\sqrt{2\pi}$$

and integrating both sides with respect to y from 0 to 1 we find

(3.47) $$\int_{0}^{1}\int_{0}^{1} \ln \Gamma(x+y)\,dx\,dy = \ln\sqrt{2\pi} - \frac{3}{4} \,.$$

Example 3.2.15

This is a solution to Monthly Problem 11329 (2007, p. 925). The problem is to prove the two equations

$$\int_0^\infty 2^{-x}\Gamma(x)\,dx = 2\int_0^1 2^{-x}\Gamma(x)\,dx - \frac{\gamma + \ln\ln 2}{\ln 2}$$

and

$$\int_0^\infty x2^{-x}\Gamma(x)\,dx$$

$$= 2\int_0^1 (x+1)2^{-x}\Gamma(x)\,dx - \frac{(\gamma + \ln\ln 2)(1 + 2\ln 2) - 1}{\ln^2 2}.$$

Our starting point for the proof is the well-known representation

$$(3.48) \qquad \ln\Gamma(x) = \int_0^\infty \left(x - 1 - \frac{1 - e^{-(x-1)t}}{1 - e^{-t}} \right) \frac{e^{-t}}{t}\,dt.$$

We multiply both sides by 2^{-x}, integrate with respect to x from zero to infinity, exchange the order of integration, and integrate by parts to obtain

$$A \equiv \int_0^\infty 2^{-x}\ln\Gamma(x)\,dx$$

$$= \int_0^\infty \left\{ -\frac{1}{\ln 2} + \frac{1}{(\ln 2)^2} - \frac{1}{1 - e^{-t}}\left(\frac{1}{\ln 2} - \frac{e^t}{t + \ln 2} \right) \right\} \frac{e^{-t}}{t}\,dt.$$

In the same way

$$B \equiv \int_0^1 2^{-x}\ln\Gamma(x)\,dx$$

$$= \int_0^\infty \left\{ -\frac{1}{\ln 2} + \frac{1}{2(\ln 2)^2} - \frac{1}{1 - e^{-t}} \left(\frac{1}{2\ln 2} - \frac{e^t - 1/2}{t + \ln 2} \right) \right\} \frac{e^{-t}}{t} dt .$$

From here

$$A - 2B = \int_0^\infty \left(\frac{e^{-t}}{\ln 2} - \frac{1}{t + \ln 2} \right) \frac{dt}{t} = \frac{1}{\ln 2} \int_0^\infty \left(e^{-t\ln 2} - \frac{1}{1+t} \right) \frac{dt}{t}$$

after rescaling $t \to t \ln 2$.

Next we use the representation

$$\int_0^\infty \left(e^{-pt} - \frac{1}{1+t} \right) \frac{dt}{t} = -\ln p - \gamma$$

(this is proved in Example 5.3.7 in Chapter 5). With $p = \ln 2$

$$A - 2B = -\frac{\gamma + \ln \ln 2}{\ln 2} .$$

This proves the first equation in the problem. The second equation is proved likewise. We multiply (3.48) by $x2^{-x}$ and integrate from zero to infinity. Then we multiply (3.48) by $(x+1)2^{-x}$ and integrate from zero to one. The computation requires integration by parts twice. Details are left to the reader.

Example 3.2.16

In association with the log-sine integrals we may want to consider the interesting integral

$$\int_0^1 \ln(\arcsin x)dx$$

which cannot be found in the tables [25] and [43]. With the substitution $x = \sin t$ followed by integration by parts this integral becomes

$$\int_0^1 \ln(\arcsin x)dx = \int_0^{\pi/2} \ln t (\cos t)dt = \int_0^{\pi/2} \ln t \, d\sin t$$

$$= \ln t \sin t \Big|_0^{\pi/2} - \int_0^{\pi/2} \frac{\sin t}{t}dt = \ln\frac{\pi}{2} - \text{Si}\left(\frac{\pi}{2}\right).$$

That is,

$$\int_0^1 \ln(\arcsin x)dx = \ln\frac{\pi}{2} - \text{Si}\left(\frac{\pi}{2}\right)$$

where

$$\text{Si}(x) = \int_0^x \frac{\sin t}{t}dt$$

is the sine integral function [40, p. 150].

Another integral with a similar structure is

$$J = \int_0^1 \ln x(\arcsin x)\, dx .$$

Integrating by parts we have

$$J = x \ln x(\arcsin x)\Big|_0^1 - \int_0^1 x\left(\frac{\arcsin x}{x} + \frac{\ln x}{\sqrt{1-x^2}}\right)dx$$

$$= -\int_0^1 \arcsin x\, dx - \int_0^1 \frac{x \ln x}{\sqrt{1-x^2}}dx .$$

The first integral we integrate by parts and in the second we set $x = \sin t$ to get

$$J = 1 - \frac{\pi}{2} - \int_0^{\pi/2} \sin t \ln(\sin t)dt .$$

This integral is known from Example 3.2.1, so finally

$$\int_0^1 \ln x (\arcsin x)\, dx = 2 - \frac{\pi}{2} - \ln 2$$

in line with entry 4.1.6 (41) in [20] and entry 4.591(1) in [25].

A simple problem for the reader. Prove that

$$\int_0^1 \ln x (\arccos x)\, dx = \ln 2 - 2$$

(entry 4.591(2) in [25]).
Hint. Do not mimic the above computation – just use the simple relation between $\arccos x$ and $\arcsin x$.

Example 3.2.17

At this point it will be very natural to look at the "first cousin" of the above integral (entry 4.593(1) in [25])

$$\int_0^1 \ln x (\arctan x)\, dx = \frac{\pi^2}{48} + \frac{\ln 2}{2} - \frac{\pi}{4}$$

where the presence of π^2 hints that $\zeta(2)$ may somehow be involved. We start the solution as above, integrating by parts. Calling this integral M we write

$$M = x \ln x (\arctan x)\Big|_0^1 - \int_0^1 x \left(\frac{\arctan x}{x} + \frac{\ln x}{1 + x^2} \right) dx$$

$$= -\int_0^1 \arctan x\, dx - \int_0^1 \frac{x \ln x}{1 + x^2}\, dx .$$

Integration by parts gives

$$\int_0^1 \arctan x\, dx = \frac{\pi}{4} - \frac{\ln 2}{2}$$

so that

$$M = \frac{\ln 2}{2} - \frac{\pi}{4} - \frac{1}{2}\int_0^1 \ln x\, d\ln(1 + x^2)$$

$$= \frac{\ln 2}{2} - \frac{\pi}{4} - \frac{1}{2}\ln x \ln(1 + x^2)\Big|_0^1 + \frac{1}{2}\int_0^1 \frac{\ln(1 + x^2)}{x}\, dx.$$

The term with the product of logarithms is zero and also

$$\int_0^1 \frac{\ln(1 + x^2)}{x}\, dx = \int_0^1 \frac{1}{x}\left\{\sum_{n=1}^\infty \frac{(-1)^{n-1}}{n} x^{2n}\right\} dx = \frac{1}{2}\sum_{n=1}^\infty \frac{(-1)^{n-1}}{n^2}$$

where (here comes π^2!)

$$\sum_{n=1}^\infty \frac{(-1)^{n-1}}{n^2} = \frac{1}{2}\zeta(2) = \frac{\pi^2}{12}.$$

Putting all these pieces together we come to the desired result.

A note about the evaluation

$$\ln x \ln(1 + x^2)\Big|_0^1 = 0.$$

At the upper limit, for $x = 1$, the value is obviously zero. Then

$$\lim_{x \to 0}(\ln x \ln(1 + x^2)) = 0$$

because $\ln(1 + x^2) \sim x^2$ near zero and $\lim_{x \to 0}(x \ln x) = 0$.

Example 3.2.18

This is a solution of Monthly Problem 11639 (2012, p. 345). The question is to evaluate the integral

$$\int_0^{\pi/2} \ln^2(2\sin t)\,dt\ .$$

For this evaluation we use again the Fourier series (3.1). With the substitution $x = 2t$ the first series can be written in the form

$$\ln\left(2\sin\frac{x}{2}\right) = -\sum_{n=1}^{\infty} \frac{\cos(nx)}{n},\ 0 < x < 2\pi\ .$$

Next we use Parseval's theorem for Fourier series (see below). According to this theorem

$$\frac{1}{\pi}\int_0^{2\pi} \ln^2\left(2\sin\frac{x}{2}\right)dx = \sum_{n=1}^{\infty}\frac{1}{n^2} = \frac{\pi^2}{6}\ .$$

Returning back to $t = x/2$ we have

$$\frac{2}{\pi}\int_0^{\pi} \ln^2(2\sin t)\,dt = \frac{\pi^2}{6}\ .$$

Using the symmetry in the graph of $\sin t$ in the first and the second quadrants we have

$$\int_0^{\pi} \ln^2(2\sin t)\,dt = 2\int_0^{\pi/2} \ln^2(2\sin t)\,dt$$

and therefore,

$$\int_0^{\pi/2} \ln^2(2\sin t)\,dt = \frac{\pi^3}{24}\ .$$

Parseval's Theorem.

If the real valued function $f(t)$ has Fourier expansion

$$f(t) = \frac{a_0}{2} + \sum_{n=1}^{\infty}(a_n\cos nt + b_n\sin nt)$$

in the interval $[0, 2\pi]$ and is square integrable in this interval, then

$$\frac{1}{\pi}\int_0^{2\pi} f^2(t)\,dt = \frac{a_0^2}{2} + \sum_{n=1}^{\infty}(a_n^2 + b_n^2)$$

(see [46, p. 119]).

Example 3.2.19

The standard logarithmic series

$$\ln(1-x) = -\sum_{n=1}^{\infty}\frac{x^n}{n} \quad (|x| < 1)$$

can be used for the evaluation of the interesting integral

$$J(\lambda) = \int_0^1 \ln x\ln(1-\lambda x)\,dx \quad (|\lambda| < 1).$$

For $0 < \lambda < 1$ we have

$$J(\lambda) = -\int_0^1 \ln x\left\{\sum_{n=1}^{\infty}\frac{\lambda^n x^n}{n}\right\}dx = -\sum_{n=1}^{\infty}\frac{\lambda^n}{n}\left\{\int_0^1 x^n \ln x\,dx\right\}$$

$$= \sum_{n=1}^{\infty}\frac{\lambda^n}{n}\left\{\frac{1}{(n+1)^2}\right\} = \sum_{n=1}^{\infty}\lambda^n\left\{\frac{1}{n} - \frac{1}{n+1} - \frac{1}{(n+1)^2}\right\}$$

$$= -\ln(1-\lambda) + \frac{1}{\lambda}\left(\ln(1-\lambda) + \lambda\right) - \frac{1}{\lambda}\left(\mathrm{Li}_2(\lambda) - \lambda\right)$$

$$= \frac{1-\lambda}{\lambda}\ln(1-\lambda) - \frac{1}{\lambda}\mathrm{Li}_2(\lambda) + 2.$$

That is,

$$\int_0^1 \ln x\ln(1-\lambda x)\,dx = \frac{1-\lambda}{\lambda}\ln(1-\lambda) - \frac{1}{\lambda}\mathrm{Li}_2(\lambda) + 2.$$

By analytic continuation this extends on the disk $|\lambda| < 1$.

With $\lambda \to 1$ we find

$$\int_0^1 \ln x \ln(1-x)\,dx = 2 - \mathrm{Li}_2(1) = 2 - \zeta(2) = 2 - \frac{\pi^2}{6}.$$

With $\lambda \to -1$

$$\int_0^1 \ln x \ln(1+x)\,dx = -2\ln 2 + \mathrm{Li}_2(-1) + 2 = -2\ln 2 + 2 - \frac{\pi^2}{12}.$$

These two integrals are entries 4.221(1) and 4.221(2) correspondingly in Gradshteyn and Ryzhik's reference table [25].

Example 3.2.20

Not all log-sine integrals should be evaluated by the Fourier series (3.1). For example, the log-sine integral

$$\int_0^{\pi/2} \frac{\ln(\sin t)}{\cos t}\,dt$$

can be approached differently, with the ordinary logarithmic series

$$\int_0^{\pi/2} \frac{\ln(\sin t)}{\cos t}\,dt = \frac{1}{2}\int_0^{\pi/2} \frac{\ln(\sin^2 t)}{\cos t}\,dt = \frac{1}{2}\int_0^{\pi/2} \frac{\ln(1-\cos^2 t)}{\cos t}\,dt$$

$$= -\frac{1}{2}\int_0^{\pi/2} \frac{1}{\cos t}\left\{\sum_{n=0}^{\infty} \frac{\cos^{2n+2}}{n+1}\right\}dt = -\frac{1}{2}\sum_{n=0}^{\infty} \frac{1}{n+1}\int_0^{\pi/2} \cos^{2n+1} t\,dt.$$

We see here the well-known Wallis integral

$$\int_0^{\pi/2} \cos^{2n+1} t\,dt = \frac{4^n}{2n+1}\binom{2n}{n}^{-1}.$$

and therefore,

$$\int_0^{\pi/2} \frac{\ln(\sin t)}{\cos t} dt = -\frac{1}{2} \sum_{n=0}^{\infty} \frac{4^n}{(n+1)(2n+1)} \binom{2n}{n}^{-1}.$$

The inverse binomial series here can be evaluated by certain means

$$\sum_{n=0}^{\infty} \frac{4^n}{(n+1)(2n+1)} \binom{2n}{n}^{-1} = \frac{\pi^2}{4}$$

and we find

$$\int_0^{\pi/2} \frac{\ln(\sin t)}{\cos t} dt = -\frac{\pi^2}{8}.$$

Example 3.2.21

We show a solution to Monthly Problem 12221 (December 2020). Prove that

$$J = \int_0^1 \frac{\log(x^6 + 1)}{x^2 + 1} dx = \frac{\pi}{2} \log 6 - 3G.$$

For the solution we make the substitution $x = \tan t$ and write

$$J = \int_0^{\pi/4} \log(1 + \tan^6 t) dt = \int_0^{\pi/4} \log\left(\frac{\cos^6 t + \sin^6 t}{\cos^6 t}\right) dt$$

$$= \int_0^{\pi/4} \log(\cos^6 t + \sin^6 t) dt - 6 \int_0^{\pi/4} \log(\cos t) dt = A - 6B$$

where A and B are the two integrals. The value of B is known from Example 3.2.1

$$B = \int_0^{\pi/4} \log(\cos t) dt = \frac{G}{2} - \frac{\pi}{4} \ln 2.$$

To evaluate A we use the trigonometric identities

$$\cos^6 t + \sin^6 t = 1 - 3\cos^2 t \sin^2 t \quad \text{and} \quad 2\cos t \sin t = \sin 2t$$

$$A = \int_0^{\pi/4} \log(\cos^6 t + \sin^6 t) dt = \int_0^{\pi/4} \log(1 - 3\cos^2 t \sin^2 t) dt$$

$$= \int_0^{\pi/4} \log(1 - \frac{3}{4}\sin^2 2t) dt = \frac{1}{2} \int_0^{\pi/2} \log(1 - \frac{3}{4}\sin^2 \theta) d\theta \quad (2t = \theta).$$

Now we expand the logarithm and integrate term by term

$$A = -\frac{1}{2} \int_0^{\pi/2} \left\{ \sum_{n=1}^{\infty} \frac{1}{n} \left(\frac{3}{4}\right)^n \sin^{2n}\theta \right\} d\theta = -\frac{1}{2} \sum_{n=1}^{\infty} \frac{1}{n} \left(\frac{3}{4}\right)^n \left\{ \int_0^{\pi/2} \sin^{2n}\theta\, d\theta \right\}$$

$$= -\frac{1}{2} \sum_{n=1}^{\infty} \frac{1}{n} \left(\frac{3}{4}\right)^n \left\{ \frac{1}{4^n} \binom{2n}{n} \frac{\pi}{2} \right\}$$

according to the Wallis formula (see Section 5.8 in Chapter 5).

The series can be evaluated by using the well-known generating function

$$\sum_{n=1}^{\infty} \binom{2n}{n} \frac{x^n}{n} = 2\log\left(\frac{1 - \sqrt{1 - 4x}}{2x}\right) \quad (|x| < 1/4)$$

(see [32]) which gives $A = \frac{\pi}{2}\log 3 - \pi\log 2$. Finally

$$J = A - 6B = \frac{\pi}{2}\ln 6 - 3G.$$

The same technique solves Problem 2107 in the Mathematics Magazine (vol. 93, December 2020). Namely,

$$\int_0^{\infty} \frac{\log(x^6 + 1)}{x^2 + 1} dx = \pi\log 6.$$

3.3　A Binet Type Formula for the Log-Gamma Function

In Example 3.2.11 we used the second Binet formula for $\ln\Gamma(\lambda)$

$$(3.49)\qquad \ln\Gamma(\lambda)=\left(\lambda-\frac{1}{2}\right)\ln\lambda-\lambda+\frac{\ln 2\pi}{2}+2\int_0^\infty\frac{\arctan(t/\lambda)}{e^{2\pi t}-1}dt$$

in order to prove the interesting evaluation

$$\int_0^\infty\frac{\arctan(x)}{e^{2\pi x}-1}dx=\frac{1}{2}-\frac{\ln 2\pi}{4}.$$

It is desirable to evaluate also the similar integral

$$\int_0^\infty\frac{\arctan(x)}{e^{2\pi x}+1}dx$$

which is not present in the popular tables [25] and [43], but appears in Brychkov's handbook [20]. For the evaluation of this integral we need a Binet type formula like (3.49) where the integral will have denominator $e^{2\pi x}+1$ instead of $e^{2\pi x}-1$. We will present this formula here.

Formula (3.49) is usually proved by using the Abel-Plana summation formula

$$\sum_{n=0}^\infty f(n)=\frac{f(0)}{2}+\int_0^\infty f(x)dx+i\int_0^\infty\frac{f(ix)-f(-ix)}{e^{2\pi x}-1}dy$$

valid for functions holomorphic on the half-plane $\mathrm{Re}\,z\geq 0$ with certain growth conditions (see Whittaker-Watson [48, pp. 145-146]). A similar Abel-Plana formula exist with $e^{2\pi y}+1$ in the denominator

$$(3.50)\qquad \sum_{n=0}^\infty f\left(n+\frac{1}{2}\right)=\int_0^\infty f(x)dx-i\int_0^\infty\frac{f(ix)-f(-ix)}{e^{2\pi x}+1}dx$$

(see, for example, Frappier's paper "A generalization of the summation formula of Plana" in the *Bulletin of the Australian Mathematical Society*,

59 (1999), 315-322). Using this formula we will prove the following proposition (a different Binet type formula).

Proposition 3.1. *For* $\operatorname{Re} z > 0$

$$(3.51) \quad \ln \Gamma \left(z + \frac{1}{2} \right) = z \ln z - z + \ln \sqrt{2\pi} - 2 \int_0^\infty \frac{\arctan(x/z)}{e^{2\pi x} + 1} dx.$$

Proof. We apply (3.50) to the function

$$f(x) = \frac{1}{(x+z)^2}$$

and first we evaluate

$$\int_0^\infty f(x)dx = -\frac{1}{x+z} \bigg|_0^\infty = \frac{1}{z}.$$

Next we compute

$$f(ix) - f(-ix) = \frac{1}{(z+ix)^2} - \frac{1}{(z-ix)^2} = -\frac{4ixz}{(z^2+x^2)^2}$$

$$-i \int_0^\infty \frac{f(ix) - f(-ix)}{e^{2\pi x} + 1} dx = -4 \int_0^\infty \frac{xz}{(z^2+x^2)^2(e^{2\pi x}+1)} dx$$

and therefore, (3.50) implies

$$(3.52) \quad \sum_{n=0}^\infty \left(n + z + \frac{1}{2} \right)^{-2} = \frac{1}{z} - 4 \int_0^\infty \frac{xz}{(z^2+x^2)^2(e^{2\pi x}+1)} dx.$$

It is easy to recognize the series on the left-hand side here. The digamma function $\psi(z)$ has the series representation

$$\psi(z) = -\gamma + \sum_{n=0}^\infty \left(\frac{1}{n+1} - \frac{1}{n+z} \right)$$

($\gamma = -\psi(1)$ is Euler's constant). From this

$$\psi'(z) = \sum_{n=0}^{\infty} \frac{1}{(n+z)^2}$$

and so (3.52) becomes

$$\psi'\left(z+\frac{1}{2}\right) = \frac{1}{z} - 4\int_0^\infty \frac{xz}{(z^2+x^2)^2(e^{2\pi x}+1)}\,dx\ .$$

Integrating this equation with respect to z provides

$$\psi\left(z+\frac{1}{2}\right) = \ln z + 2\int_0^\infty \frac{x}{(x^2+z^2)(e^{2\pi x}+1)}\,dx + C\ .$$

Setting $z \to \infty$ and using the fact that $\lim_{n\to\infty}(\psi(x+n) - \ln(n)) = 0$ (see entry 8.365(5) in [25]) we find $C = 0$. Thus we produce the interesting representation

(3.53) $$\psi\left(z+\frac{1}{2}\right) = \ln z + 2\int_0^\infty \frac{x}{(x^2+z^2)(e^{2\pi x}+1)}\,dx\ .$$

This is entry 4.1.2(17) in [20].

Remembering that $\psi(z) = (\ln\Gamma(z))'$ we integrate equation (3.53) to find

$$\ln\Gamma\left(z+\frac{1}{2}\right) = z\ln z - z + 2\int_0^\infty \frac{\arctan(z/x)}{e^{2\pi x}+1}\,dx + C$$

and setting $z \to 0$ we find $C = \ln\Gamma(1/2) = \ln\sqrt{\pi}$.

At this point we have proved the representation

(3.54) $$\ln\Gamma\left(z+\frac{1}{2}\right) = z\ln z - z + \ln\sqrt{\pi} + 2\int_0^\infty \frac{\arctan(z/x)}{e^{2\pi x}+1}\,dx$$

which is equivalent to (3.51). To obtain (3.51) from (3.54) we use the identity

$$\arctan\frac{z}{x} = \frac{\pi}{2} - \arctan\frac{x}{z}$$

so that

$$2\int_0^\infty \frac{\arctan(z/x)}{e^{2\pi x}+1}\,dx = \pi\int_0^\infty \frac{1}{e^{2\pi x}+1}\,dx - 2\int_0^\infty \frac{\arctan(x/z)}{e^{2\pi x}+1}\,dx$$

where the integral in the middle is easy to compute

$$\pi\int_0^\infty \frac{1}{e^{2\pi x}+1}\,dx = \pi\int_0^\infty \frac{e^{-2\pi x}}{1-(-e^{2\pi x}))}\,dx = \pi\int_0^\infty \left\{\sum_{n=1}^\infty (-1)^{n-1} e^{-2\pi nx}\right\}dx$$

$$= \pi\sum_{n=1}^\infty (-1)^{n-1}\int_0^\infty e^{-2\pi nx}\,dx = \pi\sum_{n=1}^\infty \frac{(-1)^{n-1}}{2\pi n} = \frac{1}{2}\ln 2 = \ln\sqrt{2}\,.$$

Replacing now the integral in (3.54) we come to (3.51). The proof is completed.

Now we set $z = 1$ in (3.51). On the left-hand side we have

$$\ln\Gamma(3/2) = \ln\left(\frac{1}{2}\Gamma\left(\frac{1}{2}\right)\right) = \ln\left(\frac{\sqrt{\pi}}{2}\right) = \ln\sqrt{\pi} - \ln 2$$

and we compute

(3.55) $$\int_0^\infty \frac{\arctan(x)}{e^{2\pi x}+1}\,dx = \frac{3}{4}\ln 2 - \frac{1}{2}$$

(entry 4.1.6(135) in [20]).

3.4 Some Theorems

Theorem A. *Let $f(x)$ be a piecewise continuous function with Fourier series*

$$\frac{a_0}{2} + \sum_{n=1}^\infty (a_n \cos nx + b_n \sin nx)$$

in the interval $(-\pi, \pi)$. *Then for any two numbers* $a \le b$ *from* $[-\pi, \pi]$
we have

$$\int_a^b f(x)dx = \frac{a_0}{2}(b-a) + \sum_{n=1}^{\infty} \int_a^b (a_n \cos nx + b_n \sin nx)dx$$

whether or not the Fourier series of $f(x)$ *converges.*

(See [23, vol. 3, Chapter 20] or [46, p. 125].)

In some examples we integrate term-wise functional series. The operation is supported by the following theorem (see [23, vol. 2, p. 436]).

Theorem B. *Let*

$$f(x) = \sum_{n=0}^{\infty} u_n(x)$$

where the functions $u_n(x)$, $n = 0,1...$ *are continuous on the interval* $[a,b]$ *and the series is uniformly convergent on that interval. In this case*

$$\int_a^b f(x)dx = \sum_{n=0}^{\infty} \int_a^b u_n(x)dx.$$

Uniform convergence is assured by the classical criterion of Weierstrass.

Theorem C. *If* $|u_n(x)| \le c_n$ *on* $[a,b]$ *for every* $n = 0,1...$, *where the series* $\sum_{n=0}^{\infty} c_n$ *is convergent, then the series* $\sum_{n=0}^{\infty} u_n(x)$ *is uniformly convergent on* $[a,b]$.

(See, for example, [23, vol. 2, p. 427].)

Proofs can be found in the corresponding references.

Chapter 4

Evaluating Integrals by Laplace and Fourier Transforms. Integrals Related to Riemann's Zeta Function

4.1 Introduction

The Laplace and Fourier transforms can be used to solve integrals directly or by means of Parseval's theorem. In this chapter the reader will find examples of both techniques.

Among other things we prove the Fresnel integrals and several important integrals involving trigonometric and hyperbolic functions. The reader will see how the functional equation for the Riemann zeta function can be derived from two special integrals. Another special integral is used to prove Euler's formula expressing $\zeta(2n)$ in terms of Bernoulli numbers. Same is done also for Euler's $L(s)$ function. The functional equation is derived and the connection between $L(2n+1)$ and Euler's numbers E_{2n} is proved again using certain integrals. At the end of the chapter we show how the functions $\zeta(s)$ and $L(s)$ can be used to evaluate integrals.

Also in this chapter the reader will meet the interesting exponential polynomials which are used for the evaluations of some nontrivial integrals with the gamma function.

4.2 Laplace Transform

We use the standard notations. The originals $f(t)$ are functions defined on $[0,\infty)$ and their images $F(s), s > 0$, are given by

$$F(s) = L\{f(t)\} = \int_0^\infty f(t)e^{-st}dt .$$

Needed are mostly the formulas

$$L\{\cos at\} = \frac{s}{s^2 + a^2}, \; L\{\sin at\} = \frac{a}{s^2 + a^2}$$

and also

$$L\{t^\beta\} = \frac{\Gamma(\beta+1)}{s^{\beta+1}}, \; L\{e^{-at}\} = \frac{1}{s+a}$$

where $-1 < \beta$ and a are constants.

Good references for the Laplace transform are the books [22] and [42]. An elementary way to use the Laplace transform for evaluation of integrals is to change one integral into another with a known value. A more powerful method is to introduce a parameter into the integral and evaluate the Laplace transform of the resulting function. This method will be illustrated by several examples.

Example 4.2.1

We want to evaluate the integral

$$\int_0^\infty \frac{x - \sin x}{x^3} dx .$$

For this purpose we define the function

$$f(t) = \int\limits_0^\infty \frac{xt - \sin(xt)}{x^3} \, dx$$

and apply the Laplace transform

$$L\{f(t)\} = \int\limits_0^\infty \left\{ \int\limits_0^\infty \frac{xt - \sin(xt)}{x^3} \, dx \right\} e^{-st} \, dt$$

$$= \int\limits_0^\infty \frac{1}{x^3} \left\{ \int\limits_0^\infty (xt - \sin(xt)) \, e^{-st} \, dt \right\} dx = \int\limits_0^\infty \frac{1}{x^3} \left\{ \frac{x}{s^2} - \frac{x}{s^2 + x^2} \right\} dx$$

$$= \frac{1}{s^2} \int\limits_0^\infty \frac{1}{s^2 + x^2} \, dx = \frac{1}{s^3} \arctan \frac{x}{s} \bigg|_0^\infty = \frac{\pi}{2s^3} = L\left\{ \frac{\pi}{4} t^2 \right\}.$$

Therefore,

$$f(t) = \frac{\pi}{4} t^2$$

and with $t = 1$

$$\int\limits_0^\infty \frac{x - \sin x}{x^3} \, dx = \frac{\pi}{4}.$$

Example 4.2.2

Here we prove the evaluation

(4.1) $$\int\limits_0^\infty \left(\frac{\sin x}{x} \right)^3 dx = \frac{3\pi}{8}$$

(this is entry 3.827(4) in [25]). Defining

$$f(t) = \int\limits_0^\infty \left(\frac{\sin xt}{x} \right)^3 dx$$

and using the decomposition

$$\sin^3 \theta = \frac{1}{4}(3\sin\theta - \sin 3\theta)$$

we compute

$$L\{f\} = \frac{1}{4}\int_0^\infty \left\{ \int_0^\infty \{3\sin(xt) - \sin(3xt)\}e^{-st}dt \right\} \frac{dx}{x^3}$$

$$= \frac{1}{4}\int_0^\infty \left\{ \frac{3x}{s^2 + x^2} - \frac{3x}{s^2 + 9x^2} \right\} \frac{dx}{x^3} = 6\int_0^\infty \frac{1}{(s^2 + x^2)(s^2 + 9x^2)} dx$$

$$= \frac{3}{4s^2}\int_0^\infty \left\{ \frac{9}{s^2 + 9x^2} - \frac{1}{s^2 + x^2} \right\} dx$$

$$= \frac{3}{4s^2}\left\{ \frac{3}{s}\arctan\frac{3x}{s} - \frac{1}{s}\arctan\frac{x}{s} \right\}\Big|_0^\infty = \frac{3}{4s^3}\left(\frac{3\pi}{2} - \frac{\pi}{2} \right) = \frac{3\pi}{4s^3}$$

$$= \frac{3\pi}{8}L\{t^2\}$$

so that

$$f(t) = \frac{3\pi}{8}t^2$$

and (4.1) is proved.

In the same way one can prove the other integrals from the group 3.827 in [25].

Example 4.2.3. The Fresnel integrals

For this example we need some preparation. First, using Euler's beta function

$$B(u,v) = \int_0^\infty \frac{t^{u-1}}{(1+t)^{u+v}} dt = \frac{\Gamma(u)\Gamma(v)}{\Gamma(u+v)}$$

we evaluate one important integral (with $v = 1 - u$)

$$\int_0^\infty \frac{t^{u-1}}{1+t} dt = \Gamma(u)\Gamma(1-u) = \frac{\pi}{\sin \pi u} \quad (0 < u < 1).$$

Here we change the variables by setting $t = x^2 / s^2$, $s > 0$, and then also $2u - 1 = q$ to get

(4.2) $$\int_0^\infty \frac{x^{-q}}{s^2 + x^2} dx = \frac{\pi}{2s^{q+1}} \sec \frac{\pi q}{2}, \quad -1 < q < 1.$$

This integral will be needed very soon.

Now we turn to Fresnel's integrals. The two integrals

$$C_\alpha = \int_0^\infty \cos(z^\alpha) dz, \quad S_\alpha = \int_0^\infty \sin(z^\alpha) dz$$

where $\alpha > 1$, are known as Fresnel integrals, named after the French mathematician and scientist Augustin-Jean Fresnel (1788-1827). The two special transcendental functions

$$C(x) = \int_0^x \cos(z^2) dz, \quad S(x) = \int_0^x \sin(z^2) dz$$

have important applications in optics.

We will evaluate C_α and S_α by using Laplace transform. First, the substitution $z^\alpha = x$ brings to

$$C_\alpha = \frac{1}{\alpha} \int_0^\infty x^{-p} \cos(x) dx, \quad S_\alpha = \frac{1}{\alpha} \int_0^\infty x^{-p} \sin(x) dx$$

where $p = 1 - \frac{1}{\alpha}$ with $0 < p < 1$. We define the two functions

$$C_\alpha(t) = \frac{1}{\alpha}\int_0^\infty x^{-p}\cos(xt)dx, \quad S_\alpha(t) = \frac{1}{\alpha}\int_0^\infty x^{-p}\sin(xt)dx$$

and compute their Laplace transforms

$$L\{C_\alpha(t)\} = \frac{1}{\alpha}\int_0^\infty x^{-p}\left\{\int_0^\infty \cos(xt)e^{-st}dt\right\}dx = \frac{s}{\alpha}\int_0^\infty \frac{x^{-p}}{x^2+s^2}dx$$

$$L\{S_\alpha(t)\} = \frac{1}{\alpha}\int_0^\infty x^{-p}\left\{\int_0^\infty \sin(xt)e^{-st}dt\right\}dx = \frac{1}{\alpha}\int_0^\infty \frac{x^{1-p}}{x^2+s^2}dx.$$

According to (4.2) we have

$$L\{C_\alpha(t)\} = \frac{\pi}{2\alpha s^p}\sec\frac{\pi p}{2} = \frac{\pi}{2\alpha}\sec\frac{\pi p}{2}\frac{L\{t^{p-1}\}}{\Gamma(p)}$$

$$L\{S_\alpha(t)\} = \frac{\pi}{2\alpha s^p}\csc\frac{\pi p}{2} = \frac{\pi}{2\alpha}\csc\frac{\pi p}{2}\frac{L\{t^{p-1}\}}{\Gamma(p)}$$

and therefore,

$$C_\alpha(t) = \frac{\pi t^{p-1}}{2\alpha\Gamma(p)}\sec\frac{\pi p}{2}, \quad S_\alpha(t) = \frac{\pi t^{p-1}}{2\alpha\Gamma(p)}\csc\frac{\pi p}{2}$$

or

$$\int_0^\infty x^{-p}\cos(xt)dx = \frac{\pi t^{p-1}}{2\Gamma(p)}\sec\frac{\pi p}{2}$$

$$\int_0^\infty x^{-p}\sin(xt)dx = \frac{\pi t^{p-1}}{2\Gamma(p)}\csc\frac{\pi p}{2}$$

(see entries 3.761 (4) and 3.761(9) in [25]).

Now since

$$\frac{1}{\Gamma(p)} = \frac{1}{\Gamma(1-1/\alpha)} = \frac{\Gamma(1/\alpha)\sin(\pi/\alpha)}{\pi}$$

we come to the evaluations

$$C_\alpha(t) = \frac{1}{\alpha}\Gamma\left(\frac{1}{\alpha}\right)\cos\frac{\pi}{2\alpha}t^{-1/\alpha}, \quad S_\alpha(t) = \frac{1}{\alpha}\Gamma\left(\frac{1}{\alpha}\right)\sin\frac{\pi}{2\alpha}t^{-1/\alpha}.$$

Setting $t = 1$ we find

$$\int_0^\infty \cos(z^\alpha)dz = \frac{1}{\alpha}\Gamma\left(\frac{1}{\alpha}\right)\cos\frac{\pi}{2\alpha}$$

$$\int_0^\infty \sin(z^\alpha)dz = \frac{1}{\alpha}\Gamma\left(\frac{1}{\alpha}\right)\sin\frac{\pi}{2\alpha}$$

which are practically entries 3.712 (1) and 3.712(2) in [25].

Problem for the reader: Show that the integral

$$\int_0^\infty x^{-p}\cos(xt)dx = \frac{\pi t^{p-1}}{2\Gamma(p)}\sec\frac{\pi p}{2}$$

can be written in the form

$$\int_0^\infty x^s \cos xt\, dx = -\frac{\Gamma(s+1)}{t^{s+1}}\sin\frac{\pi s}{2} \quad (t>0,\ -1<\mathrm{Re}\, s<0)$$

by setting $p = -s$ and using the property of the gamma function

$$\Gamma(z)\Gamma(1-z) = \frac{\pi}{\sin \pi z}.$$

(This integral will be needed later in Section 4.7.)

Remark. The import integral (4.2) appears also in disguise as entry 2.5.26(7) in [43], namely,

$$\int_0^{\pi/2} \tan^\mu x \, dx = \frac{\pi}{2} \sec\left(\frac{\pi\mu}{2}\right) \quad (|\operatorname{Re}\mu| < 1).$$

The substitution $\tan x = t, t = \arctan x$ transforms this integral into

$$\int_0^\infty \frac{t^\mu}{1+t^2} \, dt = \frac{\pi}{2} \sec\left(\frac{\pi\mu}{2}\right)$$

which is equivalent to (4.2).

A problem for the reader: Show that

$$\int_0^\infty \frac{\ln(1+\lambda t)}{t^{z+1}} \, dt = \frac{\pi\lambda^z}{z\sin(\pi z)} \quad (\lambda > 0, 0 < \operatorname{Re} z < 1).$$

Example 4.2.4

We present a solution to Monthly Problem 11650 (2012, p. 522) where Laplace transform is used. The problem is to evaluate the double integral

$$A = \int_0^\infty \int_x^\infty e^{-(x-y)^2} \sin^2(x^2 + y^2) \frac{x^2 - y^2}{(x^2 + y^2)^2} \, dy \, dx.$$

Introducing polar coordinates $x = r\cos\theta, \; y = r\sin\theta$ where

$$\left\{0 < r < \infty, \; \frac{\pi}{4} < \theta < \frac{\pi}{2}\right\}$$

the integral becomes

$$A = \int_0^\infty \int_{\pi/4}^{\pi/2} e^{-r^2(1-\sin 2\theta)} \sin^2(r^2) \frac{\cos 2\theta}{r^2} \, r \, d\theta \, dr$$

and after the substitution $t = r^2$

$$A = \frac{1}{2}\int_0^\infty e^{-t}\left(\frac{\sin t}{t}\right)^2 \left\{\int_{\pi/4}^{\pi/2} e^{t\sin 2\theta}\cos 2\theta \, d\theta\right\} dt$$

$$= \frac{1}{4} \int_0^\infty e^{-t} \left(\frac{\sin t}{t} \right)^2 \left(e^{t \sin 2\theta} \big|_{\pi/4}^{\pi/2} \right) dt = \frac{1}{4} \int_0^\infty e^{-t} \left(\frac{\sin t}{t} \right)^2 (1 - e^t) \, dt$$

$$= \frac{1}{4} \int_0^\infty e^{-t} \left(\frac{\sin t}{t} \right)^2 dt - \frac{1}{4} \int_0^\infty \left(\frac{\sin t}{t} \right)^2 dt \ .$$

The Laplace transform integral

$$J(s) = \int_0^\infty e^{-st} \left(\frac{\sin t}{t} \right)^2 dt$$

is known, this is entry 3.948 (4) in [25]

$$J(s) = \frac{s}{4} \log \frac{s^2}{s^2 + 4} - \arctan \frac{s}{2} + \frac{\pi}{2}$$

and with $s = 1$ an $s = 0$ we find $A = \frac{1}{4} \big(J(1) - J(0) \big)$, that is,

$$A = \frac{1}{16} \log \frac{1}{5} - \frac{1}{4} \arctan \frac{1}{2} \ .$$

For the convenience of the reader we give the evaluations of $J(s)$ here.

Differentiating twice and using the Laplace transform formula for the cosine

$$J''(s) = \int_0^\infty e^{-st} \sin^2 t \, dt = \frac{1}{2} \int_0^\infty e^{-st} (1 - \cos 2t) \, dt = \frac{1}{2} \left(\frac{1}{s} - \frac{s}{s^2 + 4} \right).$$

Integrating this we find

$$J'(s) = \frac{1}{2} \log s - \frac{1}{4} \log(s^2 + 4) = \frac{1}{4} \log \frac{s^2}{s^2 + 4}$$

(with $s \to \infty$ we find the constant of integration to be zero). Integrating again

$$J(s) = \frac{s}{2}\log s - \frac{s}{4}\log(s^2 + 4) - \arctan\frac{s}{2} + C$$

and with $s \to \infty$ we compute $C = \frac{\pi}{2}$. The formula is proved.

Example 4.2.5

In this example we prove two interesting integrals

$$A = \int_0^\infty t^{x-1} e^{-at} \cos bt\, dt = \frac{\Gamma(x)}{(a^2 + b^2)^{\frac{x}{2}}} \cos\left(x \arctan\frac{b}{a}\right)$$

$$B = \int_0^\infty t^{x-1} e^{-at} \sin bt\, dt = \frac{\Gamma(x)}{(a^2 + b^2)^{\frac{x}{2}}} \sin\left(x \arctan\frac{b}{a}\right)$$

with $x, a > 0$ and b an arbitrary real number. These are entries 3.944(6) and 3.944(5) correspondingly in [25].

The two integrals can be viewed as Laplace transforms of $t^{x-1}\cos bt$ and $t^{x-1}\sin bt$. For $x = 1, 2, 3$ they can be computed by differentiating with respect to the variable a the Laplace transforms

$$\int_0^\infty e^{-at} \cos bt\, dt = \frac{a}{a^2 + b^2} \quad \text{and} \quad \int_0^\infty e^{-at} \sin bt\, dt = \frac{b}{a^2 + b^2}.$$

For arbitrary $x > 0$ this method does not work. Fikhtengolt's [23, Section 539] gave a nice evaluation of A and B by using a differential equation. We will prove the integrals here by using the definition of the gamma function

$$\Gamma(x) = \int_0^\infty t^{x-1} e^{-t} dt.$$

First we make the substitution $t \to ct, c > 0$ in this integral to get

$$\Gamma(x) = c^x \int_0^\infty t^{x-1} e^{-ct} dt$$

and then we extend the above equation to the right half plane for complex numbers $c = a + ib, a > 0$. Thus

$$\Gamma(x) c^{-x} = \int_0^\infty t^{x-1} e^{-at} e^{-ibt} dt .$$

All we have to do now is separate real and imaginary parts in this equation by using Euler's formula

$$e^{-ibt} = \cos bt - i \sin bt$$

and also by using the polar representation

$$a + ib = (a^2 + b^2)^{1/2} e^{i\theta}, \ \theta = \arctan \frac{b}{a} .$$

This gives

$$c^{-x} = (a + ib)^{-x} = (a^2 + b^2)^{-x/2} e^{-ix\theta}$$

$$(a + ib)^{-x} = (a^2 + b^2)^{-x/2} \left(\cos x \arctan \frac{b}{a} + i \sin x \arctan \frac{b}{a} \right)$$

so comparing real and imaginary parts proves A and B.

With $x = \frac{1}{2}$ in A and B we find

$$\int_0^\infty e^{-at} \frac{\cos bt}{\sqrt{t}} dt = \frac{\sqrt{\pi}}{(a^2 + b^2)^{\frac{1}{4}}} \cos\left(\frac{1}{2} \arctan \frac{b}{a} \right)$$

$$\int_0^\infty e^{-at} \frac{\sin bt}{\sqrt{t}} dt = \frac{\sqrt{\pi}}{(a^2 + b^2)^{\frac{1}{4}}} \sin\left(\frac{1}{2} \arctan \frac{b}{a} \right)$$

and now we can use the identities

$$\cos^2\left(\frac{1}{2}\arctan\frac{b}{a}\right) = \frac{1}{2}\left(1 + \frac{a}{\sqrt{a^2+b^2}}\right)$$

$$\sin^2\left(\frac{1}{2}\arctan\frac{b}{a}\right) = \frac{1}{2}\left(1 - \frac{a}{\sqrt{a^2+b^2}}\right)$$

to write the curious integrals

$$\int_0^\infty e^{-ax}\frac{\cos bx}{\sqrt{x}}\,dx = \sqrt{\frac{\pi}{2}}\frac{\sqrt{\sqrt{b^2+a^2}+a}}{\sqrt{b^2+a^2}}$$

$$\int_0^\infty e^{-ax}\frac{\sin bx}{\sqrt{x}}\,dx = \sqrt{\frac{\pi}{2}}\frac{\sqrt{\sqrt{b^2+a^2}-a}}{\sqrt{b^2+a^2}}.$$

Differentiating *n*-times these integrals with respect to the variable *a* we prove the strange looking entries 3.944(13) and (14) in [25]

$$\int_0^\infty x^{n-\frac{1}{2}}e^{-ax}\cos bt\,dx = (-1)^n\sqrt{\frac{\pi}{2}}\frac{d^n}{da^n}\frac{\sqrt{\sqrt{b^2+a^2}+a}}{\sqrt{b^2+a^2}}$$

$$\int_0^\infty x^{n-\frac{1}{2}}e^{-ax}\sin bt\,dx = (-1)^n\sqrt{\frac{\pi}{2}}\frac{d^n}{da^n}\frac{\sqrt{\sqrt{b^2+a^2}-a}}{\sqrt{b^2+a^2}}.$$

Example 4.2.6

This is a solution to Monthly Problem 12260 (AMM, June-July 2021). Show that

$$J = \int_0^\infty \frac{\sin^2 x - x\sin x}{x^3}\,dx = \frac{1}{2} - \ln 2.$$

Integration by parts gives (the middle term is zero)

$$J = \frac{1}{2}\int_0^\infty \frac{\sin 2x - \sin x - x\cos x}{x^2}dx$$

and now we introduce the function

$$F(s) = \int_0^\infty e^{-sx}\frac{\sin 2x - \sin x - x\cos x}{x^2}dx \quad (s \geq 0)$$

which is the Laplace transform of the integrand. Differentiating with respect to s twice gives consecutively

$$F'(s) = -\int_0^\infty e^{-sx}\frac{\sin 2x - \sin x - x\cos x}{x}dx$$

$$F''(s) = \int_0^\infty e^{-sx}(\sin 2x - \sin x - x\cos x)dx .$$

Using the properties of the Laplace transform we can write this in the form

$$F''(s) = \frac{2}{s^2+4} - \frac{1}{s^2+1} + \frac{d}{ds}\frac{s}{s^2+1}$$

and now we can integrate back to find

$$F'(s) = \arctan\frac{s}{2} - \arctan s + \frac{s}{s^2+1}$$

(the constant of integration is zero, as $\lim_{s\to\infty} F'(s) = 0$). Integrating this again (the arctangent we integrate by parts) we come to the explicit form

$$F(s) = \ln\frac{s^2+1}{s^2+4} + s\left(\arctan\frac{s}{2} - \arctan s\right).$$

The constant of integration is zero again as $\lim_{s\to\infty} F(s) = 0$ (a simple calculus exercise for the reader). Setting $s = 0$ we compute

$$J = \frac{1}{2}F(0) = \frac{1}{2}(-\ln 4 + 1) = \frac{1}{2} - \ln 2.$$

4.3   A Tale of Two Integrals

Here we consider two interesting similar integrals. For $t \geq 0$, let

(4.3)
$$A(t) = \int_0^\infty \frac{\cos(xt)}{x^2 + 1} dx$$

(4.4)
$$B(t) = \int_0^\infty \frac{\sin(xt)}{x^2 + 1} dx.$$

Both integrals can be viewed as Fourier cosine and sine transforms of the function $(x^2 + 1)^{-1}$. They really look very much alike.

We will evaluate both integrals now by using Laplace transform and it will become clear that the two functions $A(t)$ and $B(t)$ are quite different.

Thus

$$L\{A(t)\} = \int_0^\infty \left\{ \int_0^\infty \frac{\cos(xt)}{x^2 + 1} dx \right\} e^{-st} dt = \int_0^\infty \frac{1}{x^2 + 1} \left\{ \int_0^\infty \cos(xt) e^{-st} dt \right\} dx$$

$$= \int_0^\infty \frac{1}{x^2 + 1} \left\{ \frac{s}{x^2 + s^2} \right\} dx = \frac{s}{s^2 - 1} \int_0^\infty \left\{ \frac{1}{x^2 + 1} - \frac{1}{x^2 + s^2} \right\} dx$$

$$= \frac{s}{s^2 - 1} \left\{ \arctan x - \frac{1}{s} \arctan \frac{x}{s} \right\} \Big|_0^\infty = \frac{\pi}{2} \frac{s}{s^2 - 1} \left(1 - \frac{1}{s} \right) = \frac{\pi}{2} \frac{1}{s + 1}$$

so that

$$L\{A(t)\} = \frac{\pi}{2} L\{e^{-t}\}.$$

Therefore,

$$A(t) = \frac{\pi}{2} e^{-t}.$$

(This result was obtained in Chapter 2, Example 2.3.2 by a different method.) The integral is entry 3.723(2) in [25]. The change of the order of integration above is easy to justify and we leave it to the reader.

Now we try the same technique on the second integral (4.4)

$$L\{B(t)\} = \int_0^\infty \left\{ \int_0^\infty \frac{\sin(xt)}{x^2+1} dx \right\} e^{-st} dt = \int_0^\infty \frac{1}{x^2+1} \left\{ \int_0^\infty \cos(xt) e^{-st} dt \right\} dx$$

$$= \int_0^\infty \frac{1}{x^2+1} \left\{ \frac{x}{x^2+s^2} \right\} dx = \frac{1}{s^2-1} \int_0^\infty \left\{ \frac{x}{x^2+1} - \frac{x}{x^2+s^2} \right\} dx$$

$$\frac{1}{s^2-1} \left\{ \frac{1}{2} \ln \frac{x^2+1}{x^2+s^2} \right\} \Bigg|_0^\infty = \frac{1}{s^2-1} \left(-\frac{1}{2} \ln \frac{1}{s^2} \right) = \frac{\ln s}{s^2-1}.$$

That is,

(4.5) $$L\{B(t)\} = \frac{\ln s}{s^2-1}$$

and it is not clear what is the original for this Laplace image.

Consulting the table of Fourier sine transforms [5] we find that

$$\int_0^\infty \frac{\sin(xt)}{x^2+1} dx = \frac{1}{2} \left[e^{-t} \overline{\mathrm{Ei}}(t) - e^t \, \mathrm{Ei}(-t) \right]$$

where Ei is the exponential integral function

$$\mathrm{Ei}(t) = -\int_{-x}^\infty \frac{e^{-u}}{u} du$$

and $$\overline{\mathrm{Ei}}(t) = \frac{1}{2} \big(\mathrm{Ei}(t+i0) + \mathrm{Ei}(t-i0) \big).$$

A good reference for the exponential integral is [40].

This evaluation exists also in the popular tables [25] and [43]. It can be traced back to the handbook *"Nouvelles tables d'intégrales définies"* (Amsterdam, 1858) by the Dutch mathematician David Bierens de Haan. The integral appears there as number 7 in Table 204 on p. 282. (In the improved 1867 edition of this book the integral appears as number 3 in table 160 on page 223.) This integral possibly originated in the works of Joseph Ludwig Raabe (1801-1859) and Oscar Schlömilch (1823-1901) - see the historical note by Nielsen on page 24 in his book [39]. The integral is sometimes called Raabe integral.

The exponential integral $\mathrm{Ei}(t)$ is not a very convenient function. We prefer to evaluate $B(t)$ in terms of the entire function

$$\mathrm{Ein}(t) = \sum_{n=1}^{\infty} \frac{(-1)^{n-1} t^n}{n!\, n}$$

which is related to the exponential integral by the equation

$$\mathrm{Ein}(-t) = \gamma + \ln t - \mathrm{Ei}(t), \quad t > 0 .$$

Here $\gamma = -\psi(1)$ is Euler's constant and $\psi(z) = \dfrac{d}{dz} \ln \Gamma(z)$ is the digamma function. With the function $\mathrm{Ein}(t)$ we do not have to deal with one-sided limits.

Proposition 4.1. *For every* $t > 0$

(4.6) $B(t) = \dfrac{1}{2}\left[e^t\, \mathrm{Ein}(t) - e^{-t}\, \mathrm{Ein}(-t) \right] - (\ln t + \gamma)\sinh t .$

Before proving this evaluation we need to prepare two known facts.

First let

$$H_n = 1 + \frac{1}{2} + \frac{1}{3} + \ldots + \frac{1}{n}, \quad H_0 = 0$$

be the harmonic numbers. Multiplying the two power series for e^x and $\text{Ein}(x)$ we write

(4.7)
$$e^x \text{Ein}(x) = \sum_{n=0}^{\infty} \frac{x^n}{n!} \left\{ \sum_{k=1}^{n} \binom{n}{k} \frac{(-1)^{k-1}}{k} \right\} = \sum_{n=0}^{\infty} \frac{x^n}{n!} H_n$$

according to the well-known binomial identity

$$\sum_{k=1}^{n} \binom{n}{k} \frac{(-1)^{k-1}}{k} = H_n$$

(see [8]). Thus $e^x \text{Ein}(x)$ is the exponential generation function for the harmonic numbers. Their well-known ordinary generating function is

(4.8)
$$-\frac{\ln(1-x)}{1-x} = \sum_{n=0}^{\infty} H_n x^n \quad (|x| < 1).$$

Second fact. The Laplace transform of the logarithm is

$$\int_0^{\infty} e^{-st} \ln t \, dt = -\frac{1}{s}(\ln s + \gamma).$$

This is known. For convenience we give a proof. Differentiating with respect to p the equation

$$\Gamma(p) = s^p \int_0^{\infty} t^{p-1} e^{-st} \, dt$$

we get

$$\Gamma'(p) = s^p \ln s \int_0^{\infty} t^{p-1} e^{-st} \, dt + s^p \int_0^{\infty} t^{p-1} (\ln t) e^{-st} \, dt$$

and then with $p = 1$ we find the desired result since $\Gamma'(1) = \psi(1) = -\gamma$. It follows that

$$\int_0^\infty (\ln t + \gamma) e^{-st}\, dt = -\frac{\ln s}{s}.$$

Proof of the proposition. We will compute the Laplace transform of the right-hand side in (4.6). First, for $s > 1$ we have in view of (4.7) and (4.8)

$$\frac{1}{2} L\{e^t \operatorname{Ein}(t) - e^{-t} \operatorname{Ein}(-t)\}$$

$$= \frac{1}{2}\left\{ \sum_{n=0}^\infty \frac{H_n}{n} \frac{1}{s^{n+1}} - \sum_{n=0}^\infty \frac{H_n}{n} \frac{(-1)^n}{s^{n+1}} \right\}$$

$$= \frac{1}{2s}\left\{ -\frac{\ln\left(1 - \dfrac{1}{s}\right)}{1 - \dfrac{1}{s}} + \frac{\ln\left(1 + \dfrac{1}{s}\right)}{1 + \dfrac{1}{s}} \right\}$$

$$= \frac{1}{2}\left\{ \frac{\ln(s+1) - \ln s}{s+1} - \frac{\ln(s-1) - \ln s}{s-1} \right\}$$

$$= \frac{1}{2}\left\{ \frac{\ln(s+1)}{s+1} - \frac{\ln(s-1)}{s-1} \right\} + \frac{\ln s}{s^2 - 1}.$$

Also

$$L\{(\ln t + \gamma)\sinh t\} = \frac{1}{2} L\{(\ln t + \gamma)(e^t - e^{-t})\}$$

$$= \frac{1}{2}\left\{ \frac{\ln(s+1)}{s+1} - \frac{\ln(s-1)}{s-1} \right\}$$

by using the "s-shift" property of the Laplace transform. Combining this with the previous evaluation we see that the right-hand side in (4.6) becomes

$$\frac{\ln s}{s^2 - 1}.$$

According to (4.5) the proposition is proved.

We will finish this section with three remarks.

Remark 1. The above proof shows an interesting connection between $B(t)$ and the harmonic numbers H_n. Namely, (4.6) can be written in the form

$$\int_0^\infty \frac{\sin(xt)}{x^2+1}dx + (\ln t + \gamma)\sinh t = \sum_{n=0}^\infty \frac{t^{2n+1}}{(2n+1)}H_{2n+1}.$$

Remark 2. In comparison to (4.4) the integral

$$\int_0^\infty \frac{\sin^2(xt)}{x^2+1}dx$$

has a simple evaluation. Writing $\sin^2(xt)=(1/2)(1-\cos(2xt))$ and using the evaluation of (4.3) we find

$$\int_0^\infty \frac{\sin^2(xt)}{x^2+1}dx = \frac{\pi}{4}(1-e^{-2t}).$$

More generally, using the representations 1.320 from [25]

$$\sin^{2n}x = \frac{1}{2^{2n}}\left\{\sum_{k=0}^{n-1}\binom{2n}{k}(-1)^{n-k}2\cos 2(n-k)x + \binom{2n}{n}\right\}$$

$$\cos^{2n}x = \frac{1}{2^{2n}}\left\{\sum_{k=0}^{n-1}\binom{2n}{k}2\cos 2(n-k)x + \binom{2n}{n}\right\}$$

$$\cos^{2n-1}x = \frac{1}{2^{2n-2}}\sum_{k=0}^{n-1}\binom{2n-1}{k}\cos(2n-2k-1)x$$

we prove the evaluations

$$\int_0^\infty \frac{\sin^{2n}(xt)}{x^2+1}dx = \frac{\pi}{2^{2n+1}}\left\{\sum_{k=0}^{n-1}\binom{2n}{k}(-1)^{n-k}2e^{-2(n-k)t} + \binom{2n}{n}\right\}$$

$$\int_0^\infty \frac{\cos^{2n}(xt)}{x^2+1}dx = \frac{\pi}{2^{2n+1}}\left\{\sum_{k=0}^{n-1}\binom{2n}{k}2e^{-2(n-k)t} + \binom{2n}{n}\right\}$$

$$\int_0^\infty \frac{\cos^{2n-1}(xt)}{x^2+1}dx = \frac{\pi}{2^{2n-1}}\sum_{k=0}^{n-1}\binom{2n-1}{k}e^{-(2n-2k-1)t}$$

(cf. [25], entries 3.824 (1), (3), (6), and (7)).

Remark 3. The famous Indian mathematician Srinivasa Ramanujan (1887-1920) discovered the equation

$$\frac{\pi}{2}\int_0^\infty \frac{\sin(xt)}{x^2+1}dx + \int_0^\infty \frac{\cos(xt)\ln x}{x^2+1}dx = 0$$

(this was confirmed by contour integration in the recent paper by Bruce C. Berndt and Armin Straub "Certain Integrals Arising from Ramanujan's Notebooks", SIGMA 11 (2015), 083). From this equation we learn the value of the second integral as well. With $t = 0$ we find again

$$\int_0^\infty \frac{\ln x}{x^2+1}dx = 0$$

(see Example 1.8.3).

4.4 Parseval's Theorem

In this section we will work with the Fourier cosine transform and we will evaluate interesting and challenging integrals by using Parseval's theorem.

For functions $f(x)$ defined on $[0, \infty)$ Fourier's cosine transform is defined by

$$F_c(y) = \int_0^\infty f(x)\cos(xy)dx .$$

Let $f(x)$ and $g(x)$ be two real valued functions on $[0,\infty)$ with Fourier cosine transforms $F_c(y)$ and $G_c(y)$. **Parseval's theorem** says that

(4.9) $$\int_0^\infty F_c(y)G_c(y)dy = \frac{\pi}{2}\int_0^\infty f(x)g(x)dx$$

(see [22], section 2.14). We can evaluate hard integrals by using this property. The idea is to select such pairs of transforms that one of the integrals in (4.9) will be easy to evaluate and this will provide the evaluation of the other one.

In the examples below we use Batemann's table of cosine transforms in [5].

Example 4.4.1

Consider the pair of Fourier cosine transforms

$$\pi\frac{1-e^{-ay}}{y} = \int_0^\infty \ln\left(1+\frac{a^2}{x^2}\right)\cos(xy)dx$$

$$\frac{\pi}{2b}e^{-by} = \int_0^\infty \frac{\cos(xy)}{x^2+b^2}dx$$

where $a,b > 0$. Parseval's theorem implies

$$\frac{\pi}{b}\int_0^\infty \frac{(1-e^{-ay})e^{-by}}{y}dy = \int_0^\infty \ln\left(1+\frac{a^2}{x^2}\right)\frac{dx}{x^2+b^2} .$$

The integral on the left-hand side is a Frullani integral (see Chapter 5)

$$\int_0^\infty \frac{(1-e^{-ay})e^{-by}}{y}dy = \int_0^\infty \frac{e^{-by}-e^{-(a+b)y}}{y}dy = \ln\frac{a+b}{b}$$

and we obtain the evaluation

$$\frac{\pi}{b}\ln\frac{a+b}{b} = \int_0^\infty \ln\left(1+\frac{a^2}{x^2}\right)\frac{dx}{x^2+b^2}.$$

This integral is equivalent (through a simple manipulation) to entry 4.295(7) in [25].

Example 4.4.2

Here we work with the pair of cosine transforms ($a, s > 0$)

$$\frac{s}{y^2+s^2} = \int_0^\infty e^{-sx}\cos(xy)dx$$

$$\frac{\sqrt{\pi}}{2\sqrt{a}}e^{-\frac{y^2}{4a}} = \int_0^\infty e^{-ax^2}\cos(xy)dx$$

where (4.9) implies

$$\frac{s}{\sqrt{\pi a}}\int_0^\infty e^{-\frac{y^2}{4a}}\frac{dy}{y^2+s^2} = \int_0^\infty e^{-ax^2-sx}dx.$$

Setting $a = \dfrac{1}{4}$ and $x = 2t$ we can write this in the form

$$\frac{s}{\sqrt{\pi}}\int_0^\infty \frac{e^{-y^2}}{y^2+s^2}dy = e^{s^2}\int_0^\infty e^{-(t+s)^2}dt = e^{s^2}\int_s^\infty e^{-u^2}du$$

$$= e^{s^2}\frac{\sqrt{\pi}}{2}(1-\text{erf}(s))$$

where

$$\operatorname{erf}(x) = \frac{2}{\sqrt{\pi}} \int_0^x e^{-u^2} du$$

is the error function. This way we have the evaluation

$$(4.10) \qquad \int_0^\infty \frac{e^{-y^2}}{y^2 + s^2} dy = \frac{\sqrt{\pi} \, e^{s^2}}{s} \int_s^\infty e^{-u^2} du = \frac{\pi e^{s^2}}{2s} (1 - \operatorname{erf}(s)).$$

In particular, for $s = 1$ we obtain the interesting formula

$$\int_0^\infty \frac{e^{-y^2}}{y^2 + 1} dy = \sqrt{\pi} e \int_1^\infty e^{-u^2} du.$$

Example 4.4.3

The third pair we employ is

$$\frac{1}{2} \left(\frac{1}{y^2} - \frac{\pi^2}{\sinh^2 \pi y} \right) = \int_0^\infty \frac{x \cos(xy)}{e^x - 1} dx$$

$$\frac{\pi}{2\Gamma(\beta)} \sec \frac{\pi\beta}{2} y^{\beta-1} = \int_0^\infty x^{-\beta} \cos(xy) dx$$

where $0 < \beta < 1$. (The second integral was evaluated in Example 4.2.3.) Parseval's identity gives

$$\frac{1}{2\Gamma(\beta)} \sec \frac{\pi\beta}{2} \int_0^\infty \left(\frac{1}{y^2} - \frac{\pi^2}{\sinh^2 \pi y} \right) y^{\beta-1} dy = \int_0^\infty \frac{x^{1-\beta}}{e^x - 1} dx$$

$$= \zeta(2 - \beta)\Gamma(2 - \beta).$$

Here the last equality comes from the well-known representation

$$\zeta(s)\Gamma(s) = \int_0^\infty \frac{x^{s-1}}{e^x - 1} dx \quad (\operatorname{Re} s > 1)$$

for the Riemann zeta function $\zeta(s)$ (see below Section 4.7). The result can be simplified by using the properties

$$\Gamma(\beta)\Gamma(1-\beta) = \frac{\pi}{\sin \pi\beta}, \quad \sin \pi\beta = 2\sin\frac{\pi\beta}{2}\cos\frac{\pi\beta}{2}$$

and we write

$$\int_0^\infty \left(\frac{1}{y^2} - \frac{\pi^2}{\sinh^2 \pi y}\right) y^{\beta-1} dy = \pi(1-\beta)\csc\frac{\pi\beta}{2}\zeta(2-\beta).$$

Also, with the substitution $1-\beta=\alpha$, $0<\alpha<1$ this becomes

(4.11)
$$\int_0^\infty \left(\frac{1}{y^2} - \frac{\pi^2}{\sinh^2 \pi y}\right) y^{-\alpha} dy = \pi\alpha \sec\frac{\pi\alpha}{2}\zeta(1+\alpha)$$

or

$$\int_0^\infty \frac{1}{y^2}\left[1 - \left(\frac{\pi y}{\sinh \pi y}\right)^2\right] y^{-\alpha} dy = \pi\alpha \sec\frac{\pi\alpha}{2}\zeta(1+\alpha).$$

With $\alpha \to 0$ in (4.11) we easily compute

(4.12)
$$\int_0^\infty \left(\frac{1}{y^2} - \frac{\pi^2}{\sinh^2 \pi y}\right) dy = \pi.$$

Example 4.4.4

From the pair

$$\pi\frac{1-e^{-ay}}{y} = \int_0^\infty \ln\left(1 + \frac{a^2}{x^2}\right)\cos(xy)dx$$

$$\frac{\pi}{2\Gamma(\beta)}\sec\frac{\pi\beta}{2}y^{\beta-1} = \int_0^\infty x^{-\beta}\cos(xy)dx$$

($a > 0$, $0 < \beta < 1$) we obtain the equation

$$\frac{\pi}{\Gamma(\beta)} \sec \frac{\pi\beta}{2} \int_0^\infty (1 - e^{-ay}) y^{\beta-2} dy = \int_0^\infty x^{-\beta} \ln\left(1 + \frac{a^2}{x^2}\right) dx .$$

Integrating by parts the integral on the left-hand side we find

$$\int_0^\infty (1 - e^{-ay}) y^{\beta-2} dy$$

$$= \frac{1}{\beta-1} (1 - e^{-ay}) y^{\beta-1} \Big|_0^\infty - \frac{a}{\beta-1} \int_0^\infty y^{\beta-1} e^{-ay} dy$$

$$= \frac{a}{1-\beta} \cdot \frac{\Gamma(\beta)}{a^{\beta+1}} = \frac{\Gamma(\beta)}{a^\beta (1-\beta)}$$

(the limits of $(1 - e^{-ay}) y^{\beta-1}$ at ∞ and 0 are zeros).

This way the equation takes the form

(4.13) $$\frac{\pi}{a^\beta (1-\beta)} \sec \frac{\pi\beta}{2} = \int_0^\infty x^{-\beta} \ln\left(1 + \frac{a^2}{x^2}\right) dx .$$

This integral is equivalent to entry 4.293 (10) in [25].

Example 4.4.5

Here we work again with the two integrals

$$\frac{\pi}{2\Gamma(\beta)} \sec \frac{\pi\beta}{2} y^{\beta-1} = \int_0^\infty x^{-\beta} \cos(xy) dx$$

$$\frac{\pi}{2b} e^{-by} = \int_0^\infty \frac{\cos(xy)}{x^2 + b^2} dx$$

$(0 < \beta < 1, \; b > 0)$. Parseval's theorem implies

$$\frac{\pi}{2b\Gamma(\beta)}\sec\frac{\pi\beta}{2}\int_0^\infty y^{\beta-1}e^{-by}\,dy = \int_0^\infty \frac{x^{-\beta}}{x^2+b^2}\,dx.$$

Since the first integral here equals $\Gamma(\beta)/b^\beta$, the left-hand side simplifies and we come to the important integral

$$(4.14) \qquad\qquad \frac{\pi}{2b^{\beta+1}}\sec\frac{\pi\beta}{2} = \int_0^\infty \frac{x^{-\beta}}{x^2+b^2}\,dx.$$

This integral was evaluated also in Example 4.2.3 by using Euler's beta function (cf. Example 2.3.21 on p. 116 in [34]).

Example 4.4.6

Using the pair

$$\frac{\pi}{2a}\mathrm{sech}\left(\frac{\pi y}{2a}\right) = \int_0^\infty \cos(xy)\,\mathrm{sech}(ax)\,dx$$

$$\frac{\pi}{2b}e^{-by} = \int_0^\infty \frac{\cos(xy)}{x^2+b^2}\,dx$$

$(a, b > 0)$ we find

$$\frac{\pi}{2ab}\int_0^\infty e^{-by}\,\mathrm{sech}\left(\frac{\pi y}{2a}\right)dy = \int_0^\infty \frac{\mathrm{sech}(ax)}{x^2+b^2}\,dx.$$

With $a = \pi/2$ this becomes

$$\int_0^\infty \frac{e^{-by}}{\cosh y}\,dy = b\int_0^\infty \mathrm{sech}\left(\frac{\pi x}{2}\right)\frac{dx}{x^2+b^2}.$$

The integral on the left-hand side will be evaluated in the next section in terms of Nielsen's beta function $\beta(x)$. Namely,

(4.15)
$$\int_0^\infty \frac{e^{-by}}{\cosh y}\,dy = \beta\left(\frac{b+1}{2}\right).$$

This way we have also

$$\int_0^\infty \operatorname{sech}\left(\frac{\pi x}{2}\right)\frac{dx}{x^2+b^2} = \frac{1}{b}\beta\left(\frac{b+1}{2}\right).$$

4.5 Some Important Hyperbolic Integrals

4.5.1 *Expansion of the cotangent in partial fractions*

We will need the expansions of $\cot x, \tan x, \sec x$ in partial fractions.

These classical expansions can be obtained from the famous Euler formula

$$\sin x = x\prod_{n=1}^\infty\left(1-\frac{x^2}{\pi^2 n^2}\right)$$

expressing the sine function as an infinite product.
Let $0 < x < \pi$. We take logarithms of both sides to get

$$\ln\sin x = \ln x + \sum_{n=1}^\infty \ln\left(1-\frac{x^2}{\pi^2 n^2}\right)$$

$$= \ln x + \sum_{n=1}^\infty\left\{\ln(\pi^2 n^2 - x^2) - \ln(\pi^2 n^2)\right\}.$$

Differentiating this equation we come to the expansion

$$\frac{\cos x}{\sin x} = \cot x = \frac{1}{x} - 2x\sum_{n=1}^\infty\frac{1}{\pi^2 n^2 - x^2}$$

$$= \frac{1}{x} + \sum_{n=1}^{\infty} \left(\frac{1}{x - \pi n} + \frac{1}{x + \pi n} \right).$$

Replacing here x by xi we find also the representation of the hyperbolic cotangent

$$\coth x = \frac{1}{x} + 2x \sum_{n=1}^{\infty} \frac{1}{\pi^2 n^2 + x^2} .$$

Next, replacing x by $\frac{\pi}{2} - x$ in the cotangent expansion and rearranging that series we come to

$$\tan x = -\sum_{n=1}^{\infty} \left(\frac{1}{x - (n - 1/2)\pi} + \frac{1}{x + (n - 1/2)\pi} \right)$$

which can be written in the form

$$\tan x = -8x \sum_{n=1}^{\infty} \frac{1}{4x^2 - (2n-1)^2 \pi^2} .$$

The substitution $x \to xi$ provides the expansion of the hyperbolic tangent

$$\tanh x = 8x \sum_{n=1}^{\infty} \frac{1}{4x^2 + (2n-1)^2 \pi^2} .$$

Similar expansions hold for $\sec x$ and $\csc x$. We know that

$$\sin x = \frac{2 \tan \frac{x}{2}}{1 + \tan^2 \frac{x}{2}}$$

and from here

$$\frac{1}{\sin x} = \frac{1}{2} \left(\tan \frac{x}{2} + \cot \frac{x}{2} \right).$$

Adding the expansions of $\tan\dfrac{x}{2}$ and $\cot\dfrac{x}{2}$ we obtain after a simple adjustment

$$\frac{1}{\sin x} = \frac{1}{x} + \sum_{n=1}^{\infty}(-1)^n\left(\frac{1}{x-\pi n}+\frac{1}{x+\pi n}\right) = \frac{1}{x} + 2x\sum_{n=1}^{\infty}\frac{(-1)^n}{x^2-\pi^2 n^2}.$$

Replacing x by $\dfrac{\pi}{2}-x$ and rearranging the series we find also

$$\frac{1}{\cos x} = 4\pi\sum_{n=0}^{\infty}\frac{(-1)^n(2n+1)}{(2n+1)^2\pi^2-4x^2}.$$

Here again the substitution $x \to xi$ gives

$$\frac{1}{\sinh x} = \frac{1}{x} + 2x\sum_{n=1}^{\infty}\frac{(-1)^n}{x^2+\pi^2 n^2}$$

$$\frac{1}{\cosh x} = 4\pi\sum_{n=0}^{\infty}\frac{(-1)^n(2n+1)}{(2n+1)^2\pi^2+4x^2}.$$

Example 4.5.1

We prove now the important integral

$$\int_0^{\infty}\frac{\sin xt}{\sinh\dfrac{\pi t}{2}}dt = \tanh x$$

by using the expansion of $\tanh x$ in partial fractions. We write

$$\int_0^{\infty}\frac{\sin xt}{\sinh\dfrac{\pi t}{2}}dt = 2\int_0^{\infty}\frac{\sin xt}{e^{\pi t/2}-e^{-\pi t/2}}dt = 2\int_0^{\infty}\sin xt\,\frac{e^{-\pi t/2}}{1-e^{-\pi t}}dt$$

$$= 2\int_0^\infty \sin xt \left\{ \sum_{n=0}^\infty e^{-\pi\left(n+\frac{1}{2}\right)t} \right\} dt = 2\sum_{n=0}^\infty \left\{ \int_0^\infty \sin xt \, e^{-\pi\left(n+\frac{1}{2}\right)t} \right\} dt$$

$$2\sum_{n=0}^\infty \frac{x}{x^2 + \pi^2 (n+1/2)^2} = 8x \sum_{n=0}^\infty \frac{1}{4x^2 + \pi^2 (2n+1)^2} = \tanh x$$

according to the expansion of $\tanh x$ obtained above.

This integral will be used later in Section 4.7.

Example 4.5.2

Using the expansion of $\coth x$ in partial fractions we will prove here one very interesting integral, namely,

$$\int_0^\infty \frac{\sin xt}{e^t - 1} dt = \frac{\pi}{2} + \frac{\pi}{e^{2\pi x} - 1} - \frac{1}{2x}.$$

For $x > 0$ we write

$$\int_0^\infty \frac{\sin xt}{e^t - 1} dt = \int_0^\infty \sin xt \frac{e^{-t}}{1 - e^{-t}} dt = \int_0^\infty \sin xt \left\{ \sum_{n=1}^\infty e^{-nt} \right\} dt$$

$$= \sum_{n=1}^\infty \left\{ \int_0^\infty \sin xt \, e^{-nt} dt \right\} = \sum_{n=1}^\infty \frac{x}{x^2 + n^2} = \frac{\pi}{2} \coth \pi x - \frac{1}{2x}.$$

At the same time

$$\frac{\pi}{2} \coth \pi x = \frac{\pi}{2} \frac{e^{\pi x} + e^{-\pi x}}{e^{\pi x} - e^{-\pi x}} = \frac{\pi}{2} \frac{(e^{\pi x} - e^{-\pi x} + 2e^{-\pi x})}{e^{\pi x} - e^{-\pi x}}$$

$$= \frac{\pi}{2} + \frac{\pi e^{-\pi x}}{e^{\pi x} - e^{-\pi x}} = \frac{\pi}{2} + \frac{\pi}{e^{2\pi x} - 1},$$

and the proof is completed.
This integral will be used later in Section 4.9.

The integral

$$\int_0^\infty \frac{\sin xt}{e^t - 1} dt = \frac{\pi}{2} \coth \pi x - \frac{1}{2x}$$

appears as entry 3.911(2) in [25] and as entry 2.534(4) in [43].

With rescaling we can write the above integral in the form

$$\int_0^\infty \frac{\sin \lambda t}{e^{2\pi t} - 1} dt = \frac{1}{4}\left(1 + 2\frac{e^{-\lambda}}{1 - e^{-\lambda}}\right) - \frac{1}{2\lambda}$$

$$= \frac{1}{4}\left(1 + 2\sum_{k=1}^\infty e^{-k\lambda}\right) - \frac{1}{2\lambda}$$

or

$$\sum_{k=1}^\infty e^{-k\lambda} = -\frac{1}{2} + \frac{1}{\lambda} + 2\int_0^\infty \frac{\sin \lambda t}{e^{2\pi t} - 1} dt \; .$$

Differentiating n times with respect to λ we find

$$\sum_{k=1}^\infty k^n e^{-k\lambda} = \frac{n!}{\lambda^{n+1}} + 2(-1)^n \int_0^\infty \sin\left(\lambda t + \frac{n\pi}{2}\right) \frac{t^n}{e^{2\pi t} - 1} dt \; .$$

This representation can be used to solve the *American Mathematical Monthly* Problem 12075 (2018). The problem is this:

Let $x_n = \sum_{k=1}^\infty k^n / e^k$. Prove that $\lim_{n \to \infty} (x_n / n!) = 1$, but the sequence $(x_n - n!)_{n \geq 1}$ is unbounded.

Indeed, we have with $\lambda = 1$

$$\frac{x_n}{n!} - 1 = \frac{2(-1)^n}{n!} \int_0^\infty \sin\left(t + \frac{n\pi}{2}\right) \frac{t^n}{e^{2\pi t} - 1} dt$$

and $\lim\limits_{n\to\infty}\left(\dfrac{x_n}{n!}-1\right)=0$ since the right-hand side approaches zero. We can estimate this way

$$|(x_n/n!)-1|\le\frac{2}{n!}\int_0^\infty\frac{t^n}{e^{2\pi t}-1}\,dt$$

$$=\frac{2}{n!(2\pi)^{n+1}}\int_0^\infty\frac{t^n}{e^t-1}\,dt=\frac{2\zeta(n+1)}{(2\pi)^{n+1}}$$

and the sequence $\zeta(n+1)$ is bounded. For the second part of the problem we can write

$$x_{4m}-(4m)!=2\int_0^\infty\sin t\,\frac{t^{4m}}{e^{2\pi t}-1}\,dt$$

which is an unbounded subsequence of $(x_n-n!)_{n\ge1}$.

Problem for the reader: Prove that for $x>0$

$$\int_0^\infty\frac{\sin xt}{e^t+1}\,dt=\frac{1}{2x}-\frac{\pi}{2}\operatorname{csch}\pi x$$

(entry 3.911(1) in [25] and entry 2.5.34(2) in [43]).

4.5.2 *Evaluation of important hyperbolic integrals*

Consider the function

$$(4.16)\qquad\beta(x)=\sum_{n=0}^\infty\frac{(-1)^n}{n+x}=\int_0^1\frac{u^{x-1}}{u+1}\,du=\int_0^\infty\frac{e^{-xt}e^t}{e^t+1}\,dt$$

for $\operatorname{Re}(x)\ne0,-1,-2,\dots$ This function was studied and used extensively by Nielsen and is sometimes called Nielsen's beta function. It was called the incomplete beta function in [16] (see also [35]).

The first equality above follows from expanding $(1+u)^{-1}$ in geometric series inside the integral and then integrating term by term. For the second equality we use the substitution $u = e^{-t}$.

The function $\beta(x)$ is closely related to the digamma function $\psi(x) = (d/dx)\ln\Gamma(x)$ which has the series representation

$$\psi(x) = -\gamma + \sum_{n=0}^{\infty}\left(\frac{1}{n+1} - \frac{1}{n+x}\right).$$

Comparing this series to (4.16) we see that

$$\beta(x) = \frac{1}{2}\left[\psi\left(\frac{x+1}{2}\right) - \psi\left(\frac{x}{2}\right)\right].$$

Now we prove the evaluation (4.15)

$$\int_0^{\infty}\frac{e^{-by}}{\cosh y}\,dy = 2\int_0^{\infty}\frac{e^{-by}}{e^y + e^{-y}}\,dy = 2\int_0^{\infty}\frac{e^{-(b+1)y}}{1+e^{-2y}}\,dy$$

$$= 2\int_0^{\infty}e^{-(b+1)y}\left\{\sum_{n=0}^{\infty}(-1)^n e^{-2ny}\right\}dy = 2\sum_{n=0}^{\infty}(-1)^n\int_0^{\infty}e^{-(b+1+2n)y}\,dy$$

$$= 2\sum_{n=0}^{\infty}\frac{(-1)^n}{b+1+2n}.$$

That is, for every $b > 0$

(4.17)
$$\int_0^{\infty}\frac{e^{-by}}{\cosh y}\,dy = 2\sum_{n=0}^{\infty}\frac{(-1)^n}{b+1+2n}.$$

Comparing this to (4.16) proves (4.15). We can also write

$$\int_0^{\infty}\frac{e^{-by}}{\cosh y}\,dy = \frac{1}{2}\left[\psi\left(\frac{b+3}{4}\right) - \psi\left(\frac{b+1}{4}\right)\right].$$

Now let $|y|<1$. We will prove the important formula

(4.18)
$$\int_0^\infty \frac{\cos(xy)}{\cosh x}\,dx = \frac{\pi}{2}\operatorname{sech}\left(\frac{\pi y}{2}\right)$$

known as Ramanujan's formula. It shows the invariance of the hyperbolic secant under the cosine transform (entry 3.981(3) in [25]). We write

$$\int_0^\infty \frac{\cos(xy)}{\cosh x}\,dx = \frac{1}{2}\int_0^\infty \frac{e^{ity}+e^{-ity}}{\cosh t}\,dt$$

$$= 2\sum_{n=0}^\infty \left\{\frac{1}{2n+1-yi}+\frac{1}{2n+1+yi}\right\}$$

$$= 2\sum_{n=0}^\infty (-1)^n \frac{2n+1}{(2n+1)^2+y^2}\,.$$

In the last series we recognize the expansion of $\dfrac{\pi}{2}\operatorname{sech}\left(\dfrac{\pi y}{2}\right)$ in partial fractions obtained above and (4.18) follows. (See also entry 6.1.62 in Hansen's table [26].)

We can write (4.18) in the form

$$\int_0^\infty \frac{\cos(xy)}{\cosh x}\,dx = \frac{\pi}{e^{\pi y/2}+e^{-\pi y/2}} = \frac{\pi e^{\pi y/2}}{1+e^{\pi y}}$$

and integrating both sides with respect to y we find

$$\int_0^\infty \frac{\sin(xy)}{x\cosh x}\,dx = \int \frac{\pi e^{\pi y/2}\,dy}{1+e^{\pi y}} = 2\int \frac{e^{\pi y/2}}{1+e^{\pi y}}\,d\frac{\pi y}{2}$$

$$= 2\int \frac{d\,e^{\pi y/2}}{1+e^{\pi y}} = 2\arctan\left(e^{\frac{\pi y}{2}}\right)+C\,.$$

With $y = 0$ we compute $C = -\dfrac{\pi}{2}$ so that

(4.19) $\displaystyle\int_0^\infty \frac{\sin(xy)}{x\cosh x}\,dx = 2\arctan\!\left(e^{\frac{\pi y}{2}}\right) - \frac{\pi}{2}$

which is entry 4.111(7) in [25] and entry 2.5.47(6) in [43].

Next we evaluate the integral

$$\int_0^\infty \frac{\sin(xy)}{\cosh x}\,dx$$

which looks simpler, but in fact, is not. We will evaluate it this way:

$$\beta\!\left(\frac{1-iy}{2}\right) = \int_0^\infty \frac{e^{ixy}}{\cosh x}\,dx = \int_0^\infty \frac{\cos(xy)}{\cosh x}\,dx + i\int_0^\infty \frac{\sin(xy)}{\cosh x}\,dx$$

so that

(4.20) $\displaystyle\int_0^\infty \frac{\sin(xy)}{\cosh x}\,dx = \operatorname{Im}\beta\!\left(\frac{1-iy}{2}\right) = 2y\sum_{n=0}^\infty \frac{(-1)^n}{(2n+1)^2 + y^2}$

where we used (4.16) for the series representation.
Prudnikov et al in [43, entry 2.5.46(4)], and Gradshteyn and Ryzhik [25, 3.981(2)], give the evaluation

$$\int_0^\infty \frac{\sin(xy)}{\cosh x}\,dx = \frac{i}{2}\left[\psi\!\left(\frac{1-iy}{4}\right) - \psi\!\left(\frac{1+iy}{4}\right)\right] - \frac{\pi}{2}\tanh\frac{\pi y}{2}.$$

The same evaluation can be found also in [5].

Some other interesting hyperbolic integrals can be found in [17].

Problems for the reader: Show that

$$\frac{1}{2}\beta\left(\frac{\mu+1}{2}\right) = \int_0^1 \frac{t^\mu}{1+t^2}\,dt = \int_0^{\pi/4} \tan^\mu x\,dx$$

for $\operatorname{Re}\mu > -1$ (entry 2.5.26(1) in [43]).

$$\int_0^\infty \frac{\sin^2(xy)}{x\sinh x}\,dx = \frac{1}{2}\ln\cosh(\pi y)$$

(entry 2.5.47(3) in [43]). Also, for $|a| < 1$, b real

$$\int_0^\infty \frac{\cosh(ax)\sin(bx)}{\sinh(x)}\,dx = \frac{\pi}{2}\frac{\sinh(\pi b)}{\cosh(\pi b) + \cos(\pi a)}$$

$$\int_0^\infty \frac{\sinh(ax)\cos(bx)}{\sinh(x)}\,dx = \frac{\pi}{2}\frac{\sinh(\pi a)}{\cosh(\pi b) + \cosh(\pi a)}$$

(entries 2.5.46(9) and 2.5.46(13) in [43]).

Another problem for the reader.

Using the expansion of $\operatorname{sech}(x)$ from section 4.5.1 and also equation (2.24) from Example 2.3.2 prove the integral (entry 4.113(12) in [25])

$$y(\lambda) = \int_0^\infty \frac{\cos\lambda x}{\cosh(\pi x/2)} \cdot \frac{dx}{1+x^2} = \lambda e^{-\lambda} + \cosh(\lambda)\ln(1 + e^{-2\lambda}).$$

Another possible solution: compute $y'' - y$ and using the integral from (4.18) solve the resulting second order differential equation by variation of parameters under initial conditions $y(0) = \ln 2$, $y'(0) = 0$. In this case the result comes in the form

$$y(\lambda) = \cosh(\lambda)\ln(2\cosh(\lambda)) - \lambda\sinh(\lambda)$$

as evaluated by Ramanujan (see equation (12) on p. 61 in the *Collected Papers of Srinivasa Ramanujan*, Chelsea, 1962).

4.6 Exponential Polynomials and Gamma Integrals

In this section we will evaluate some integrals containing the gamma function. We will use Parseval's theorem for the Fourier transform and also a special class of polynomials, the exponential polynomials.

Let $D = \dfrac{d}{dx}$. The exponential polynomials $\varphi_n(x)$ are defined by the equation

(4.21) $$(xD)^n e^x = \varphi_n(x) e^x$$

for $n = 0, 1, \ldots$. Thus we have

$$\varphi_0(x) = 1$$

$$\varphi_1(x) = x$$

$$\varphi_2(x) = x^2 + x$$

$$\varphi_3(x) = x^3 + 2x^2 + x$$

$$\varphi_4(x) = x^4 + 6x^3 + 7x^2 + x$$

$$\varphi_5(x) = x^5 + 10x^4 + 25x^3 + 15x^2 + x$$

etc. From the definition it follows immediately that

$$\varphi_{n+1}(x) = x(\varphi_n'(x) + \varphi_n(x)).$$

Using the Taylor series of the exponential function

$$e^x = \sum_{k=0}^{\infty} \frac{x^k}{k!}$$

equation (4.21) can be written in the form (with the agreement $0^0 = 1$)

$$\varphi_n(x)e^x = \sum_{k=0}^{\infty} \frac{k^n x^k}{k!} .$$

Another important property is this

(4.22)
$$\varphi_{n+1}(x) = x\sum_{k=0}^{n}\binom{n}{k}\varphi_k(x) .$$

Here is the proof. Starting from

$$\varphi_k(x)e^x = \sum_{j=0}^{\infty} \frac{j^k x^j}{j!}$$

we compute

$$\sum_{k=0}^{n}\binom{n}{k}\varphi_k(x)e^x = \sum_{j=0}^{\infty}\frac{x^j}{j!}\left\{\sum_{k=0}^{n}\binom{n}{k}j^k\right\} = \sum_{j=0}^{\infty}\frac{x^j}{j!}(j+1)^n$$

$$=\frac{1}{x}\sum_{j=0}^{\infty}\frac{(j+1)^{n+1}x^{j+1}}{(j+1)!} = \frac{1}{x}\varphi_{n+1}(x)$$

as needed.

The generating function for the polynomials φ_n justifies their name

$$e^{x(e^t-1)} = \sum_{n=0}^{\infty}\frac{\varphi_n(x)t^n}{n!} .$$

Their coefficients are the Stirling numbers of the second kind $S(n,k)$

$$\varphi_n(x) = \sum_{k=0}^{n}S(n,k)x^k .$$

The exponential polynomials appeared in the works of Ramanujan, E. T. Bell, J. Touchard, Gean-Carlo Rota and many others. As indicated in [14], they appeared as early as 1843 in the works of the prominent German mathematician Johann August Grunert.

We can write

$$x\frac{d}{dx} = ax\frac{d}{d(ax)}$$

for any nonzero number a, so equation (4.21) can be written in the more flexible form

$$(xD)^n e^{ax} = \varphi_n(ax)e^{ax}.$$

With the substitution $x = e^t$ this becomes

$$\text{(4.23)} \qquad \left(\frac{d}{dt}\right)^n e^{ae^t} = \varphi_n(ae^t)e^{ae^t}.$$

This formula describes the higher order derivatives of the function e^{ae^t}.

Next we recall the definition of Fourier transform. For functions $f(t)$ defined on $(-\infty, +\infty)$ the Fourier transform $F(x)$ is defined by

$$F(x) = \int_{-\infty}^{\infty} e^{-ixt} f(t)\, dt$$

with inversion

$$f(t) = \frac{1}{2\pi} \int_{-\infty}^{\infty} e^{ixt} F(x)\, dx.$$

We will use **Parseval's theorem**. If $g(t)$, $G(x)$ is another pair original-image, it is true that

$$\int_{-\infty}^{\infty} F(x)\overline{G(x)}\, dx = \int_{-\infty}^{\infty} f(t)\overline{g(t)}\, dt$$

where the bar on the top indicates complex conjugate.

In the following examples we will use Parseval's theorem in the same way we used it in Section 4.4.

Example 4.6.1

First we represent the gamma function as a Fourier transform.
The usual integral representation of the gamma function

$$\Gamma(z) = \int_0^\infty x^{z-1} e^{-x} dx, \quad \text{Re}(z) > 0$$

is a Mellin transform formula. It turns into a Fourier transform by the substitution $x = e^\lambda$ and also by setting $z = a + it, a > 0, -\infty < t < \infty$

$$\Gamma(a + it) = \int_{-\infty}^\infty e^{i\lambda t} e^{a\lambda} e^{-e^\lambda} d\lambda$$

From this equation by inversion

(4.24) $$2\pi e^{a\lambda} e^{-e^\lambda} = \int_{-\infty}^\infty e^{-i\lambda t} \Gamma(a + it) dt .$$

In particular, with $a = 1$ we can write

$$2\pi(-e^\lambda) e^{-e^\lambda} = 2\pi \frac{d}{d\lambda} e^{-e^\lambda} = -\int_{-\infty}^\infty e^{-i\lambda t} \Gamma(1 + it) dt .$$

Differentiating n times with respect to λ we find in accordance with (4.23) the formula

$$2\pi \varphi_{n+1}(-e^{-\lambda}) e^{-e^\lambda} = -(-i)^n \int_{-\infty}^\infty e^{-i\lambda t} t^n \Gamma(1 + it) dt$$

$(n = 0,1,...)$ or

(4.25) $$\int_{-\infty}^\infty e^{-i\lambda t} t^n \Gamma(1 + it) dt = -2\pi i^n \varphi_{n+1}(-e^{-\lambda}) e^{-e^\lambda}$$

which represents the Fourier transform of $t^n \Gamma(1+it)$. From here by setting $\lambda = 0$ we obtain the evaluation

$$(4.26) \qquad \int_{-\infty}^{\infty} t^n \Gamma(1+it)dt = -2\pi e^{-1} i^n \varphi_{n+1}(-1)$$

with

$$\varphi_{n+1}(-1) = \sum_{k=0}^{n+1} S(n+1,k)(-1)^k .$$

If $p(t) = c_0 + c_1 t + ... + c_m t$ is a polynomial we can write

$$\int_{-\infty}^{\infty} p(t)\, \Gamma(1+it)dt = -2\pi e^{-1} \sum_{k=0}^{m} c_k i^k\, \varphi_{k+1}(-1) .$$

Differentiating n-times equation (4.24) we have

$$2\pi \left(\frac{d}{d\lambda} \right)^n e^{a\lambda} e^{-e^\lambda} = (-i)^n \int_{-\infty}^{\infty} e^{-i\lambda t} t^n \Gamma(a+it)dt .$$

On the left-hand side we use the Leibniz rule to compute

$$2\pi \left(\frac{d}{d\lambda} \right)^n e^{a\lambda} e^{-e^\lambda} = \sum_{k=0}^{n} \binom{n}{k} \left(\frac{d}{d\lambda} \right)^k e^{-e^\lambda} \left(\frac{d}{d\lambda} \right)^{n-k} e^{a\lambda}$$

$$= \sum_{k=0}^{n} \binom{n}{k} \varphi_k(-e^\lambda) e^{-e^\lambda} a^{n-k} e^{a\lambda}$$

$$= e^{a\lambda} e^{-e^\lambda} \sum_{k=0}^{n} \binom{n}{k} \varphi_k(-e^\lambda) a^{n-k} .$$

This provides the Fourier transform formula $(n = 0,1,2,...)$

$$(4.27) \qquad \int_{-\infty}^{\infty} e^{-i\lambda t} t^n \Gamma(a+it)dt = 2\pi i^n e^{a\lambda} e^{-e^\lambda} \sum_{k=0}^{n} \binom{n}{k} \varphi_k(-e^\lambda) a^{n-k} .$$

When $a = 1$ this turns into (4.25) in view of (4.22). When $\lambda = 0$ we have

(4.28) $$\int_{-\infty}^{\infty} t^n \Gamma(a+it)dt = 2\pi i^n e^{-1} \sum_{k=0}^{n} \binom{n}{k} \varphi_k(-1)a^{n-k}.$$

Example 4.6.2

We are going to obtain a new integral formula with the product of two gamma functions. We replace a by $b > 0$ and λ by $\lambda - \mu$ in (4.24) to write

$$2\pi e^{b\lambda} e^{-b\mu} e^{-e^{\lambda} e^{-\mu}} = \int_{-\infty}^{\infty} e^{-i\lambda t} e^{i\mu t} \Gamma(b+it)dt.$$

Now we apply Parseval's theorem for this Fourier transform and the one in (4.27). The result is

$$2\pi i^n e^{-b\mu} \sum_{k=0}^{n} \binom{n}{k} a^{n-k} \int_{-\infty}^{\infty} e^{(a+b)\lambda} e^{-e^{\lambda}(1+e^{-\mu})} \varphi_k(-e^{\lambda})d\lambda$$

$$= \int_{-\infty}^{\infty} e^{-i\mu t} t^n \Gamma(a+it)\Gamma(b-it)dt.$$

Returning to the variable $x = e^{\lambda}$ we write this in the form

$$\int_{-\infty}^{\infty} e^{-i\mu t} t^n \Gamma(a+it)\Gamma(b-it)\,dt$$

$$= 2\pi i^n e^{-b\mu} \sum_{k=0}^{n} \binom{n}{k} a^{n-k} \int_{0}^{\infty} \varphi_k(-x)x^{a+b-1} e^{-x(1+e^{-\mu})}\,dx$$

$$= 2\pi i^n e^{-b\mu} \sum_{k=0}^{n} \binom{n}{k} a^{n-k} \sum_{j=0}^{k} S(k,j)\,(-1)^j \int_{0}^{\infty} x^{a+b+j-1} e^{-x(1+e^{-\mu})}\,dx$$

$$= 2\pi i^n e^{-b\mu} \sum_{k=0}^{n} \binom{n}{k} a^{n-k} \sum_{j=0}^{k} S(k,j) (-1)^j \frac{\Gamma(a+b+j)}{(1+e^{-\mu})^{a+b+j}}.$$

This proves the evaluation

(4.29) $$\int_{-\infty}^{\infty} e^{-i\mu t} t^n \Gamma(a+it)\Gamma(b-it) dt$$

$$= 2\pi i^n e^{-b\mu} \sum_{k=0}^{n} \binom{n}{k} a^{n-k} \sum_{j=0}^{k} S(k,j) (-1)^j \frac{\Gamma(a+b+j)}{(1+e^{-\mu})^{a+b+j}}$$

for any $n = 0,1,2,...$ and any $a,b > 0$. In particular, when $\mu = 0$

$$\int_{-\infty}^{\infty} t^n \Gamma(a+it)\Gamma(b-it) dt$$

$$= \pi i^n \sum_{k=0}^{n} \binom{n}{k} a^{n-k} \sum_{j=0}^{k} S(k,j) (-1)^j \frac{\Gamma(a+b+j)}{2^{a+b+j-1}}.$$

For $n = 0$ we have the Fourier transform formula.

(4.30) $$\int_{-\infty}^{\infty} e^{-i\mu t} \Gamma(a+it)\Gamma(b-it) dt = 2\pi \Gamma(a+b) \frac{e^{-b\mu}}{(1+e^{-\mu})^{a+b}}.$$

Note that all gamma integrals above are absolutely convergent because of the well-known estimate (M - a constant)

$$|\Gamma(a+it)| \le M |t|^{a-\frac{1}{2}} e^{-\frac{\pi}{2}|t|} \quad (|t| \to \infty).$$

Example 4.6.3

Let $0 < a \le b$. We apply Parseval's theorem to the pair of Fourier transforms

$$2\pi e^{b\lambda}e^{-b\mu}e^{-e^{\lambda}e^{-\mu}} = \int_{-\infty}^{\infty} e^{-i\lambda t}e^{i\mu t}\Gamma(b+it)dt$$

$$\frac{\pi}{a}e^{-a|\lambda|} = \int_{-\infty}^{\infty} e^{-i\lambda t}\frac{1}{a^2+t^2}dt$$

(the first formula is from the previous example). The result is

$$\frac{2\pi^2}{a}\int_{-\infty}^{\infty} e^{-a|\lambda|}e^{b\lambda}e^{-b\mu}e^{-e^{\lambda}e^{-\mu}}d\lambda = \int_{-\infty}^{\infty} e^{i\mu t}\frac{\Gamma(b+it)}{a^2+t^2}dt\ .$$

We change here μ to $-\mu$ and in the first integral we make the substitution $x = e^{\lambda}$ to get

$$\int_{-\infty}^{\infty} e^{-i\mu t}\frac{\Gamma(b+it)}{a^2+t^2}dt = \frac{2\pi^2}{a}\int_0^{\infty} e^{-a|\ln x|}x^{b-1}e^{b\mu}e^{-xe^{\mu}}dx$$

$$= \frac{2\pi^2 e^{b\mu}}{a}\left[\int_0^1 x^{a+b-1}e^{-xe^{\mu}}dx + \int_1^{\infty} x^{b-a-1}e^{-xe^{\mu}}dx\right]$$

$$= \frac{2\pi^2 e^{b\mu}}{a}\left[\frac{1}{e^{\mu(a+b)}}\int_0^{e^{\mu}} u^{a+b-1}e^{-u}du + \frac{1}{e^{\mu(b-a)}}\int_{e^{\mu}}^{\infty} u^{b-a-1}e^{-u}du\right](u = xe^{\mu})$$

$$= \frac{2\pi^2}{a}\left[e^{-a\mu}\gamma(a+b,e^{\mu}) + e^{a\mu}\Gamma(b-a,e^{\mu})\right].$$

That is,

(4.31) $$\int_{-\infty}^{\infty} e^{-i\mu t}\frac{\Gamma(b+it)}{a^2+t^2}dt$$

$$= \frac{2\pi^2}{a}\left[e^{-a\mu}\gamma(a+b,e^{\mu}) + e^{a\mu}\Gamma(b-a,e^{\mu})\right]$$

where

$$\gamma(s,p) = \int_0^p x^{s-1} e^{-x} dx$$

and

$$\Gamma(s,p) = \int_p^\infty x^{s-1} e^{-x} dx$$

are the lower and upper incomplete gamma functions. When $a = b$

$$\int_{-\infty}^\infty e^{-i\mu t} \frac{\Gamma(a+it)}{a^2+t^2} dt = \frac{2\pi^2}{a} \left[e^{-a\mu} \gamma(2a, e^\mu) - e^{a\mu} \operatorname{Ei}(-e^\mu) \right]$$

where

$$\operatorname{Ei}(x) = \int_{-x}^\infty \frac{e^{-t}}{t} dt$$

is the exponential integral function. When $\mu = 0$ we have

$$\int_{-\infty}^\infty \frac{\Gamma(b+it)}{a^2+t^2} dt = \frac{2\pi^2}{a} \left(\gamma(a+b,1) + \Gamma(b-a,1) \right).$$

4.7 The Functional Equation of the Riemann Zeta Function

The Riemann zeta function $\zeta(s)$ is defined for $\operatorname{Re} s > 1$ by the series

$$\zeta(s) = \sum_{n=1}^\infty \frac{1}{n^s}$$

Riemann's zeta function has very important applications in number theory (the distribution of prime numbers) and in analysis. Its theory originates in the works of Leonhard Euler, who often used the function

$$\eta(s) = \sum_{n=1}^\infty \frac{(-1)^{n-1}}{n^s}$$

defined for $\operatorname{Re} s > 0$. This function is known as Euler's eta function, or as the alternating zeta function. Clearly,

$$\eta(1) = \ln 2.$$

For $\operatorname{Re} s > 1$ we have

$$\eta(s) = 1 - \frac{1}{2^s} + \frac{1}{3^s} - \frac{1}{4^s} + \dots$$

$$= \left(1 + \frac{1}{2^s} + \frac{1}{3^s} + \frac{1}{4^s} + \dots\right) - 2\left(\frac{1}{2^s} + \frac{1}{4^s} + \dots\right)$$

$$= \zeta(s) - \frac{2}{2^s}\zeta(s).$$

That is, $\eta(s) = (1 - 2^{1-s})\zeta(s)$, or

$$\zeta(s) = \frac{1}{1 - 2^{1-s}}\eta(s).$$

By means of $\eta(s)$ Riemann's zeta function $\zeta(s)$ extends to the half plane $\operatorname{Re} s > 0$ with a simple pole at $s = 1$.

A fundamental property of Riemann's zeta function is the functional equation

(4.32) $$\zeta(s) = 2^s \pi^{s-1} \sin\frac{\pi s}{2}\Gamma(1-s)\zeta(1-s)$$

which helps to extend $\zeta(s)$ also for $\operatorname{Re} s \le 0$. For example,

$$\zeta(-1) = -\frac{1}{12}.$$

Titchmarsh [45, pp.12-27] gave seven proofs of the functional equation (4.32) and several others exist in the literature. Here we present yet another proof of this equation based on some results in this chapter. Our point is to show that the functional equation follows from the two integrals

(4.33)
$$\int_0^\infty \frac{\sin xt}{\sinh \dfrac{\pi t}{2}}\,dt = \tanh x$$

and

(4.34)
$$\int_0^\infty x^s \cos xt\,dx = -\frac{\Gamma(s+1)}{t^{s+1}}\sin\frac{\pi s}{2}$$

where $t > 0,\ -1 < \mathrm{Re}\,s < 0$.

The first integral can be viewed as a sine Fourier transform. It was evaluated above in Example 4.5.1. The second integral is a well-known Fresnel integral. It can be found at the end of Example 4.2.3.

Here are some integral representations of the functions $\zeta(s)$ and $\eta(s)$.

Proposition 4.2. *For* $\mathrm{Re}\,s > 1$ *and* $\mathrm{Re}\,b > 0$

(4.35)
$$\int_0^\infty \frac{x^{s-1}}{e^{bx}-1}\,dx = b^{-s}\Gamma(s)\zeta(s)$$

(4.36)
$$\int_0^\infty \frac{x^{s-1}}{e^{bx}+1}\,dx = b^{-s}\Gamma(s)\eta(s)$$

(4.37)
$$\int_0^\infty \frac{x^{s-1}}{\sinh bx}\,dx = 2b^{-s}(1-2^{-s})\Gamma(s)\zeta(s)$$

(4.38)
$$\int_0^\infty x^{s-1}(1-\tanh bx)\,dx = 2(2b)^{-s}\Gamma(s)\eta(s)$$

(4.39)
$$\int_0^\infty x^{s-1}(\coth bx - 1)\,dx = 2(2b)^{-s}\Gamma(s)\zeta(s).$$

In particular, from (4.36) and (4.38)

$$\int_0^\infty \frac{1}{e^{bx}+1}\,dx = \frac{\eta(1)}{b} = \frac{\ln 2}{b}$$

and

$$\int_0^\infty (1 - \tanh bx)\,dx = \frac{\ln 2}{b}.$$

Proof. For the proof we can take $b = 1$. Then the general form comes after the substitution $x \to bx$.

Using geometric series

$$\int_0^\infty \frac{x^{s-1}}{e^x - 1}\,dx = \int_0^\infty x^{s-1}\frac{e^{-x}}{1-e^{-x}}\,dx = \int_0^\infty x^{s-1}\left\{\sum_{n=1}^\infty e^{-nx}\right\}dx.$$

After changing the order of integration and summation this equals

$$= \sum_{n=1}^\infty \left\{\int_0^\infty x^{s-1}e^{-nx}\,dx\right\} = \Gamma(s)\sum_{n=1}^\infty \frac{1}{n^s} = \Gamma(s)\zeta(s)$$

and (4.35) is proved. Equation (4.36) is proved the same way. For (4.37) we write

$$\int_0^\infty \frac{x^{s-1}}{\sinh bx}\,dx = 2\int_0^\infty x^{s-1}\frac{e^{-x}}{1-e^{-2x}}\,dx = \int_0^\infty x^{s-1}\left\{\sum_{n=0}^\infty e^{-(2n+1)x}\right\}dx$$

$$= \sum_{n=0}^\infty \left\{\int_0^\infty x^{s-1}e^{-(2n+1)x}\,dx\right\} = \Gamma(s)\sum_{n=0}^\infty \frac{1}{(2n+1)^s}$$

and (4.37) follows, because

$$\sum_{n=0}^\infty \frac{1}{(2n+1)^s} = \zeta(s) - \frac{1}{2^s}\zeta(s).$$

To prove (4.38) we write the function $1 - \tanh x$ in terms of exponentials and expand in geometric series

$$1 - \tanh x = 1 - \frac{e^x - e^{-x}}{e^x + e^{-x}} = \frac{2e^{-x}}{e^x + e^{-x}}$$

$$= \frac{2e^{-2x}}{1 + e^{-2x}} = 2\sum_{n=1}^{\infty} (-1)^{n-1} e^{-2nx}.$$

Multiplying this by x^{s-1} and integrating we obtain

$$\int_0^\infty x^{s-1}(1 - \tanh x)\,dx = 2\Gamma(s)\sum_{n=1}^{\infty} \frac{(-1)^{n-1}}{(2n)^s} = 2(2)^{-s}\Gamma(s)\eta(s).$$

The representation (4.39) is proved in a similar way.

Differentiating (4.38) and (4.39) with respect to the variable b we find two more representations

(4.40)
$$\int_0^\infty \frac{x^s}{\cosh^2 bx}\,dx = \frac{4}{(2b)^{s+1}}\Gamma(s+1)\eta(s)$$

(4.41)
$$\int_0^\infty \frac{x^s}{\sinh^2 bx}\,dx = \frac{4}{(2b)^{s+1}}\Gamma(s+1)\zeta(s).$$

The left-hand side in (4.40) is well defined for $\mathrm{Re}\,s > -1$ and this representation provides the extension of $\eta(s)$ and $\zeta(s)$ for $\mathrm{Re}\,s > -1$, since we have

(4.42) $\quad \eta(s) = \dfrac{(2b)^{s+1}}{4\Gamma(s+1)} \displaystyle\int_0^\infty \frac{x^s}{\cosh^2 bx}\,dx, \quad \zeta(s) = \dfrac{1}{1 - 2^{1-s}}\eta(s).$

Differentiating equation (4.33) we find also

(4.43)
$$\frac{1}{\cosh^2 x} = \int_0^\infty \frac{t\cos xt}{\sinh \dfrac{\pi t}{2}}\,dt.$$

This integral will be used immediately in the following proof.

Proof of the functional equation.

For $-1 < \operatorname{Re} s < 0$ we compute

$$\int_0^\infty \frac{x^s}{\cosh^2 bx} dx = \int_0^\infty x^s \left\{ \int_0^\infty \frac{t \cos xt}{\sinh \frac{\pi t}{2}} dt \right\} dx$$

$$= \int_0^\infty \left\{ \int_0^\infty x^s \cos xt \, dx \right\} \frac{t}{\sinh \frac{\pi t}{2}} dt$$

$$= -\Gamma(s+1) \sin \frac{\pi s}{2} \left\{ \int_0^\infty \frac{t^{-s}}{\sinh \frac{\pi t}{2}} dt \right\}$$

$$= -\Gamma(s+1) \sin \frac{\pi s}{2} \left\{ 2 \left(\frac{\pi}{2} \right)^{s-1} \Gamma(1-s)(1-2^{s-1})\zeta(1-s) \right\}$$

by using (4.34) and (4.37). Combining this result with (4.40) we find

$$\frac{4}{2^{s+1}} \Gamma(s+1)(1-2^{1-s})\zeta(s)$$

$$= -\Gamma(s+1) \sin \frac{\pi s}{2} \left\{ 2 \left(\frac{\pi}{2} \right)^{s-1} \Gamma(1-s)(1-2^{s-1})\zeta(1-s) \right\}.$$

Simplifying this equation gives the functional equation (4.32). The restriction $-1 < \operatorname{Re} s < 0$ can now be removed by analytic continuation.

The functional equation can be written also in the form

$$(4.44) \qquad \zeta(1-s) = 2(2\pi)^{-s} \cos \frac{\pi s}{2} \Gamma(s)\zeta(s)$$

by using the property

$$\Gamma(1-s)\Gamma(s) = \frac{\pi}{\sin \pi s}.$$

We come to the classical result:

Proposition 4.3. *The Riemann zeta function* $\zeta(s)$ *can be extended as analytical function for all complex values* $s \neq 1$. *At that,* $\zeta(0) = -1/2$ *and* $\zeta(-2k) = 0, \ k = 1, 2, \dots$.

Proof. The property $\zeta(-2k) = 0, \ k = 1, 2, \dots$ follows immediately from (4.43) with $s = 1 + 2k$. Setting $s = 0, b = 1$ in (4.42) we find also

$$2\eta(0) = \int_0^\infty \frac{1}{\cosh^2 x} dx = \tanh x \Big|_0^\infty = 1, \quad \eta(0) = \frac{1}{2}$$

and

$$\zeta(0) = -\eta(0) = -\frac{1}{2}.$$

Notice that for the proof of the functional equation we used only the Fresnel integral (4.34), the derivative of (4.33) (which is the integral (4.43)), and the representation (4.37). The functional equation is "built into" the structure if the integral (4.33).

Example 4.7.1

The integral representations of the functions $\eta(s)$ and $\zeta(s)$ can be used to evaluate some difficult integrals. For example, differentiating with respect to s equation (4.36), that is,

$$\int_0^\infty \frac{x^{s-1}}{e^{bx}+1} dx = b^{-s} \Gamma(s) \eta(s)$$

we come to the equation

$$\int_0^\infty \frac{x^{s-1}\ln x}{e^{bx}+1}dx$$

$$= -b^{-s}\ln b\,\Gamma(s)\eta(s)+b^{-s}\Gamma'(s)\eta(s)+b^{-s}\Gamma(s)\eta'(s)$$

and with $s=1$

$$\int_0^\infty \frac{\ln x}{e^{bx}+1}dx = -\frac{\ln b}{b}\eta(1)+\frac{1}{b}\Gamma'(1)\eta(1)+\frac{1}{b}\eta'(1).$$

Here $\eta(1)=\ln 2$, $\Gamma'(1)=\Gamma'(1)/\Gamma(1)=\psi(1)=-\gamma$ and we write

$$\int_0^\infty \frac{\ln x}{e^{bx}+1}dx = -\frac{1}{b}\ln b\ln 2 -\frac{1}{b}\gamma\ln 2 +\frac{1}{b}\eta'(1).$$

The value of $\eta'(1)$ can be found from the equation

$$\eta(s)=(1-2^{1-s})\zeta(s)$$

and the well-known series representation

$$\zeta(s)-\frac{1}{s-1}=\gamma+\sum_{n=1}^\infty \frac{(-1)^n}{n!}\gamma_n(s-1)^n$$

(the coefficients γ_n here are known as the Stieltjes constants). A simple computation gives

$$\eta'(1)=\lim_{s\to 1}\frac{\eta(s)-\ln 2}{s-1}=\gamma\ln 2 -\frac{1}{2}\ln^2 2\,.$$

Therefore,

$$\int_0^\infty \frac{\ln x}{e^{bx}+1}dx = -\frac{\ln 2}{2b}(\ln 2 + 2\ln b)\,.$$

Problem for the reader (easy!). Prove that

$$\int_0^\infty \ln x (1 - \tanh bx)\, dx = -\frac{\ln 2}{2b}(3\ln 2 + 2\ln b) \ .$$

Example 4.7.2

Differentiating equation (4.35)

$$\int_0^\infty \frac{x^{s-1}}{e^{bx}-1}\, dx = b^{-s}\Gamma(s)\zeta(s)$$

with respect to s and then setting $s = 2$ we obtain

$$\int_0^\infty \frac{x\ln x}{e^{bx}-1}\, dx = \frac{1}{b^2}(-\zeta(2)\ln b + \psi(2)\zeta(2) + \zeta'(2)) \ .$$

Now we have the values $\zeta(2) = \dfrac{\pi^2}{6}$, $\psi(2) = 1 - \gamma$ and

$$\zeta'(2) = \frac{\pi^2}{6}(\gamma + \ln(2\pi) - 12\ln A)$$

where A is the Glaisher-Kinkelin constant. We conclude that

$$\int_0^\infty \frac{x\ln x}{e^{bx}-1}\, dx = \frac{\pi^2}{6b^2}\left(\ln\frac{2\pi}{b} + 1 - 12\ln A\right) \ .$$

For the Glaisher-Kinkelin constant it is known that

$$1 - 12\ln A = 12\zeta'(-1)$$

and when $b = 2\pi$ we have

$$\int_0^\infty \frac{x\ln x}{e^{2\pi x}-1}\, dx = \frac{1}{2}\zeta'(-1) \ .$$

4.8   The Functional Equation for Euler's $L(s)$ Function

Euler worked with the function

(4.45)
$$L(s) = \sum_{n=0}^{\infty} \frac{(-1)^n}{(2n+1)^s} \quad (\operatorname{Re} s > 0)$$

which is somewhat similar to the zeta function. Here

$$L(1) = 1 - \frac{1}{3} + \frac{1}{5} - \frac{1}{7} + \ldots = \frac{\pi}{4}$$

is the Gregory series. Another important value is

$$L(2) = \sum_{n=0}^{\infty} \frac{(-1)^n}{(2n+1)^2} = G$$

the Catalan constant (see Section 5.5).

The function $L(s)$ is also known as Dirichlet's $L(s)$ function, or Dirichlet's beta function $\beta(s)$.

It is easy to show that

(4.46)
$$2\Gamma(s)L(s) = \int_0^{\infty} \frac{x^{s-1}}{\cosh x} dx .$$

The proof is exactly like that of (4.37).

Proposition 4.4. *The function $L(s)$ satisfies the functional equation*

(4.47)
$$\left(\frac{2}{\pi}\right)^s \sin\frac{\pi s}{2}\Gamma(s)L(s) = L(1-s)$$

or, in equivalent form

(4.48)
$$L(s) = \left(\frac{\pi}{2}\right)^{s-1} \cos\frac{\pi s}{2}\Gamma(1-s)L(1-s) .$$

Proof. We use the integral

(4.49)
$$\frac{1}{\cosh x} = \int_0^\infty \frac{\cos xt}{\cosh \dfrac{\pi t}{2}}\, dt$$

(this is a modification of (4.18) from Section 4.5.2).

For $\operatorname{Re} s > 0$ we write

$$\int_0^\infty \frac{x^{s-1}}{\cosh x}\, dx = \int_0^\infty \frac{1}{\cosh \dfrac{\pi t}{2}} \left\{ \int_0^\infty x^{s-1} \cos xt\, dx \right\} dt$$

$$= \int_0^\infty \frac{1}{\cosh \dfrac{\pi t}{2}} \left\{ \frac{\Gamma(s)}{t^s} \cos \frac{\pi s}{2} \right\} dt = \Gamma(s) \cos \frac{\pi s}{2} \left\{ \int_0^\infty \frac{t^{-s}}{\cosh \dfrac{\pi t}{2}}\, dt \right\}$$

$$= \Gamma(s) \cos \frac{\pi s}{2} \left\{ 2 \left(\frac{\pi}{2} \right)^{s-1} \Gamma(1-s) L(1-s) \right\}.$$

Comparing this to (4.46) we come to the functional equation (4.48). The equivalent form (4.47) follows in view of the gamma function property

$$\Gamma(s)\Gamma(1-s) = \frac{\pi}{\sin \pi s}.$$

In the proof we have used again the Fresnel integral (4.34) in a slightly different form.

The functional equation shows that $L(s)$ extends as analytic function for all complex s. From (4.48) we find

$$L(0) = \frac{2}{\pi} L(1) = \frac{1}{2}$$

and from (4.47) the property

$$L(1-2k) = 0, \quad k = 1, 2, \ldots .$$

Like in the previous section we can say that the functional equation is built into the structure of the integral (4.49).

4.9 Euler's Formula for Zeta(2*n*)

In this section we present two important formulas relating the values of Riemann's zeta function $\zeta(2n)$ and $\zeta(1-2n)$, $n = 1, 2, \dots$ to the Bernoulli numbers B_n. We also give two formulas relating the values $L(2n+1)$ and $L(-2n)$ of Euler's $L(s)$-function to Euler's numbers E_n. Following the ideas of the previous two sections these formulas will be based on two special integrals.

4.9.1 *Bernoulli numbers*

The popular Bernoulli numbers B_k appeared in the works of the Swiss mathematician Jakob Bernoulli (1654-1705) who evaluated sums of powers of consecutive integers

$$1^n + 2^n + \dots + (m-1)^n = \frac{1}{n+1} \sum_{k=0}^{n} \binom{n+1}{k} m^{n+1-k} B_k .$$

The Bernoulli numbers have numerous applications in mathematics: in combinatorics, analysis, and probability (see, for instance, [40, Chapter 24] and [44, Section 1.6]).

The exponential generating function for the Bernoulli numbers is

(4.50) $\qquad \dfrac{x}{e^x - 1} = \sum_{n=0}^{\infty} \dfrac{B_n}{n!} x^n = 1 - \dfrac{x}{2} + \sum_{n=2}^{\infty} \dfrac{B_n}{n!} x^n \quad (|x| < 2\pi) .$

We have

$$B_0 = 1, \ B_1 = -\frac{1}{2}, \ B_2 = \frac{1}{6}, \ B_3 = 0, \ B_4 = -\frac{1}{30}, \dots .$$

Equation (4.50) can be put in the form

(4.51)
$$\frac{x}{e^x - 1} + \frac{x}{2} - 1 = \sum_{n=2}^{\infty} \frac{B_n}{n!} x^n$$

where the function on the left-hand side is even. Indeed, setting

$$f(x) = \frac{x}{e^x - 1} + \frac{x}{2}$$

we have

$$f(-x) = \frac{-x}{e^{-x} - 1} - \frac{x}{2} = \frac{-xe^x}{1 - e^x} - \frac{x}{2} = \frac{-xe^x - x}{2(1 - e^x)}$$

$$= \frac{-x(e^x - 1 + 2)}{2(1 - e^x)} = \frac{x}{2} + \frac{x}{e^x - 1} = f(x).$$

It follows that $B_{2n+1} = 0$, $n = 1, 2, \dots$.

Most interestingly, the nonzero numbers B_{2n} are related to the values $\zeta(2n)$ of Riemann's zeta function. Euler discovered the formula

(4.52)
$$\zeta(2n) = \frac{(-1)^{n-1}(2\pi)^{2n}}{2(2n)!} B_{2n} \quad (n = 1, 2, \dots)$$

bearing his name. In particular

$$\zeta(2) = \frac{\pi^2}{6}, \ \zeta(4) = \frac{\pi^4}{90}, \ \zeta(6) = \frac{\pi^6}{945}, \dots.$$

There are many proofs of this formula. We will present here a proof based on the integral

(4.53)
$$\int_0^{\infty} \frac{\sin xt}{e^t - 1} dt = \frac{\pi}{2} + \frac{\pi}{e^{2\pi x} - 1} - \frac{1}{2x} \quad (x > 0)$$

which was evaluated in Example 4.5.2. We will see that Euler's formula is hidden in the structure of this integral. We need also the integral representation of the zeta function

(4.54) $$\zeta(s) = \frac{1}{\Gamma(s)} \int_0^\infty \frac{t^{s-1}}{e^t - 1} dt \quad (\text{Re}\, s > 1)$$

(see equation (4.35) in the previous section).

Proof of Euler's formula.

We look at the right-hand side in equation (4.53)

$$G(x) = \frac{\pi}{2} + \frac{\pi}{e^{2\pi x} - 1} - \frac{1}{2x}.$$

It is easy to see that $\lim_{x \to 0} G(x) = 0$ and defining $G(0) = 0$ this function becomes analytic in the disk $|x| < 1$. According to (4.51)

$$2xG(x) = \frac{2\pi x}{2} + \frac{2\pi x}{e^{2\pi x} - 1} - 1 = \sum_{n=1}^\infty \frac{(2\pi)^{2n} B_{2n}}{(2n)!} x^{2n}$$

and therefore,

$$G(x) = \sum_{n=1}^\infty \frac{(2\pi)^{2n} B_{2n}}{2(2n)!} x^{2n-1}.$$

which is obviously the Taylor series for this function centered at $x = 0$. Using the formula for the Taylor coefficients we conclude that

(4.55) $$\frac{G^{(2n-1)}(0)}{(2n-1)!} = \frac{(2\pi)^{2n} B_{2n}}{2(2n)!}.$$

At the same time from (4.53)

$$G(x) = \int_0^\infty \frac{\sin xt}{e^t - 1} dt, \quad G^{(2n-1)}(x) = (-1)^{n-1} \int_0^\infty \frac{t^{2n-1} \cos xt}{e^t - 1} dt$$

and

$$\frac{G^{(2n-1)}(0)}{(2n-1)!} = \frac{(-1)^{n-1}}{(2n-1)!} \int_0^\infty \frac{t^{2n-1}}{e^t - 1} dt = (-1)^{n-1} \zeta(2n) .$$

Comparing this to (4.55) we find

$$(-1)^{n-1} \zeta(2n) = \frac{(2\pi)^{2n} B_{2n}}{2(2n)!}$$

and Euler's formula is proved. Since $\zeta(0) = -\frac{1}{2}$ and $B_0 = 1$ the formula is true also for $n = 0$.

Note that this proof can be used in reverse order and from Euler's formula we can derive equation (4.53). Therefore, the two equations (4.52) and (4.53) are equivalent.

Combining Euler's formula (4.52) with the functional equation

$$\zeta(1 - s) = 2(2\pi)^{-s} \cos\frac{\pi s}{2} \Gamma(s)\zeta(s)$$

and setting here $s = 2n$ we come to another remarkable formula

(4.56) $$\zeta(1 - 2n) = -\frac{B_{2n}}{2n} \quad (n = 1, 2, \ldots)$$

also due to Euler. In particular,

$$\zeta(-1) = -\frac{1}{12}, \ \zeta(-3) = \frac{1}{120} .$$

Remark. As we know from Example 4.5.2 equation (4.53) can be written also in the form

$$\int_0^\infty \frac{\sin xt}{e^t - 1} dt = \frac{\pi}{2} \coth \pi x - \frac{1}{2x}$$

and it follows from the representation of $G(x)$ that

$$\frac{\pi}{2}\coth \pi x - \frac{1}{2x} = \sum_{n=1}^{\infty} \frac{(2\pi)^{2n} B_{2n}}{2(2n)!} x^{2n-1} .$$

This shows the interesting Taylor series for the function $\pi x \coth \pi x$

$$\pi x \coth \pi x = \sum_{n=0}^{\infty} \frac{(2\pi)^{2n} B_{2n}}{(2n)!} x^{2n} .$$

In terms of zeta values

$$\pi x \coth \pi x = 2\sum_{n=0}^{\infty} (-1)^{n-1} \zeta(2n) x^{2n} .$$

If we replace x by xi we come to the representation of the trigonometric cotangent

$$\pi x \coth \pi x = -2\sum_{n=0}^{\infty} \zeta(2n) x^{2n} \quad (|x| < 1)$$

4.9.2   *Euler numbers and Euler's formula for L(2n+1)*

The Euler numbers E_n are usually defined by their exponential generating function

(4.57) $$\frac{1}{\cosh x} = \sum_{n=0}^{\infty} \frac{E_n}{n!} x^n .$$

This function is even, so $E_{2n-1} = 0$, $n = 1, 2, \dots$. For even indices we have in particular,

$$E_0 = 1, \, E_2 = -1, \, E_4 = 5, \, E_6 = -61, \, E_8 = 1385, \dots$$

(see [40, Chapter 24]).

We will prove a formula similar to Euler's formula (4.52) involving Euler's $L(s)$ function

$$L(s) = \sum_{n=0}^{\infty} \frac{(-1)^n}{(2n+1)^s} \quad (\operatorname{Re} s > 0)$$

with integral representation

$$L(s) = \frac{1}{2\Gamma(s)} \int_0^{\infty} \frac{t^{s-1}}{\cosh t} dt$$

(see Section 4.7).

Proposition 4.5. *For every* $n = 0, 1, 2, \ldots$ *we have*

(4.58) $$L(2n+1) = \frac{(-1)^n \pi^{2n+1}}{2^{2n+2}(2n)!} E_{2n}.$$

For example

$$L(1) = \frac{\pi}{4}, \quad L(3) = \frac{\pi^3}{32}, \quad L(5) = \frac{5\pi^5}{1536}, \ldots.$$

This formula also comes from Euler. We give here a proof based on the special integral

(4.59) $$\int_0^{\infty} \frac{\cos xt}{\cosh t} dt = \frac{\pi}{2} \frac{1}{\cos \dfrac{\pi x}{2}}$$

evaluated in Section 4.5.2 (this is integral (4.18)).

The argument is this. First we write

$$\frac{\pi}{2} \frac{1}{\cos \dfrac{\pi x}{2}} = \sum_{n=0}^{\infty} \frac{\pi^{2n+1} E_{2n}}{2^{2n+2}(2n)!} x^{2n}$$

as a result from (4.57). Then equation (4.59) says that

$$(4.60) \qquad \int_0^\infty \frac{\cos xt}{\cosh t}\, dt = \sum_{n=0}^\infty \frac{\pi^{2n+1} E_{2n}}{2^{2n+2}(2n)!} x^{2n}$$

giving explicitly the Taylor coefficients of the function

$$F(x) = \int_0^\infty \frac{\cos xt}{\cosh t}\, dt .$$

At the same time we can find the Taylor coefficients of this even function by repeated differentiation with respect to x. That is,

$$\frac{F^{(2n)}(0)}{(2n)!} = \frac{(-1)^n}{(2n)!} \int_0^\infty \frac{t^{2n}}{\cosh t}\, dt = (-1)^n L(2n+1)$$

for $n = 0, 1, 2,...$ This way

$$F(x) = \sum_{n=0}^\infty (-1)^n L(2n+1) x^{2n} .$$

Comparing this equation to equation (4.60) completes the proof.

The functional equation for $L(s)$ proved in Section 4.8

$$L(s) = \left(\frac{\pi}{2}\right)^{s-1} \cos \frac{\pi s}{2} \Gamma(1-s) L(1-s)$$

in view of the representation

$$L(2n+1) = \frac{(-1)^n \pi^{2n+1}}{2^{2n+2}(2n)!} E_{2n}$$

implies a formula similar to (4.56), namely,

$$(4.61) \qquad L(-2n) = \frac{1}{2} E_{2n} \quad (n = 0, 1, 2,...) .$$

Example 4.9.1

In Gradshteyn and Ryzhik's reference book [25], on page 532, there sits one monster integral, a visitor from the bad dream of some calculus student

$$(4.62) \qquad J = \int_{\pi/4}^{\pi/2} \ln(\ln \tan x)\, dx = \frac{\pi}{2}\ln\left(\frac{\Gamma(3/4)}{\Gamma(1/4)}\sqrt{2\pi}\right).$$

This is entry 4.229(7) in [25]. It is difficult to understand how such an integral has come into life. Closer examinations shows, however, that the integral has several faces, one of which is quite civilized.

First, the substitution $u = \tan x$ ($x = \arctan u$) turns this integral into

$$(4.63) \qquad J = \int_{1}^{\infty} \frac{\ln(\ln u)}{1 + u^2}\, du$$

(entry 4.325(4)). The substitution $x = \arctan\dfrac{1}{u}$ gives it another look, an integral with finite limits

$$(4.64) \qquad J = \int_{0}^{1} \frac{1}{1+u^2}\ln\left(\ln\frac{1}{u}\right) du = \int_{0}^{1} \frac{\ln(-\ln u)}{1+u^2}\, du.$$

Finally, the substitution $u = \ln(\tan x)$ or $x = \arctan e^u$ makes the logarithm simpler

$$(4.65) \qquad J = \int_{0}^{\infty} \ln u\, \frac{e^u}{1 + e^{2u}}\, du = \frac{1}{2}\int_{0}^{\infty} \frac{\ln u}{\cosh u}\, du$$

(entry 4.371(1)).
This last one we will evaluate using Euler's $L(s)$ function. More precisely, we look at equation (4.46)

$$\int_{0}^{\infty} \frac{x^{s-1}}{\cosh x}\, dx = 2\Gamma(s)L(s)$$

and realize that the logarithm will appear inside after differentiation. Indeed, differentiating and setting $s = 1$ we find

$$\frac{1}{2}\int_0^\infty \frac{\ln x}{\cosh x}dx = \Gamma'(1)L(1) + \Gamma(1)L'(1)$$

and since $\Gamma'(1) = \psi(1) = -\gamma$, $L(1) = \dfrac{\pi}{4}$ we have

$$J = L'(1) - \frac{\pi}{4}\gamma.$$

Now we need to find $L'(1)$. From the functional equation

$$\left(\frac{2}{\pi}\right)^s \sin\frac{\pi s}{2}\Gamma(s)L(s) = L(1-s)$$

differentiating and setting $s = 1$ we find

(4.66) $$L'(1) - \frac{\pi}{4}\gamma = \frac{\pi}{2}\left(\frac{1}{2}\ln\frac{\pi}{2} - L'(0)\right).$$

We see now that we need $L'(0)$. To compute this value we will follow the idea of Ilan Vardi [54]. Vardi involved the Hurwitz zeta function

$$\zeta(s,a) = \sum_{n=0}^\infty \frac{1}{(n+a)^s} \quad (a > 0, \operatorname{Re} s > 1).$$

This function can be continued analytically as a function of s on the whole complex plane with a simple pole at $s = 1$. For its analytical continuation we have a very important property, the formula of Lerch (see [48, p. 273])

(4.67) $$\zeta'(0, a) = \ln\frac{\Gamma(a)}{\sqrt{2\pi}}.$$

We can express $L(s)$ in terms of $\zeta(s,a)$

$$\varsigma\left(s,\frac{1}{4}\right)-\varsigma\left(s,\frac{3}{4}\right)=\left(\sum_{n=0}^{\infty}\frac{4^s}{(4n+1)^s}-\sum_{n=0}^{\infty}\frac{4^s}{(4n+3)^s}\right)$$

$$=4^s\left(1-\frac{1}{3^s}+\frac{1}{5^s}-\frac{1}{7^s}+...\right)$$

that is,

(4.68) $$\varsigma\left(s,\frac{1}{4}\right)-\varsigma\left(s,\frac{3}{4}\right)=4^s L(s).$$

Differentiating this equation and then setting $s = 0$ we find in view of Lerch's formula (4.67) and $L(0) = 1/2$

$$\ln\frac{\Gamma(1/4)}{\Gamma(3/4)}=\ln 2 + L'(0).$$

Placing this value of $L'(0)$ into equation (4.66) we find

$$L'(1)-\frac{\pi}{4}\gamma=\frac{\pi}{2}\left(\ln\frac{\Gamma(3/4)}{\Gamma(1/4)}+\ln\sqrt{2\pi}\right)$$

which proves (4.62).

A good survey on this integral and several similar integrals can be found in Iaroslav V. Blagouchine's paper "Rediscovery of Malmsten's integrals, their evaluation by contour integration methods and some related results" (*Ramanujan J.*, 35 (2014), 21–110).

Example 4.9.2

The integral from the previous example has an interesting neighbor in Gradshteyn and Ryzhik's table [25] with a similar structure

(4.69) $$\int_0^1\left(\ln\frac{1}{x}\right)^{\mu-1}\ln\left(\ln\frac{1}{x}\right)dx=\psi(\mu)\Gamma(\mu)\quad(\mathrm{Re}\,\mu>0)$$

(entry 4.229(4)). This integral is much easier to prove. The substitution $t = \ln\dfrac{1}{x}$ or $x = -\ln t$ brings the integral to something familiar

$$\int_0^1 \left(\ln\frac{1}{x}\right)^{\mu-1} \ln\left(\ln\frac{1}{x}\right) dx = \int_0^\infty t^{\mu-1}(\ln t)e^{-t} dt .$$

This is the derivative of Euler's gamma function

$$\Gamma(\mu) = \int_0^\infty t^{\mu-1}e^{-t} dt .$$

Since $\Gamma'(\mu) = \psi(\mu)\Gamma(\mu)$ our proof is done.

When $\mu = 1$ we have

(4.70) $$\int_0^1 \ln\left(\ln\frac{1}{x}\right) dx = \psi(1) = -\gamma$$

which is entry 4.229(1). With $\mu = \dfrac{1}{2}$ in (4.69)

$$\int_0^1 \left(\ln\frac{1}{x}\right)^{-1/2} \ln\left(\ln\frac{1}{x}\right) dx = \psi\left(\frac{1}{2}\right)\Gamma\left(\frac{1}{2}\right)$$

and as

$$\Gamma\left(\frac{1}{2}\right) = \sqrt{\pi}, \; \psi\left(\frac{1}{2}\right) = -2\ln 2 - \gamma$$

we find also

(4.71) $$\int_0^1 \left(\ln\frac{1}{x}\right)^{-1/2} \ln\left(\ln\frac{1}{x}\right) dx = -(\gamma + 2\ln 2)\sqrt{\pi} .$$

This is entry 4.229(3) in [25].

Example 4.9.3

Euler's numbers E_n are well-known and together with formulas (4.46) and (4.58), namely,

$$\int_0^\infty \frac{x^{s-1}}{\cosh x}dx = 2\Gamma(s)L(s), \quad L(2n+1) = \frac{(-1)^n \pi^{2n+1}}{2^{2n+2}(2n)!}E_{2n}$$

they can be used for the evaluation of a group of integrals from section 3.523 in [25]. The two formulas above together yield

$$(4.72) \qquad \int_0^\infty \frac{x^{2n}}{\cosh x}dx = \frac{(-1)^n \pi^{2n+1}}{2^{2n+1}}E_{2n} = (-1)^n \left(\frac{\pi}{2}\right)^{2n+1} E_{2n}$$

which also comes directly from (4.60). This is entry 3.523 (4) in [25].

For $n = 1, 2, 3$ we have

$$(4.73) \qquad \int_0^\infty \frac{x^2}{\cosh x}dx = \frac{\pi^3}{8}$$

$$(4.74) \qquad \int_0^\infty \frac{x^4}{\cosh x}dx = \frac{5\pi^5}{32}$$

$$(4.75) \qquad \int_0^\infty \frac{x^6}{\cosh x}dx = \frac{61\pi^7}{128}.$$

These are entries 3.523(5), 3.523(7), and 3.352(9) in the same order.

For $s = \frac{3}{2}$ we find from (4.46) and the definition of $L(s)$

$$(4.76) \qquad \int_0^\infty \frac{\sqrt{x}}{\cosh x}dx = 2\Gamma\left(\frac{3}{2}\right)L\left(\frac{3}{2}\right) = \sqrt{\pi}\sum_{n=0}^\infty \frac{(-1)^n}{(2n+1)^{3/2}}$$

(entry 3.523(11)).

In the same way with $s = \dfrac{1}{2}$ we obtain

(4.77) $$\int_0^\infty \frac{1}{\sqrt{x}\cosh x}\,dx = 2\Gamma\left(\frac{1}{2}\right)L\left(\frac{1}{2}\right) = 2\sqrt{\pi}\sum_{n=0}^{\infty}\frac{(-1)^n}{\sqrt{2n+1}}$$

(entry 3.523(12) in [25]). Etc.

Example 4.9.4

We will show now a solution to the *American Mathematical Monthly* Problem 11973 (Vol. 124, No. 4, (2017), p. 369). Our solution is based on the representation

(4.78) $$\pi x \coth \pi x = -2\sum_{n=0}^{\infty}\zeta(2n)x^{2n} \quad (|x|<1)$$

from Section 4.9.1.

The problem is to prove that

(4.79) $$\frac{\pi}{2}\sum_{n=0}^{\infty}\frac{\zeta(2n)}{(2n+1)4^n}\left(1-\frac{2}{4^n}\right) = G$$

where G is Catalan's constant.
For the solution we integrate (4.78)

$$\int_0^x t\frac{\cos \pi t}{\sin \pi t}\,d\pi t = \int_0^x td\ln(\sin \pi t)$$

$$= x\ln(\sin \pi x) - \int_0^x \ln(\sin \pi t)\,dt = -2\sum_{n=0}^{\infty}\frac{\zeta(2n)}{2n+1}x^{2n+1}$$

so that

$$-x\ln(\sin \pi x) + \int_0^x \ln(\sin \pi t)\,dt = 2\sum_{n=0}^{\infty}\frac{\zeta(2n)}{2n+1}x^{2n+1}.$$

Here we set $x = \dfrac{1}{2}$ and rescale the variable in the integral to get

$$(4.80) \qquad \int_0^{\pi/2} \ln(\sin t)\,dt = \pi \sum_{n=0}^{\infty} \frac{\zeta(2n)}{(2n+1)4^n}.$$

In the same way setting $x = \dfrac{1}{4}$ we find

$$(4.81) \qquad \frac{\pi}{4}\ln 2 + 2\int_0^{\pi/4} \ln(\sin t)\,dt = \pi \sum_{n=0}^{\infty} \frac{\zeta(2n)}{(2n+1)16^n}.$$

The log-sine integrals are known (see Example 3.2.1 in Chapter 3)

$$\int_0^{\pi/2} \ln(\sin t)\,dt = -\frac{\pi}{2}\ln 2$$

$$\int_0^{\pi/4} \ln(\sin t)\,dt = \frac{-\pi}{4}\ln 2 - \frac{1}{2}G.$$

Finally, from (4.80) and (4.81) we find

$$\frac{\pi}{2}\sum_{n=0}^{\infty} \frac{\zeta(2n)}{(2n+1)4^n}\left(1 - \frac{2}{4^n}\right)$$

$$= \frac{\pi}{2}\left(\sum_{n=0}^{\infty} \frac{\zeta(2n)}{(2n+1)4^n} - 2\sum_{n=0}^{\infty} \frac{\zeta(2n)}{(2n+1)16^n}\right)$$

$$= -\frac{\pi}{4}\ln 2 - \frac{\pi}{4}\ln 2 + \frac{\pi}{2}\ln 2 + G = G$$

as desired.

As an added bonus we have the two interesting series evaluations

$$(4.82) \qquad \sum_{n=0}^{\infty} \frac{\zeta(2n)}{(2n+1)4^n} = -\ln\sqrt{2}$$

(4.83) $$\sum_{n=0}^{\infty} \frac{\zeta(2n)}{(2n+1)16^n} = -\frac{1}{4}\ln 2 - \frac{G}{\pi}.$$

In the last one three interesting constants appear together.

Chapter 5

Various Techniques

5.1 The Formula of Poisson

Many challenging integrals can be evaluated by using the residue theorem and contour integration. This popular method is well represented in most books on complex variables and will not be discussed here. However, in the theory of analytic functions there is one nice and efficient formula which is useful for integral evaluation and does not require special contours.

Suppose $f(z)$ is a function bounded and analytic on the right half plane $\mathrm{Re}\, z > 0$. Setting $z = x + iy$ we have the important integral representation (see chapter 6 in [30])

$$(5.1) \qquad f(x+iy) = \frac{1}{\pi} \int_{-\infty}^{\infty} f(it) \frac{x}{x^2 + (y-t)^2} dt$$

where $f(it), -\infty < t < \infty$ are the boundary values on the y-axis.

For real valued integrals we use the two representations

$$(5.2) \qquad \mathrm{Re}\, f(x+iy) = \frac{1}{\pi} \int_{-\infty}^{\infty} \mathrm{Re}\, f(it) \frac{x}{x^2 + (y-t)^2} dt$$

$$(5.3) \qquad \mathrm{Im}\, f(x+iy) = \frac{1}{\pi} \int_{-\infty}^{\infty} \mathrm{Im}\, f(it) \frac{x}{x^2 + (y-t)^2} dt \; .$$

The same representation is true for harmonic functions bounded on the closed right half plane. The formula has numerous applications in harmonic analysis, potential theory and partial differential equations.

Example 5.1.1

Let $a > 0$ and for $z = x + iy$ consider the function

$$f(z) = e^{-az} = e^{-ax}e^{-iay} = e^{-ax}(\cos ay - i\sin ay)$$

which is bounded and analytic for $x \geq 0$. By Poisson's formula

$$e^{-ax}(\cos ay - i\sin ay) = \frac{x}{\pi} \int_{-\infty}^{\infty} \frac{\cos at - i\sin at}{x^2 + (y-t)^2} dt .$$

Separating real and imaginary parts we find for $x > 0$

$$(5.4) \qquad \int_{-\infty}^{\infty} \frac{\cos at}{x^2 + (y-t)^2} dt = \frac{\pi}{x} e^{-ax} \cos ay$$

$$(5.5) \qquad \int_{-\infty}^{\infty} \frac{\sin at}{x^2 + (y-t)^2} dt = \frac{\pi}{x} e^{-ax} \sin ay .$$

For $y = 0$ in (5.4) we find

$$\int_{0}^{\infty} \frac{\cos at}{x^2 + t^2} dt = \frac{1}{2} \int_{-\infty}^{\infty} \frac{\cos at}{x^2 + t^2} dt = \frac{\pi}{2x} e^{-ax} .$$

This integral was evaluated in Example 2.3.2 (Chapter 2) by differentiation with respect to the variable a.

Example 5.1.2

Now consider $f(z) = e^{-bz} \cosh az$ where $0 < a \leq b$. Then

$$f(it) = \cos(bt)\cos(at) - i\sin(bt)\cos(at)$$

We apply formula (5.2) and then set $y = 0$ to obtain for $x > 0$

(5.6) $$\int_0^\infty \frac{\cos bt \cos at}{x^2 + t^2} dt = \frac{\pi}{2x} e^{-bx} \cosh ax$$

as the integrand is an even function. This is entry 3.743(3) from [25] in a slightly different form.

Example 5.1.3

Let $\alpha > 0$ and consider the function $f(z) = \dfrac{\log(1 + \alpha z)}{z}$ with the principal

value of the logarithm. Then $f(it) = \dfrac{-i}{t}\log(1 + i\alpha t)$ and

$$\operatorname{Re} f(it) = \frac{1}{t}\operatorname{Arg}(1 + i\alpha t) = \frac{\arctan \alpha t}{t}.$$

With $x > 0$, $y = 0$ we find from (5.2)

(5.7) $$\frac{\log(1 + \alpha x)}{x^2} = \frac{1}{\pi}\int_{-\infty}^\infty \frac{\arctan \alpha t}{t} \frac{dt}{x^2 + t^2}$$

$$= \frac{2}{\pi}\int_0^\infty \frac{\arctan \alpha t}{t} \frac{dt}{x^2 + t^2}$$

that is,

$$\int_0^\infty \frac{\arctan \alpha t}{t(x^2 + t^2)} dt = \frac{\pi}{2} \frac{\log(1 + \alpha x)}{x^2}.$$

The proofs of the next two examples are left to the reader.

Example 5.1.4

Show that for every $x > 0$ and $0 < \alpha < \beta$

(5.8) $$\int_0^\infty \frac{\sin \alpha t \cos \beta t}{t} \frac{dt}{x^2 + t^2} = \frac{\pi}{2x^2} e^{-\beta x} \sinh \alpha x.$$

This is entry 3.725 (3) in [25]. The function to be used is

$$f(z) = e^{-\beta z} \frac{\sinh \alpha z}{z}.$$

Example 5.1.5

Prove that for every $x > 0, a > 0, b > 0$

(5.9) $$\int_0^\infty \frac{\cos bt}{b^2 + t^2} \frac{dt}{x^2 + t^2} = \frac{\pi}{2bx} \frac{b e^{-ax} - x e^{-ab}}{b^2 - x^2}$$

(entry 3.728 in [25]). Computing the limit when $x \to b$ show that

$$\int_0^\infty \frac{\cos bt}{(b^2 + t^2)^2} dt = \frac{\pi e^{-ab}(ab + 1)}{4b^3}.$$

The function to be used for this integral is

$$f(z) = \frac{b e^{-az} - z e^{-ab}}{b^2 - z^2}.$$

Example 5.1.6

Here we use this technique to solve Problem 2116 from the *Mathematics Magazine* (94 (2021), 150). The problem is to evaluate

$$\int_0^\infty \frac{e^{\cos t} \cos(\sin t + \alpha t)}{x^2 + t^2} dt \quad (x, \alpha > 0).$$

For the solution we consider the entire function $f(z) = e^{e^{-z}-\alpha z}$, $z = x + iy$ which is bounded on the closed right half plane $x \geq 0$. We have $\operatorname{Re} f(it) = e^{\cos t} \cos(\sin t + \alpha t)$ and Poisson's formula (5.2) implies (with $y = 0$)

$$\int_0^\infty \frac{e^{\cos t} \cos(\sin t + \alpha t)}{x^2 + t^2} dt = \frac{\pi}{2x} e^{e^{-x} - \alpha x}$$

(as the integrand is an even function).

5.2 Frullani Integrals

Frullani's formula says that for $a, b > 0$

$$(5.10) \qquad \int_0^\infty \frac{f(bx) - f(ax)}{x} dx = [f(\infty) - f(0)] \ln \frac{b}{a}$$

where $f(\infty) = \lim_{x \to \infty} f(x)$. For the validity of this formula it is sufficient to assume that $f(x)$ is continuous on $[0, \infty)$, and its derivative $f'(x)$ exists and is continuous on $(0, \infty)$. We also assume the limit $f(\infty)$ exists. These conditions are too strong, but quite appropriate for most applications and for the short proof below.

Giuliano Frullani (1795-1834) was an Italian mathematician. The above formula carries his name because of his contributions to integral solving techniques.

Here is a simple proof of the formula.

$$\int_0^\infty \frac{f(bx) - f(ax)}{x} dx = \int_0^\infty \left\{ \int_a^b f'(xy) dy \right\} dx$$

$$= \int_a^b \left\{ \int_0^\infty f'(xy) dx \right\} dy = \int_a^b \left\{ \frac{f(xy)}{y} \Big|_0^\infty \right\} dy$$

$$= \int_a^b \{f(\infty) - f(0)\} \frac{dy}{y} = \{f(\infty) - f(0)\} \int_a^b \frac{dy}{y}$$

$$= (f(\infty) - f(0)) \ln \frac{b}{a}.$$

A popular example follows. For $a, b > 0$

$$\int_0^\infty \frac{e^{-ax} - e^{-bx}}{x} dx = \ln \frac{b}{a}$$

with $f(x) = e^{-x}$, $\lim\limits_{x \to \infty} f(x) = 0$, $f(0) = 1$.

More essential examples are given below.

In all following examples until the end of the section we assume that a and b are two positive numbers.

Example 5.2.1

Taking $f(x) = e^{-\lambda x} \cos x$, $\lambda > 0$ in (5.10) we have

$$\int_0^\infty \frac{e^{-\lambda ax} \cos ax - e^{-\lambda bx} \cos bx}{x} dx = \ln \frac{b}{a}.$$

In this case $\lim\limits_{x \to \infty} f(x) = 0$, $f(0) = 1$.

A similar integral was evaluated in Example 2.2.9 by differentiation with respect to a parameter.

It is most remarkable that the integral does not depend on λ. Even more, we can use the function $f(x) = e^{-\lambda x} \cos^p(x)$ where λ is any positive number and p is any positive integer. We still have

(5.11) $$\int_0^\infty \frac{e^{-\lambda ax} \cos^p(ax) - e^{-\lambda bx} \cos^p(bx)}{x} dx = \ln \frac{b}{a}$$

and the integral is independent of λ and p.

Similarly, we have for any $\lambda > 0$ and any integer $p \geq 1$

$$\int_0^\infty \frac{e^{-\lambda ax}\sin^p(ax) - e^{-\lambda bx}\sin^p(bx)}{x}\,dx = 0$$

with $f(x) = e^{-\lambda x}\sin^p(x)$, $\lim\limits_{x\to\infty} f(x) = 0$, $f(0) = 0$.

Example 5.2.2

In this example we will evaluate the integral

$$\int_0^\infty \frac{\cos^p(ax) - \cos^p(bx)}{x}\,dx \quad (a, b > 0)$$

which is not a Frullani integral, because $\lim\limits_{x\to\infty} \cos^p(x)$ does not exist. However, this integral is interesting and resembles (5.11). The computation is based on the case $p = 1$

$$\int_0^\infty \frac{\cos(ax) - \cos(bx)}{x}\,dx = \ln\frac{b}{a}$$

evaluated in Chapter 2, in Example 2.2.13. For any integer $p \geq 1$ we consider two cases: when $p = 2n$ is even, and when $p = 2n - 1$ is odd. We also need the two representations (entries 1.320(5) and 1.320(7) from [25])

$$\cos^{2n}(x) = \frac{1}{4^n}\left\{2\sum_{k=0}^{n-1}\binom{2n}{k}\cos 2(n-k)x + \binom{2n}{n}\right\}$$

$$\cos^{2n-1}(x) = \frac{1}{4^{n-1}}\sum_{k=0}^{n-1}\binom{2n-1}{k}\cos(2n-2k-1)x.$$

Then we compute

$$\int_0^\infty \frac{\cos^{2n}(ax) - \cos^{2n}(bx)}{x} \, dx$$

$$= \frac{2}{4^n} \sum_{k=0}^{n-1} \binom{2n}{k} \int_0^\infty \frac{\cos(2(2n-k)ax) - \cos(2(2n-k)bx)}{x} \, dx$$

$$= \frac{2}{4^n} \ln\frac{b}{a} \sum_{k=0}^{n-1} \binom{2n}{k}.$$

That is,

(5.12) $$\int_0^\infty \frac{\cos^{2n}(ax) - \cos^{2n}(bx)}{x} \, dx = \frac{2}{4^n} \ln\frac{b}{a} \sum_{k=0}^{n-1} \binom{2n}{k}.$$

Similarly

$$\int_0^\infty \frac{\cos^{2n-1}(ax) - \cos^{2n-1}(bx)}{x} \, dx$$

$$= \frac{1}{4^{n-1}} \sum_{k=0}^{n-1} \binom{2n-1}{k} \int_0^\infty \frac{\cos((2n-2k-1)ax) - \cos((2n-2k-1)bx)}{x} \, dx$$

$$= \frac{1}{4^{n-1}} \ln\frac{b}{a} \left\{ \sum_{k=0}^{n-1} \binom{2n-1}{k} \right\} = \frac{1}{4^{n-1}} \ln\frac{b}{a} \{2^{2n-2}\} = \ln\frac{b}{a}.$$

Finally

(5.13) $$\int_0^\infty \frac{\cos^{2n-1}(ax) - \cos^{2n-1}(bx)}{x} \, dx = \ln\frac{b}{a}$$

(see entry 2,5,29(18) in [43]).

Example 5.2.3

Let p be any positive number. Applying Frullani's formula to the function $f(x) = \arctan^p(x)$ with $\lim_{x\to\infty} f(x) = \left(\dfrac{\pi}{2}\right)^p$ and $f(0) = 0$ we find

$$(5.14) \qquad \int_0^\infty \frac{\arctan^p(ax) - \arctan^p(bx)}{x}\, dx = \left(\frac{\pi}{2}\right)^p \ln\frac{a}{b}.$$

Example 5.2.4

For any $p > 0$ we have

$$(5.15) \qquad \int_0^\infty \frac{\operatorname{sech}^p(ax) - \operatorname{sech}^p(bx)}{x}\, dx = \ln\frac{b}{a}.$$

Here $f(x) = \operatorname{sech}^p(x)$, $\lim_{x\to\infty} f(x) = 0$, $f(0) = 1$.

Example 5.2.5

Now take the function $f(x) = e^{-\lambda x} \ln^p(1 + x)$ with arbitrary $\lambda > 0$, $p > 0$. Frullani's formula provides

$$(5.16) \qquad \int_0^\infty \frac{e^{-\lambda a x} \ln^p(1 + ax) - e^{-\lambda b x} \ln^p(1 + bx)}{x}\, dx = 0$$

since $\lim_{x\to\infty} f(x) = 0$ and also $f(0) = 0$.

We cannot pass to limits with $\lambda \to 0$ here because the integral

$$\int_0^\infty \frac{\ln^p(1 + ax) - \ln^p(1 + bx)}{x}\, dx$$

is divergent.

Example 5.2.6

At first sight the integral

$$\int_0^\infty \frac{e^{-ax^2} - e^{-bx^2}}{x}\, dx$$

(where $a, b > 0$) does not look like a Frullani integral. However, replacing a and b by $\sqrt{a}$ and $\sqrt{b}$ we apply (5.10) to the function $f(x) = e^{-x^2}$ to get

$$\int_0^\infty \frac{e^{-ax^2} - e^{-bx^2}}{x}\, dx = \ln\frac{\sqrt{b}}{\sqrt{a}} = \frac{1}{2}\ln\frac{b}{a}.$$

In the same way, for any $p > 0$

(5.17) $$\int_0^\infty \frac{e^{-ax^p} - e^{-bx^p}}{x}\, dx = \ln\frac{b^{1/p}}{a^{1/p}} = \frac{1}{p}\ln\frac{b}{a}.$$

For further examples of Frullani integrals and more theory see Albano et al. [4], Boros and Moll [7], and Moll [35].

Frullani's formula has many extensions and ramifications. Sergio Bravo et al. have connected it to the special method of brackets [19]. Hardy [27] has provided very interesting extensions and examples.

5.3 A Special Formula

An interesting extension of Frullani's formula was found by Ramanujan. His result is discussed by Bruce Brendt in his book [1, pp. 313-317]. Here we give a similar formula which quickly evaluates integrals of the form

$$\int_0^\infty \frac{f(x)}{x}\,dx$$

for appropriate functions $f(x)$.

Suppose the function $f(x)$ is defined on $[0.\infty)$ and is also analytic in a neighborhood of the origin with Taylor series

(5.18)
$$f(x) = \sum_{n=0}^\infty (-1)^{n-1} A(n) x^n$$

where $A(0) = 0$ and the coefficients $A(n)$, $n = 0,1,2,...$ extend to a "good enough" differentiable function $A(t)$ on the interval $[0, \infty)$. Then we have the remarkable formula

(5.19)
$$\int_0^\infty \frac{f(x)}{x}\,dx = A'(0).$$

At the end of the section we will sketch a justification of this formula.

Note that the summation in (5.18) can be started from $n = 1$

$$f(x) = \sum_{n=1}^\infty (-1)^{n-1} A(n) x^n.$$

Example 5.3.1

We start with a simple example. Consider the function

$$f(x) = \frac{\ln(1+x)}{1+x}$$

which is defined on $[0, \infty)$ and has the Taylor expansion

$$\frac{\ln(1+x)}{1+x} = \sum_{n=1}^\infty (-1)^{n-1} H_n x^n$$

for $|x| < 1$. Here H_n are the harmonic numbers (they appeared before in Section 4.2). The harmonic numbers have the representation

$$H_n = \psi(n+1) + \gamma$$

where $\psi(t) = \Gamma'(t)/\Gamma(t)$ is the digamma function and $\gamma = -\psi(1)$ is Euler's constant. This way the harmonic numbers extend to the function

$$H(t) = \psi(t+1) + \gamma$$

defined and differentiable on $(-1, \infty)$. The digamma function has the series representation

$$\psi(x+1) = -\gamma + \sum_{n=1}^{\infty}\left(\frac{1}{n} - \frac{1}{n+x}\right)$$

so that

$$H'(0) = \psi'(1) = \sum_{n=1}^{\infty}\frac{1}{n^2} = \frac{\pi^2}{6}$$

and we come to the evaluation

$$\int_0^{\infty} \frac{\ln(1+x)}{(1+x)x}\,dx = \frac{\pi^2}{6}$$

This integral was evaluated in Example 2.5.4 in Chapter 2. It is a good illustration of how formula (5.19) works.

A natural extension is the integral

$$\int_0^{\infty} \frac{\ln(1+x)}{(1+x)^p x}\,dx, \ p > 0.$$

We have the representation

$$\frac{\ln(1+x)}{(1+x)^p} = \sum_{n=1}^{\infty}(-1)^{n-1}\binom{n+p-1}{n}\big(\psi(n+p) - \psi(p)\big)x^n$$

(see, for instance, [13]) and the coefficients

$$A(n) = \binom{n+p-1}{n} \big(\psi(n+p) - \psi(p)\big)$$

extend to

$$A(t) = \binom{t+p-1}{t} \big(\psi(t+p) - \psi(p)\big)$$

$$= \frac{\Gamma(t+p)}{\Gamma(p)\Gamma(t+1)} \big(\psi(t+p) - \psi(p)\big).$$

It is easy to compute that

$$A'(0) = \psi'(p) = \zeta(2, p)$$

where

$$\zeta(s, p) = \sum_{n=0}^{\infty} \frac{1}{(n+p)^s}, \quad \mathrm{Re}\, s > 1$$

is the Hurwitz zeta function. Thus

$$\int_0^{\infty} \frac{\ln(1+x)}{(1+x)^p x}\, dx = \zeta(2, p) = \sum_{n=0}^{\infty} \frac{1}{(n+p)^2}.$$

This integral appears as entry 2.6.10 (52) in [43].

Example 5.3.2

We have

$$\int_0^{\infty} \left(\frac{1 - e^{-x}}{x} - e^{-x} \right) \frac{dx}{x} = 1.$$

To prove this we notice that

$$\frac{1-e^{-x}}{x} - e^{-x} = \sum_{n=1}^{\infty} (-1)^{n-1} \left(\frac{1}{n!} - \frac{1}{(n+1)!} \right) x^n$$

and the coefficients

$$A(n) = \frac{1}{n!} - \frac{1}{(n+1)!} \quad \text{extend to} \quad A(t) = \frac{1}{\Gamma(t+1)} - \frac{1}{\Gamma(t+2)}$$

with

$$A'(t) = -\frac{\Gamma'(t+1)}{\Gamma^2(t+1)} + \frac{\Gamma'(t+2)}{\Gamma^2(t+2)}$$

so that $A'(0) = \psi(2) - \psi(1) = 1$. This proves the value of the integral.

Example 5.3.3

For any $q > 0$ we have

(5.20) $$\int_0^{\infty} \left(\frac{\sin\sqrt{x}}{\sqrt{x}} - \frac{1}{1+qx} \right) \frac{dx}{x} = 2(1-\gamma) + \ln q .$$

Here

$$\frac{\sin\sqrt{x}}{\sqrt{x}} - \frac{1}{1+qx} = \sum_{n=1}^{\infty} (-1)^{n-1} \left(q^n - \frac{1}{(2n+1)!} \right) x^n$$

and the coefficients $A(n) = q^n - 1/(2n+1)!$ extend to

$$A(t) = q^t - \frac{1}{\Gamma(2t+2)} .$$

Therefore,

$$A'(t) = q^t \ln q + \frac{2\Gamma'(2t+2)}{\Gamma^2(2t+2)}$$

$$A'(0) = \ln q + 2\psi(2) = \ln q + 2(1 - \gamma).$$

The evaluation is proved.

With the substitution $x = t^2$ the integral (5.20) turns into

$$\int_0^\infty \left(\frac{\sin t}{t} - \frac{1}{1 + qt^2} \right) \frac{dt}{t} = 1 - \gamma + \frac{1}{2} \ln q$$

which is similar to entry 2.5.29 (8) in [43].

Remark. From (5.20) we know that when $q = 1$

$$\int_0^\infty \left(\frac{\sin \sqrt{x}}{\sqrt{x}} - \frac{1}{1 + x} \right) \frac{dx}{x} = 2(1 - \gamma).$$

It is most interesting that for any $p > 0$ we have also

(5.21)
$$\int_0^\infty \left(\frac{\sin \sqrt{x}}{\sqrt{x}} - \frac{1}{1 + x^p} \right) \frac{dx}{x} = 2(1 - \gamma)$$

that is, the integral does not depend on the parameter p!

Here is a proof. We can write

$$\int_0^\infty \left(\frac{\sin \sqrt{x}}{\sqrt{x}} - \frac{1}{1 + x^p} \right) \frac{dx}{x}$$

$$= \int_0^\infty \left(\frac{\sin \sqrt{x}}{\sqrt{x}} - \frac{1}{1 + x} + \frac{1}{1 + x} - \frac{1}{1 + x^p} \right) \frac{dx}{x}$$

$$= 2(1 - \gamma) + \int_0^\infty \left(\frac{1}{1 + x} - \frac{1}{1 + x^p} \right) \frac{dx}{x}$$

and now we will show that the last integral equals zero

(5.22)
$$\int_0^\infty \left(\frac{1}{1+x} - \frac{1}{1+x^p} \right) \frac{dx}{x} = 0 .$$

Proof.

$$\int_0^\infty \left(\frac{1}{1+x} - \frac{1}{1+x^p} \right) \frac{dx}{x} = \int_0^\infty \frac{x^p - x}{(1+x)(1+x^p)} \frac{dx}{x}$$

$$= \int_0^\infty \frac{x^{p-1} - 1}{(1+x)(1+x^p)} dx$$

$$= \int_0^1 \frac{x^{p-1} - 1}{(1+x)(1+x^p)} dx + \int_1^\infty \frac{x^{p-1} - 1}{(1+x)(1+x^p)} dx .$$

These two integrals annihilate each other. Making the substitution $x = 1/t$ in the first one we get

$$\int_0^1 \frac{x^{p-1} - 1}{(1+x)(1+x^p)} dx = \int_\infty^1 \frac{1 - t^{p-1}}{(t+1)(t^p+1)} \frac{t^{p+1}}{t^{p-1}} \left(-\frac{dt}{t^2} \right)$$

$$= -\int_1^\infty \frac{t^{p-1} - 1}{(t+1)(t^p+1)} dt$$

so that

(5.23)
$$\int_0^\infty \frac{x^{p-1} - 1}{(1+x)(1+x^p)} dx = 0$$

for any $p > 0$.

With the substitution $x = t^2$ we find also that for any $p > 0$

(5.24)
$$\int_0^\infty \left(\frac{\sin t}{t} - \frac{1}{1+t^p} \right) \frac{dt}{t} = 1 - \gamma .$$

In general, we can state the proposition.

Proposition 5.1. *Suppose $h(x)$ is a function defined on $[0,\infty)$ and such that the following integral is convergent and has value M*

$$\int_0^\infty \left(h(x) - \frac{1}{1+t} \right) \frac{dt}{t} = M .$$

Then for every $p > 0$ we have also

$$\int_0^\infty \left(h(x) - \frac{1}{1+t^p} \right) \frac{dt}{t} = M .$$

For the proof we just need to add the zero integral (5.22).

Example 5.3.4

For any $q > 0$ we have

(5.25) $$\int_0^\infty \left(\cos\sqrt{x} - \frac{1}{1+qx} \right) \frac{dx}{x} = -2\gamma + \ln q .$$

Here

$$\cos\sqrt{x} - \frac{1}{1+qx} = \sum_{n=1}^\infty (-1)^{n-1} \left(q^n - \frac{1}{(2n)!} \right) x^n$$

and the coefficients $A(n) = q^n - 1/(2n)!$ extend to

$$A(t) = q^t - \frac{1}{\Gamma(2t+1)}$$

with $A'(0) = \ln q + 2\psi(1) = \ln q - 2\gamma$. The evaluation is proved.

With the substitution $x = t^2$ we find also (cf. 2.5.29 (9) in [43])

$$\int_0^\infty \left(\cos t - \frac{1}{1+qt^2} \right) \frac{dt}{t} = -\gamma + \frac{1}{2}\ln q .$$

In the same way as above we find also

(5.26) $$\int_0^\infty \left(\cos \sqrt{x} - \frac{1}{1+x^p} \right) \frac{dx}{x} = -2\gamma$$

$$\int_0^\infty \left(\cos t - \frac{1}{1+t^p} \right) \frac{dt}{t} = -\gamma$$

or,

$$\gamma = \int_0^\infty \left(\frac{1}{1+t^p} - \cos t \right) \frac{dt}{t}$$

for any $p > 0$. This extend the well-known (for $p = 1$) representation of Euler's constant γ.

Example 5.3.5

(5.27) $$\int_0^\infty \left(\frac{1-\cos \sqrt{x}}{x} - \frac{1}{2(1+qx)} \right) \frac{dx}{x} = \frac{1}{2} \ln q + \frac{3}{2} - \gamma$$

for any $q > 0$. With $x = t^2$

$$\int_0^\infty \left(\frac{1-\cos t}{t} - \frac{1}{2(1+qt^2)} \right) \frac{dt}{t} = \frac{1}{4} \left(\ln q + 3 - 2\gamma \right).$$

Now

$$\frac{1-\cos \sqrt{x}}{x} - \frac{1}{2(1+qx)} = \sum_{n=1}^\infty (-1)^{n-1} \left(\frac{q^n}{2} - \frac{1}{(2n+2)!} \right) x^n$$

and the coefficients $A(n)$ extend to

$$A(t) = \frac{q^t}{2} - \frac{1}{\Gamma(2t+3)}$$

($t \geq -1$) with

$$A'(0) = \psi(3) + \frac{1}{2}\ln q = \frac{3}{2} - \gamma + \frac{1}{2}\ln q .$$

Again, as in Example 5.3.3 we have for every $p > 0$

(5.28)
$$\int_0^\infty \left(\frac{1 - \cos t}{t^2} - \frac{1}{2(1 + t^p)} \right) \frac{dt}{t} = \frac{3}{4} - \frac{\gamma}{2}$$

which extends entry 2.5.29 (7) from [43] (given there for $p = 1$).

Example 5.3.6

For any $q > 0$ we have

(5.29)
$$\int_0^\infty \left(\frac{\arctan \sqrt{x}}{\sqrt{x}} - \frac{1}{1 + qx} \right) \frac{dx}{x} = 2 + \ln q$$

Here

$$\frac{\arctan \sqrt{x}}{\sqrt{x}} - \frac{1}{1 + qx} = \sum_{n=1}^\infty (-1)^{n-1} \left(q^n - \frac{1}{2n + 1} \right) x^n$$

and the coefficients $A(n)$ extend to $A(t) = q^t - \dfrac{1}{2t + 1}$ with $A'(0) = \ln q + 2$. This proves (5.29).

Example 5.3.7

For any $q, p > 0$

(5.30)
$$\int_0^\infty \left(e^{-px} - \frac{1}{1 + qx} \right) \frac{dx}{x} = \ln \frac{q}{p} - \gamma .$$

Special Techniques for Solving Integrals

This is entry 2.3.19 (7) in [43]. We have

$$e^{-px} - \frac{1}{1+qx} = \sum_{n=1}^{\infty} (-1)^{n-1} \left(q^n - \frac{p^n}{n!} \right) x^n$$

and the coefficients $A(n) = q^n - p^n / n!$ extend to $A(t) = q^t - p^t / \Gamma(t+1)$ with

$$A'(t) = q^t \ln q - \frac{\Gamma(t+1) \ln p - \Gamma'(t+1)}{\Gamma^2(t+1)} p^t$$

and $A'(0) = \ln q - \ln p + \psi(1) = \ln \dfrac{q}{p} - \gamma$.

Using the special integral (5.22) we can also write for any $p, r > 0$

$$(5.31) \qquad \int_0^{\infty} \left(e^{-px} - \frac{1}{1+x^r} \right) \frac{dx}{x} = -\ln p - \gamma$$

Example 5.3.8

For every $q > 0$

$$(5.32) \qquad \int_0^{\infty} \left(\beta(x+1) - \frac{\ln 2}{1+qx} \right) \frac{dx}{x} = \left(\frac{1}{2} \ln 2 - \gamma + \ln q \right) \ln 2$$

where $\beta(x)$ is the Nielsen beta function used before in Section 4.5 (see (4.16)). Recall that

$$\beta(x) = \sum_{n=0}^{\infty} \frac{(-1)^n}{n+x}.$$

We will use also Euler's eta function

$$\eta(s) = \sum_{n=1}^{\infty} \frac{(-1)^{n-1}}{n^s} = \frac{1}{\Gamma(s)} \int_0^{\infty} \frac{x^{s-1}}{e^x + 1} dx, \quad \text{Re}\, s > 0$$

with $\eta(1) = \ln 2$. It is easy to prove the expansion

$$\beta(x+1) = \sum_{n=0}^{\infty} (-1)^n \eta(n+1) x^n.$$

This is the Taylor series for $\beta(x+1)$ centered at the origin, as

$$\beta^{(n)}(1) = (-1)^n n! \sum_{k=1}^{\infty} \frac{(-1)^{k-1}}{k^{n+1}} = (-1)^n n! \, \eta(n+1).$$

Now

$$\beta(x+1) - \frac{\ln 2}{1+qx} = \sum_{n=0}^{\infty} (-1)^n (\eta(n+1) - q^n \ln 2) x^n$$

and $A(n) = \eta(n+1) - q^n \ln 2$ extend to $A(t) = \eta(t+1) - q^t \ln 2$ with $A'(0) = \eta'(1) - \ln q \ln 2$. It is known that $\eta'(1) = \gamma \ln 2 - (\ln^2 2)/2$ and (5.32) follows.

To understand this integral better we will show that it is absolutely convergent. It is sufficient to show that the integral

$$\int_{1}^{\infty} \frac{\beta(x+1)}{x} dx$$

is convergent. This follows from the estimate ($M > 0$ a constant)

$$0 < \frac{\beta(x+1)}{x} \le \frac{M}{x^2}$$

for $x > 0$. We have from (4.16)

$$x\beta(x+1) = \int_{0}^{\infty} \frac{xe^{-xt}}{e^t + 1} dt = -\int_{0}^{\infty} \frac{de^{-xt}}{e^t + 1} dt = \frac{1}{2} - \int_{0}^{\infty} \frac{e^t e^{-xt}}{(e^t + 1)^2} dt$$

and therefore, $x\beta(x+1)$ is a bounded function on $[0, \infty)$. When $q = 1$ we have for every $p > 0$

(5.33) $$\int_0^\infty \left(\beta(x+1) - \frac{\ln 2}{1+x^p} \right) \frac{dx}{x} = \left(\frac{1}{2}\ln 2 - \gamma \right) \ln 2 .$$

Example 5.3.9

Let $\zeta(s)$ be Riemann's zeta function defined for $\operatorname{Re}(s) > 1$ by

$$\zeta(s) = \sum_{n=1}^\infty \frac{1}{n^s} = \frac{1}{\Gamma(s)} \int_0^\infty \frac{x^{s-1}}{e^x - 1} dx .$$

Then for every $q > 0$ we have

(5.34) $$\int_0^\infty \left(\frac{\psi(x+1) + \gamma}{x} - \frac{\zeta(2)}{1+qx} \right) \frac{dx}{x} = \zeta(2)\ln q - \zeta'(2) .$$

We will use the well-known expansion (Taylor series centered at the origin – see 8.363(1) in [25])

$$\psi(x+1) + \gamma = \sum_{k=1}^\infty (-1)^{k-1} \zeta(k+1) x^k$$

for $|x| < 1$. Setting $k - 1 = n$ we have

$$\frac{\psi(x+1) + \gamma}{x} = \sum_{n=0}^\infty (-1)^n \zeta(n+2) x^n$$

so that

$$\frac{\psi(x+1) + \gamma}{x} - \frac{\zeta(2)}{1+qx} = \sum_{n=1}^\infty (-1)^{n-1} (q^n \zeta(2) - \zeta(n+2)) x^n$$

(here $\zeta(2)$ is needed to assure $A(0) = 0$). Now $A(n) = q^n \zeta(2) - \zeta(n+2)$ extend to $A(t) = q^t \zeta(2) - \zeta(t+2)$ with $A'(0) = \zeta(2)\ln q - \zeta'(2)$. The evaluation (5.34) follows.

Since $\zeta(2) = \dfrac{\pi^2}{6}$ and $\zeta'(2) = \zeta(2)(\gamma + \ln(2\pi)) - 12\ln A$, where A is Glaisher-Kinkelin's constant, we can write the value of the integral as

$$\frac{\pi^2}{6}(\ln q - \gamma - \ln(2\pi)) + 12\ln A .$$

It is not difficult to see that this integral is also absolutely convergent.

Explaining formula (5.19)

First we recall Euler's transformation formula for series (see [8], [13]). Let

$$g(t) = a_0 + a_1 t + a_2 t^2 + \ldots$$

be a power series. Then

$$\frac{1}{1-t} g\left(\frac{t}{1-t}\right) = \sum_{n=0}^{\infty} t^n \left\{ \sum_{k=0}^{n} \binom{n}{k} a_k \right\} .$$

Now in the integral in (5.19) we make the substitution $x = \dfrac{t}{1-t}$ to get

$$\int_0^{\infty} \frac{f(x)}{x} dx = \int_0^1 \frac{1}{1-t} f\left(\frac{t}{1-t}\right) \frac{dt}{t}$$

$$= \int_0^1 \left\{ \sum_{n=0}^{\infty} t^n \sum_{k=0}^{n} \binom{n}{k} (-1)^{k-1} A(k) \right\} \frac{dt}{t}$$

$$= \sum_{n=1}^{\infty} \left\{ \sum_{k=1}^{n} \binom{n}{k} (-1)^{k-1} A(k) \right\} \int_0^1 t^{n-1} dt$$

$$= \sum_{n=1}^{\infty} \left\{ \sum_{k=1}^{n} \binom{n}{k} (-1)^{k-1} A(k) \right\} \frac{1}{n} .$$

Note that the summation starts now from $n = 1$ because $A(0) = 0$. It was shown in [9] that for a large class of function $A(t)$ defined on $[0, \infty)$ we have

(5.35)
$$\sum_{n=1}^{\infty} \left\{ \sum_{k=1}^{n} \binom{n}{k} (-1)^{k-1} A(k) \right\} \frac{1}{n} = A'(0).$$

This is true for all functions $A(t)$ appearing in the above examples.

To give the reader an idea how (5.35) works we apply it to the function $A(t) = q^t - 1$, $|1 - q| < 1$. Note that $A(0) = 0$ and we can start the summation in the interior sum from $n = 0$

$$\sum_{n=1}^{\infty} \left\{ \sum_{k=1}^{n} \binom{n}{k} (-1)^{k-1} (q^k - 1) \right\} \frac{1}{n}$$

$$= \sum_{n=1}^{\infty} \left\{ \sum_{k=0}^{n} \binom{n}{k} (-1)^{k-1} (q^k - 1) \right\} \frac{1}{n}$$

$$= -\sum_{n=1}^{\infty} \left\{ \sum_{k=0}^{n} \binom{n}{k} (-1)^{k} q^k + \sum_{k=0}^{n} \binom{n}{k} (-1)^{k} \right\} \frac{1}{n}$$

$$= -\sum_{n=1}^{\infty} \left\{ \sum_{k=0}^{n} \binom{n}{k} (-1)^{k} q^k \right\} \frac{1}{n}$$

$$= -\sum_{n=1}^{\infty} \frac{(1-q)^n}{n} = \ln(1 - (1-q)) = \ln q = A'(0)$$

using the fact that

$$\sum_{k=0}^{n} \binom{n}{k} (-1)^{k} = 0.$$

One more example. Let us take the function

$$A(t) = \frac{1}{\Gamma(t+1)} - 1 = \frac{1}{\Gamma(t+1)} - \frac{1}{\Gamma(1)}$$

where for the reciprocal gamma function we will use an integral representation based on a positively oriented Hankel contour L extending on both sides of the negative half axis $(-\infty, 0)$ with a small loop around the origin

$$\frac{1}{\Gamma(t+1)} = \frac{1}{2\pi i} \int_L u^{-t-1} e^u du .$$

We have

$$A(t) = \frac{1}{2\pi i} \int_L (u^{-t} - 1) u^{-1} e^u du$$

and (5.35) works the same way (as it applies to the expression $u^{-t} - 1$).

A new proof of formula (5.35) appeared in the recent paper by the author "New series identities with Cauchy, Stirling, and harmonic numbers, and Laguerre polynomials" (https://arxiv.org/abs/1911.00186).

Other interesting formulas for evaluating integrals were presented by M. Laurence Glasser and Michael Milgram in their paper "Master Theorems for a Family of Integrals" (*Integral Transforms Spec. Func.* 25 (2014), 805-820).

5.4 Miscellaneous Selected Integrals

In this section we present a collection of interesting integrals some of which appear as problems in the *American Mathematical Monthly*, the *College Mathematics Journal*, or the *Mathematics Magazine*.
The solutions of the problems presented here were found independently by the author. They may differ from the published solutions.

Example 5.4.1
We start with a simple and nice integral. For any $a, b > 0$

$$(5.36) \qquad \int_0^\infty \ln \frac{x^2 + b^2}{x^2 + a^2} \, dx = \pi(b - a)$$

This is entry 2.6.15(10) in [43]. The proof is very short. We will write this integral as a double integral and then change the order of integration.

$$\int_0^\infty \ln \frac{x^2 + b^2}{x^2 + a^2} \, dx = \int_0^\infty \{ \ln(x^2 + b^2) - \ln(x^2 + a^2) \} \, dx$$

$$= \int_0^\infty \left\{ \ln(x^2 + t^2) \Big|_a^b \right\} dx = \int_0^\infty \left\{ \int_a^b \frac{2t}{x^2 + t^2} \, dt \right\} dx$$

$$= \int_a^b \left\{ \int_0^\infty \frac{2t}{x^2 + t^2} \, dx \right\} dt = 2 \int_a^b \left\{ \arctan \frac{x}{t} \Big|_0^\infty \right\} dt$$

$$= 2 \int_a^b \frac{\pi}{2} \, dt = \pi(b - a).$$

The integral can be evaluated without using double integrals, but first computing the antiderivative and then using the limits. Thus

$$\int \ln \frac{x^2 + b^2}{x^2 + a^2} \, dx = \int \ln(x^2 + b^2) \, dx - \int \ln(x^2 + a^2) \, dx$$

(and after integration by part and simplifying)

$$= x \ln \frac{x^2 + b^2}{x^2 + a^2} + 2b \arctan \frac{x}{b} - 2a \arctan \frac{x}{a}.$$

Now evaluation between the given limits brings to our result. For the limit

$$\lim_{x \to \infty} x \ln \frac{x^2 + b^2}{x^2 + a^2} = 0$$

the substitution $x = 1/t$ would be appropriate.

Example 5.4.2

Solution to Monthly Problem 11796 (2015, p.738). The problem is to evaluate the integral

$$\int_0^\infty \frac{\sin(2n+1)x}{\sin x} e^{-\alpha x} x^{m-1} dx$$

where $\alpha > 0$ is arbitrary and $m \geq 1, n \geq 0$ are integers.

We start with the well-known representation of the Dirichlet kernel

$$\frac{\sin(2n+1)x}{\sin x} = \sum_{k=-n}^n e^{2ikx}$$

(this is computed easily by summing the finite geometric series on the right-hand side). Next we multiply this equation by $e^{-\alpha x}$ and integrate

$$\int_0^\infty \frac{\sin(2n+1)x}{\sin x} e^{-\alpha x} dx = \sum_{k=-n}^n \int_0^\infty e^{-(\alpha-2ik)x} dx = \sum_{k=-n}^n \frac{1}{\alpha-2ik}.$$

All we need to do now is to differentiate both sides of this equation $m-1$ times with respect to α. The result is

(5.37)
$$\int_0^\infty \frac{\sin(2n+1)x}{\sin x} e^{-\alpha x} x^{m-1} dx = (m-1)! \sum_{k=-n}^n \frac{1}{(\alpha-2ik)^m}.$$

This evaluation is compact enough. If we want to avoid complex numbers in the answer we can separate the real part of the sum to get

$$\int_0^\infty \frac{\sin(2n+1)x}{\sin x} e^{-\alpha x} x^{m-1} dx$$

$$= (m-1)! \left(\frac{1}{\alpha^m} + 2 \sum_{k=1}^n \frac{\cos(m \arctan(2k/\alpha))}{(\alpha^2+4k^2)^{m/2}} \right).$$

Example 5.4.3

Here we prove entry 4.236(1) from [25]. For any $p > 0$

$$(5.38) \quad J = \int_0^1 \left\{ \frac{1 + (p-1)\ln x}{1-x} + \frac{x \ln x}{(1-x)^2} \right\} x^{p-1} dx = \psi'(p) - 1.$$

This is a very tricky integral, a real puzzle!

The integral is a combination of two independent integrals, $J = J_1 + J_2$ where

$$J_1 = \int_0^1 \left\{ \frac{1 + p \ln x}{1-x} + \frac{x \ln x}{(1-x)^2} \right\} x^{p-1} dx = -1$$

and

$$J_2 = \int_0^1 \frac{-\ln x}{1-x} x^{p-1} dx = \psi'(p).$$

This combination is not obvious!
First we evaluate

$$J_1 = \int_0^1 \left\{ \frac{x^{p-1} + p x^{p-1} \ln x}{1-x} + \frac{x^p \ln x}{(1-x)^2} \right\} dx$$

$$= \int_0^1 \left\{ \frac{d}{dx}(x^p \ln x) \frac{1}{1-x} + (x^p \ln x) \frac{d}{dx} \frac{1}{1-x} \right\}$$

$$= \int_0^1 \frac{d}{dx} \left(\frac{x^p \ln x}{1-x} \right) dx = \frac{x^p \ln x}{1-x} \bigg|_0^1 = -1$$

(the values at the endpoint are found by using limits).

Next we use the well-known integral representation of the digamma function (entry 8.361 (7) in [25])

$$\psi(p) = -\gamma + \int_0^1 \frac{1 - x^{p-1}}{1-x} dx$$

where γ is Euler's constant. Differentiating with respect to p brings to J_2 immediately. The formula is proved.

Example 5.4.4

In this example we give an extension to Monthly Problem 11506 (2010, p. 459). We will prove that for any two complex numbers λ, μ with $|\operatorname{Re}\lambda| < 1, |\operatorname{Re}\mu| < 1, |\operatorname{Re}(\lambda + \mu)| < 1$ and any two positive number $a, b > 0$ we have

$$(5.39) \qquad \sin\lambda\pi \int_0^\infty \frac{x^\lambda}{x+a} \frac{b^\mu - x^\mu}{b-x} dx = \sin\mu\pi \int_0^\infty \frac{x^\mu}{x+b} \frac{a^\lambda - x^\lambda}{a-x} dx$$

$$= \frac{\sin\lambda\pi \sin\mu\pi}{\pi} \int_0^\infty \int_0^\infty \frac{x^\mu t^\lambda}{(x+b)(t+a)(t+x)} dx dt .$$

The proof starts with the well-known Euler integral

$$\int_0^\infty \frac{t^\beta}{t+a} dt = \frac{-\pi a^\beta}{\sin\beta\pi}, \quad -1 < \beta < 0$$

(see Example 4.2.3). Using partial fractions we write

$$\int_0^\infty \frac{t^\lambda}{(t+a)(t+x)} dt = \frac{\pi}{\sin\lambda\pi} \frac{a^\lambda - x^\lambda}{a-x}$$

which extends to $-1 < \operatorname{Re}\lambda < 1$. Then we rewrite this in the form

$$\frac{a^\lambda - x^\lambda}{a-x} = \frac{\sin\lambda\pi}{\pi} \int_0^\infty \frac{t^\lambda}{(t+a)(t+x)} dt .$$

Here we multiply both sides by $\sin\mu\pi \dfrac{x^\mu}{x+b}$ and integrate with respect to x from zero to infinity. The result is

(5.40a)
$$\sin\mu\pi\int_0^\infty \frac{x^\mu}{x+b}\frac{a^\lambda - x^\lambda}{a-x}dx$$

$$=\frac{\sin\lambda\pi\sin\mu\pi}{\pi}\int_0^\infty\int_0^\infty \frac{x^\mu t^\lambda}{(x+b)(t+a)(t+x)}dtdx.$$

In the same way, from

$$\frac{b^\mu - x^\mu}{b-x} = \frac{\sin\mu\pi}{\pi}\int_0^\infty \frac{t^\mu}{(t+b)(t+x)}dt$$

we obtain

(5.40b)
$$\sin\lambda\pi\int_0^\infty \frac{x^\lambda}{x+a}\frac{b^\mu - x^\mu}{b-x}dx$$

$$=\frac{\sin\lambda\pi\sin\mu\pi}{\pi}\int_0^\infty\int_0^\infty \frac{x^\lambda t^\mu}{(x+a)(t+b)(t+x)}dtdx.$$

The double integrals in (5.40a) and (5.40b) are the same, so the left-hand sides are the same and the identity (5.39) is proved.

Example 5.4.5

Here we prove a very nice integral. This is entry 4.242(1) from [25]. For any $0 < b < a$

(5.41)
$$J = \int_0^\infty \frac{\ln x\, dx}{\sqrt{(a^2+x^2)(b^2+x^2)}} = \frac{\ln ab}{2a}K\left(\frac{\sqrt{a^2-b^2}}{a}\right)$$

where

$$K(k) = \int_0^{\pi/2} \frac{1}{\sqrt{1-k^2\sin^2\theta}}d\theta$$

is the complete elliptical integral of the first kind (see 8.112 (1) in [25]).

This evaluation involves a special trick – we use two independent substitutions in the same integral. First we set $x = a\tan\theta$ in (5.41) to get

$$J = \int_0^{\pi/2} \frac{\ln a + \ln\tan\theta}{\sqrt{a^2\cos^2\theta + b^2\sin^2\theta}} d\theta .$$

The second substitution (in the original integral) is $x = b\cot\theta$ and it transforms the integral into

$$J = \int_0^{\pi/2} \frac{\ln b - \ln\tan\theta}{\sqrt{a^2\cos^2\theta + b^2\sin^2\theta}} d\theta .$$

Adding these two integrals we find

$$(5.42) \qquad 2J = \ln ab \int_0^{\pi/2} \frac{1}{\sqrt{a^2\cos^2\theta + b^2\sin^2\theta}} d\theta .$$

Here $a^2\cos^2\theta + b^2\sin^2\theta = a^2\left(1 - \frac{a^2 - b^2}{a^2}\sin^2\theta\right)$ and setting $k = \frac{\sqrt{a^2 - b^2}}{a}$

we can write the above result as

$$2J = \frac{\ln ab}{a} \int_0^{\pi/2} \frac{1}{\sqrt{1 - k^2\sin^2\theta}} d\theta = \frac{\ln ab}{a} K(k)$$

which is (5.41).

An interesting equation follows from (5.41) and (5.42)

$$\int_0^\infty \frac{\ln x\, dx}{\sqrt{(a^2 + x^2)(b^2 + x^2)}} = \frac{\ln ab}{2} \int_0^{\pi/2} \frac{1}{\sqrt{a^2\cos^2\theta + b^2\sin^2\theta}} d\theta .$$

The second integral here is directly related to the Arithmetic-Geometric Mean of the numbers a and b - see Section 1.7 in Chapter 1.

Example 5.4.6

Solution to Monthly Problem 11966 (March 2017). The problem is to prove that

$$(5.43) \qquad \int_0^1 \frac{x\ln(1+x)}{1+x^2}dx = \frac{\pi^2}{96} + \frac{(\ln 2)^2}{8}.$$

We present two solutions using the method from Chapter 2, that is, differentiation with respect to a parameter.

For the first solution we define the function

$$J(\lambda) = \int_0^1 \frac{x\ln(1+\lambda x)}{1+x^2}dx, \quad 0 \le \lambda \le 1.$$

Differentiation gives

$$J'(\lambda) = \int_0^1 \frac{x^2}{(1+\lambda x)(1+x^2)}dx$$

$$= \frac{1}{1+\lambda^2}\left[\int_0^1 \frac{dx}{1+\lambda x} + \lambda\int_0^1 \frac{xdx}{1+x^2} - \int_0^1 \frac{dx}{1+x^2}\right]$$

$$= \frac{1}{1+\lambda^2}\left[\frac{\ln(1+\lambda)}{\lambda} + \lambda\frac{\ln 2}{2} - \frac{\pi}{4}\right]$$

$$= \frac{\ln(1+\lambda)}{\lambda(1+\lambda^2)} + \frac{\ln 2}{2}\frac{\lambda}{1+\lambda^2} - \frac{\pi}{4}\frac{1}{1+\lambda^2}.$$

Integrating this between 0 and 1 we find

$$J(1) = \int_0^1 \frac{\ln(1+\lambda)}{\lambda(1+\lambda^2)}d\lambda + \frac{(\ln 2)^2}{4} - \frac{\pi^2}{16}.$$

Now

$$\int_0^1 \frac{\ln(1+\lambda)}{\lambda(1+\lambda^2)}d\lambda = \int_0^1 \frac{\ln(1+\lambda)}{\lambda}d\lambda - \int_0^1 \frac{\lambda\ln(1+\lambda)}{1+\lambda^2}d\lambda = \frac{\pi^2}{12} - J(1).$$

because

$$\int_0^1 \frac{\ln(1+\lambda)}{\lambda} d\lambda = \int_0^1 \left\{ \sum_{n=1}^{\infty} \frac{(-1)^{n-1}\lambda^{n-1}}{n} \right\} d\lambda = \sum_{n=1}^{\infty} \frac{(-1)^{n-1}}{n^2} = \frac{\pi^2}{12}.$$

This way we come to the equation

$$2J(1) = \frac{(\ln 2)^2}{4} + \frac{\pi^2}{48}$$

and (5.43) follows.

For the second solution we first integrate by parts

$$\int_0^1 \frac{x\ln(1+x)}{1+x^2} dx = \frac{1}{2}\ln(1+x)\ln(1+x^2)\Big|_0^1 - \frac{1}{2}\int_0^1 \frac{\ln(1+x^2)}{1+x} dx$$

$$= \frac{(\ln 2)^2}{2} - \frac{1}{2}\int_0^1 \frac{\ln(1+x^2)}{1+x} dx$$

and then we solve the last integral in this equation. For $0 \le \alpha \le 1$ we work with the function

$$F(\alpha) = \int_0^1 \frac{\ln(1+\alpha^2 x^2)}{1+x} dx.$$

Differentiating with respect to α we find

$$F'(\alpha) = \int_0^1 \frac{2\alpha x^2}{(1+x)(1+\alpha^2 x^2)} dx$$

$$= \frac{2\alpha}{1+\alpha^2} \left[\int_0^1 \frac{dx}{1+x} + \int_0^1 \frac{x dx}{1+\alpha^2 x^2} - \int_0^1 \frac{dx}{1+\alpha^2 x^2} \right]$$

$$= \frac{2\alpha}{1+\alpha^2} \left[\ln 2 + \frac{\ln(1+\alpha^2)}{2\alpha^2} - \frac{\arctan\alpha}{\alpha} \right]$$

$$= \frac{2\alpha}{1+\alpha^2}\ln 2 + \frac{\ln(1+\alpha^2)}{\alpha(1+\alpha^2)} - \frac{2\arctan\alpha}{1+\alpha^2}.$$

Integrating this between 0 and 1 we find

$$F(1) = \int_0^1 \frac{\ln(1+x^2)}{1+x}dx = (\ln 2)^2 + \int_0^1 \frac{\ln(1+\alpha^2)}{\alpha(1+\alpha^2)}d\alpha - \frac{\pi^2}{16}.$$

In this integration the first and the third integrals are trivial. In the second integral we make the substitution $t = \alpha^2$ to get

$$\int_0^1 \frac{\ln(1+\alpha^2)}{\alpha(1+\alpha^2)}d\alpha = \frac{1}{2}\int_0^1 \frac{\ln(1+\alpha^2)2\alpha}{\alpha^2(1+\alpha^2)}d\alpha = \frac{1}{2}\int_0^1 \frac{\ln(1+t)}{t(1+t)}dt$$

$$= \frac{1}{2}\int_0^1 \frac{\ln(1+t)}{t}dt - \frac{1}{2}\int_0^1 \frac{\ln(1+t)}{1+t}dt = \frac{\pi^2}{24} - \frac{(\ln 2)^2}{4}$$

(the first integral appeared above). Finally,

$$F(1) = (\ln 2)^2 + \left(\frac{\pi^2}{24} - \frac{(\ln 2)^2}{4}\right) - \frac{\pi^2}{16} = \frac{3}{4}(\ln 2)^2 - \frac{\pi^2}{48}$$

and a simple computation gives

$$\int_0^1 \frac{x\ln(1+x)}{1+x^2}dx = \frac{\pi^2}{96} + \frac{(\ln 2)^2}{8}.$$

The integral

(5.44) $$\int_0^1 \frac{\ln(1+x^2)}{1+x}dx = \frac{3}{4}(\ln 2)^2 - \frac{\pi^2}{48}$$

will be used later in Section 5.5.

Example 5.4.7

We present here a very challenging integral. We want to evaluate

$$J = \int_0^1 \left(\frac{\pi^2}{18} - 2\arcsin^2 \frac{x}{2} \right) \frac{dx}{1-x} .$$

Our starting point is the expansion (Maclaurin series)

$$2\arcsin^2 \frac{x}{2} = \sum_{n=1}^{\infty} \binom{2n}{n}^{-1} \frac{x^{2n}}{n^2}, \quad |x| < 2$$

(for a simple derivation of this expansion see [32]). When $x = 1$ we have

$$\frac{\pi^2}{18} = \sum_{n=1}^{\infty} \binom{2n}{n}^{-1} \frac{1}{n^2}$$

and therefore,

$$\frac{\pi^2}{18} - 2\arcsin^2 \frac{x}{2} = \sum_{n=1}^{\infty} \binom{2n}{n}^{-1} \frac{1}{n^2}(1 - x^{2n})$$

$$J = \sum_{n=1}^{\infty} \binom{2n}{n}^{-1} \frac{1}{n^2} \int_0^1 \frac{1 - x^{2n}}{1-x} dx$$

$$= \sum_{n=1}^{\infty} \binom{2n}{n}^{-1} \frac{1}{n^2} \int_0^1 (1 + x + x^2 + \dots + x^{2n-1}) dx$$

$$= \sum_{n=1}^{\infty} \binom{2n}{n}^{-1} \frac{1}{n^2} H_{2n}$$

where $H_k = 1 + 1/2 + \dots + 1/k$ are the harmonic numbers.

We have the evaluation

$$\sum_{n=1}^{\infty} \binom{2n}{n}^{-1} \frac{H_{2n}}{n^2} = \zeta(3) - \frac{2}{3} \sum_{n=1}^{\infty} \binom{2n}{n}^{-1} \frac{1}{n^3}$$

(see equation (117) on p. 22 in [2]). From the same place (equation (61))

$$\sum_{n=1}^{\infty}\binom{2n}{n}^{-1}\frac{1}{n^3}=\pi\sqrt{3}(c_1+1)-\frac{4}{3}\zeta(3)-\frac{2\pi^3}{9\sqrt{3}}$$

where

$$c_1+1=\sum_{n=0}^{\infty}\frac{1}{(3n+1)^2}$$

(c_1 is a special constant used in [2]). Now we finally have

$$J=\sum_{n=1}^{\infty}\binom{2n}{n}^{-1}\frac{H_{2n}}{n^2}=\frac{17}{9}\zeta(3)+\frac{4\pi^3}{27\sqrt{3}}-\frac{2\pi}{\sqrt{3}}(c_1+1).$$

Remark. In the next two examples we will use the Stirling numbers of the first kind $s(n,k)$. They are dual to the Stirling numbers of the second kind $S(n,k)$ which appeared in Section 4.6

$$\sum_{k=0}^{n}S(n,k)s(k,m)=\delta_{nm},\quad \sum_{k=0}^{n}s(n,k)S(k,m)=\delta_{nm}.$$

The ordinary and exponential generating functions for $s(n,k)$ are

(5.45) $$x(x-1)(x-2)...(x-n+1)=\sum_{k=0}^{n}s(n,k)x^k$$

or

$$x(x+1)(x+2)...(x+n-1)=(-1)^n\sum_{k=0}^{n}(-1)^k s(n,k)x^k$$

(5.46) $$\frac{1}{n!}(\ln(1+x))^n=\sum_{p=0}^{\infty}s(p,n)\frac{x^p}{p!}.$$

Example 5.4.8

Solution to problem 1139 from the *College Mathematics Journal* (November 2018, p. 371).

(a) Prove that for any integer $n > 0$

$$\int_0^1 \frac{(\ln(1-t))^n}{t} dt = (-1)^n n! \zeta(n+1).$$

(b) Evaluate the series

$$\sum_{p=0}^{\infty} (-1)^p \frac{s(p,n)}{p! p}.$$

Solution. From the expansion (5.46) after setting $x = -t$, dividing by t, and integrating between 0 and 1 we find

(5.47) $$\int_0^1 \frac{(\ln(1-t))^n}{t} dt = n! \sum_{p=0}^{\infty} (-1)^p \frac{s(p,n)}{p! p}.$$

With the substitution $1 - t = e^{-x}$ this integral transforms to

$$\int_0^1 \frac{(\ln(1-t))^n}{t} dt = (-1)^n \int_0^{\infty} \frac{x^n e^{-x}}{1 - e^{-x}} dx$$

$$= (-1)^n \int_0^{\infty} \frac{x^n}{e^x - 1} dx = (-1)^n n! \zeta(n+1)$$

by using the well-known integral representation of the Riemann zeta function

$$\zeta(s) = \frac{1}{\Gamma(s)} \int_0^{\infty} \frac{x^{s-1}}{e^x - 1} dx, \quad \text{Re}(s) > 0.$$

Comparing the above two evaluations we come to the most interesting representation

(5.48) $$\zeta(n+1) = (-1)^n \sum_{p=0}^{\infty} (-1)^p \frac{s(p,n)}{p! p}.$$

This solves problem 1139 completely.

The representation (5.48) is not new. A good exposition of this formula, informative comments, and more general results can be found in Adamchik's paper [3].

Example 5.4.9

Here we solve the similar Problem 1117 from the *College Mathematics Journal* (January 2018, p. 60). Prove that for $n \geq 1$

(5.49) $\qquad \int_0^1 \left(\frac{\ln(1-t)}{t} \right)^n dt = n \sum_{k=0}^{n-1} (-1)^{k-1} s(n-1,k) \zeta(n+1-k)$.

For the proof we will transform both sides of this equation to one and the same expression.

First the left-hand side.

With the substitution $1 - t = e^{-x}$ we write

$$\int_0^1 \left(\frac{\ln(1-t)}{t} \right)^n dt = (-1)^n \int_0^\infty \frac{x^n e^{-x}}{(1 - e^{-x})^n} dx$$

$$= (-1)^n \int_0^\infty x^n e^{-x} \left\{ \sum_{m=0}^\infty \binom{-n}{m} (-1)^m e^{-mx} \right\} dx$$

$$= (-1)^n \sum_{m=0}^\infty \binom{-n}{m} (-1)^m \int_0^\infty x^n e^{-(m+1)x} dx$$

$$= (-1)^n \sum_{m=0}^\infty \binom{-n}{m} (-1)^m \frac{n!}{(m+1)^{n+1}}$$

$$= (-1)^n n! \sum_{m=0}^\infty \binom{n-1+m}{m} \frac{1}{(m+1)^{n+1}}.$$

The exchange of integration and summation is justified as the binomial series is absolutely convergent.

Now the right-hand side.

$$n\sum_{k=0}^{n-1}(-1)^{k-1}s(n-1,k)\zeta(n+1-k)$$

$$=n\sum_{k=0}^{n-1}(-1)^{k-1}s(n-1,k)\left\{\sum_{m=1}^{\infty}\frac{m^k}{m^{n+1}}\right\}$$

$$=n\sum_{m=1}^{\infty}\frac{1}{m^{n+1}}\left\{\sum_{k=0}^{n-1}(-1)^{k-1}s(n-1,k)m^k\right\}$$

$$=n\sum_{m=1}^{\infty}\frac{(-1)^n}{m^{n+1}}\left\{\sum_{k=0}^{n-1}(-1)^{n-1+k}s(n-1,k)m^k\right\}$$

$$=(-1)^n n\sum_{m=1}^{\infty}\frac{m(m+1)...(m+n-2)}{m^{n+1}}$$

$$=(-1)^n n\sum_{m=0}^{\infty}\frac{(m+1)...(m+n-1)}{(m+1)^{n+1}}$$

(by using the generating function (5.45) for the numbers $s(n-1,k)$ and changing the index of summation $m \to m+1$). Multiplying top and bottom of the terms of this series by $(n-1)!\,m!$ we continue

$$=(-1)^n n(n-1)!\sum_{m=0}^{\infty}\frac{(m+n-1)!}{(n-1)!\,m!\,(m+1)^{n+1}}$$

$$=(-1)^n n!\sum_{m=0}^{\infty}\binom{n-1+m}{m}\frac{1}{(m+1)^{n+1}}$$

which equals the left-hand side. Thus (5.49) is proved.

Example 5.4.10

In this example we give our solution to Monthly Problem 11418 (2009, p. 276). The problem is to evaluate the strange integral

$$J(a) = \int_{-\infty}^{\infty} \frac{x^2 \operatorname{sech}^2 x}{a - \tanh x} dx$$

where $a > 1$. The published solution (by Omran Kouba, 2010, p. 652) uses contour integration. We will give a solution without complex integration.

The integral is very tricky! Here $\operatorname{sech}^2 x$ is the derivative of $\tanh x$ and the substitution $u = a - \tanh x$ is very tempting! However, this substitution leads to nowhere.

After playing with $J(a)$ for some time, the reader may come to a natural decision – express the hyperbolic functions in terms of exponentials, simplify, and see what happens. With some simple algebra we can write $J(a)$ in the form

$$J(a) = \frac{4}{a-1} \int_{-\infty}^{\infty} \frac{x^2 e^{2x}}{(e^{2x}+1)(e^{2x}+(a+1)(a-1)^{-1})} dx .$$

Setting $b = \dfrac{a+1}{a-1} > 0$ and rescaling $x \to x/2$ we come to a more convenient form

$$J(a) = \frac{1}{2(a-1)} \int_{-\infty}^{\infty} \frac{x^2 e^x}{(e^x+1)(e^x+b)} dx .$$

In order to evaluate this integral we define the function

$$F(\lambda) = \int_{-\infty}^{\infty} \frac{e^{\lambda x}}{(e^x+1)(e^x+b)} dx$$

for $0 < \lambda < 2$. It is clear that

$$J(a) = \frac{1}{2(a-1)} F''(1) .$$

so we proceed to find $F''(1)$. The substitution $e^x = t, x = \ln t$ transforms $F(\lambda)$ into the integral

$$F(\lambda) = \int_0^\infty \frac{t^{\lambda-1}}{(t+1)(t+b)} dx.$$

Notice that $b > 1$. For the moment we assume that $0 < \lambda < 1$. Using partial fractions we write

$$F(\lambda) = \frac{1}{b-1} \left(\int_0^\infty \frac{t^{\lambda-1}}{t+1} dt - \int_0^\infty \frac{t^{\lambda-1}}{t+b} dt \right)$$

$$= \frac{1}{b-1} \left(\int_0^\infty \frac{t^{\lambda-1}}{t+1} dt - b^{\lambda-1} \int_0^\infty \frac{t^{\lambda-1}}{t+b} dt \right) = \frac{1-b^{\lambda-1}}{b-1} \int_0^\infty \frac{t^{\lambda-1}}{t+1} dt.$$

The last integral here we evaluate by using Euler's beta function

$$B(u,v) = \int_0^\infty \frac{t^{u-1}}{(1+t)^{u+v}} = \frac{\Gamma(u)\Gamma(v)}{\Gamma(u+v)}, \quad u,v > 0.$$

We set $u = \lambda, v = 1 - \lambda$ to find

$$F(\lambda) = \frac{1-b^{\lambda-1}}{b-1} \Gamma(\lambda)\Gamma(1-\lambda) = \frac{1-b^{\lambda-1}}{b-1} \frac{\pi}{\sin(\pi\lambda)}.$$

It is easy to see (using the rule of L'Hospital) that

$$\lim_{\lambda \to 1} F(\lambda) = \frac{\ln b}{b-1}$$

and defining $F(1)$ to be this value we see that $F(\lambda)$ is analytic in the disc $|\lambda - 1| < 1$. Its Taylor series centered at $\lambda - 1$ starts with

$$F(\lambda) = \frac{\ln b}{b-1} + \frac{\ln^2 b}{2(b-1)} (\lambda-1) + \frac{\ln b (\pi^2 + \ln^2 b)}{6(b-1)} (\lambda-1)^2 + \ldots$$

so that

$$F''(1) = \frac{\ln b\,(\pi^2 + \ln^2 b)}{3(b-1)}.$$

Finally,

$$J(a) = \frac{1}{12}\ln b\,(\pi^2 + \ln^2 b) = \frac{1}{12}\ln\frac{a+1}{a-1}\left(\pi^2 + \ln^2\frac{a+1}{a-1}\right).$$

Done!

Example 5.4.11

Here we will work with one excellent symmetric integral!
This is Monthly Problem 12127 (August 2019). Evaluate

$$J = \int_0^1 \left(\frac{\mathrm{Li}_2(1) - \mathrm{Li}_2(x)}{1-x}\right)^2 dx$$

where

$$\mathrm{Li}_2(x) = \sum_{n=1}^{\infty} \frac{x^n}{n^2}$$

is Euler's dilogarithm ($|x| \le 1$).

In the following work we will use the simple fact that

$$\frac{d}{dx}\mathrm{Li}_2(x) = -\frac{\ln(1-x)}{x}.$$

Integration by parts gives

$$J = \int_0^1 \left(\mathrm{Li}_2(1) - \mathrm{Li}_2(x)\right)^2 \frac{1}{(1-x)^2}\,dx$$

$$= \int_0^1 \left(\mathrm{Li}_2(1) - \mathrm{Li}_2(x)\right)^2 d\frac{1}{1-x}$$

$$= \frac{\left(\mathrm{Li}_2(1) - \mathrm{Li}_2(x)\right)^2}{1-x}\Bigg|_0^1 - 2\int_0^1 \frac{\left(\mathrm{Li}_2(1) - \mathrm{Li}_2(x)\right)}{1-x} \frac{\ln(1-x)}{x}\,dx$$

$$= A - 2\int_0^1 \left(\frac{1}{1-x} + \frac{1}{x}\right)\left(\mathrm{Li}_2(1) - \mathrm{Li}_2(x)\right)\ln(1-x)\,dx$$

(where A, the first term, will be evaluated later)

$$= A - 2\int_0^1 \left(\mathrm{Li}_2(1) - \mathrm{Li}_2(x)\right)\frac{\ln(1-x)}{x}\,dx$$

$$-2\int_0^1 \left(\mathrm{Li}_2(1) - \mathrm{Li}_2(x)\right)\frac{\ln(1-x)}{1-x}\,dx$$

$$= A - 2\int_0^1 \left(\mathrm{Li}_2(1) - \mathrm{Li}_2(x)\right) d\left(\mathrm{Li}_2(1) - \mathrm{Li}_2(x)\right)$$

$$+\int_0^1 \left(\mathrm{Li}_2(1) - \mathrm{Li}_2(x)\right) d\left(\ln(1-x)\right)^2$$

$$= A - \left(\mathrm{Li}_2(1) - \mathrm{Li}_2(x)\right)^2\Bigg|_0^1 + \left(\mathrm{Li}_2(1) - \mathrm{Li}_2(x)\right)\ln^2(1-x)\Bigg|_0^1$$

$$-\int_0^1 \frac{\ln^3(1-x)}{x}\,dx$$

$$= A + (\zeta(2))^2 + B + 6\zeta(4)$$

where B is the third term and the value

$$\int_0^1 \frac{\ln^3(1-x)}{x}\,dx = \int_0^1 \frac{\ln^3 t}{1-t}\,dt = -6\zeta(4)$$

is well known (see Example 5.4 8 above). Also

$$A = \lim_{x \to 1} \frac{\left(\mathrm{Li}_2(1) - \mathrm{Li}_2(x)\right)^2}{1-x} - \left(\mathrm{Li}_2(1)\right)^2 = -\left(\zeta(2)\right)^2.$$

The limit is zero. By the Mean Value Theorem, when $x \to 1$ the function $\mathrm{Li}_2(1) - \mathrm{Li}_2(x)$ behaves like $(1-x)\ln(1-x)/x$ and the limit is the same as

$$\lim_{x \to 1}(1-x)\ln^2(1-x) = 0$$

(It is easy to see that $\lim_{t \to 0}(t \ln^m t) = 0$ for any m.) For the same reason

$$B = \left(\mathrm{Li}_2(1) - \mathrm{Li}_2(x)\right)\ln^2(1-x)\Big|_0^1 = 0.$$

Finally,

$$J = 6\zeta(4) = \frac{\pi^4}{15}.$$

Example 5.4.12

Yet another Monthly Problem! Another impossible integral! The Monthly Problem 12184 (127, May 2020) asks for the proof of the strange integral

$$J = \int_1^\infty \frac{\ln(x^4 - 2x^2 + 2)}{x\sqrt{x^2 - 1}}\,dx = \pi \ln(2 + \sqrt{2}).$$

Here is the proof. We write $x^4 - 2x^2 + 2 = (x^2 - 1)^2 + 1$ and make the substitution $x = \sec t$. Since $\sec^2 t - 1 = \tan^2 t$ the integral becomes

$$J = \int_0^{\pi/2} \ln(1 + \tan^4 t)\,dt = \int_0^{\pi/2} \ln\left(\frac{\cos^4 t + \sin^4 t}{\cos^4 t}\right)dt$$

$$= \int_0^{\pi/2} \ln(\cos^4 t + \sin^4 t)\,dt - 4\int_0^{\pi/2} \ln(\cos t)\,dt.$$

The last integral is known from Section 3.2

$$\int_0^{\pi/2} \ln(\cos t)dt = -\frac{\pi}{2}\ln 2.$$

We now use the trigonometric identities

$$\cos^4 t = \frac{1}{8}(\cos 4t + 4\cos 2t + 3)$$

$$\sin^4 t = \frac{1}{8}(\cos 4t - 4\cos 2t + 3)$$

from which

$$\cos^4 t + \sin^4 t = \frac{1}{4}(\cos 4t + 3)$$

and therefore,

$$\int_0^{\pi/2} \ln(\cos^4 t + \sin^4 t)dt = \int_0^{\pi/2} \ln(\cos 4t + 3)dt - \int_0^{\pi/2} \ln 4 \, dt$$

$$= \int_0^{\pi/2} \ln(\cos 4t + 3)dt - \pi \ln 2.$$

The last integral here is known

$$\int_0^{\pi/2} \ln(\cos 4t + 3)dt = \frac{\pi}{2}\ln(3 + \sqrt{2}).$$

This evaluation comes from entry 2.6.36 (2) in the table of Prudnikov et al. [43]

$$\int_0^{2\pi} \ln(a + b\cos x)dx = 2\pi \ln \frac{a + \sqrt{a^2 - b^2}}{2} \quad (0 \le b < a)$$

where we take $a = 3$, $b = 1$, and make the substitution $x = 4t$ to get our particular case. Putting all pieces together we find

$$J = \frac{\pi}{2}\ln(6 + 4\sqrt{2}) = \pi \ln \sqrt{6 + 4\sqrt{2}}.$$

Simple algebra shows that $\sqrt{6 + 4\sqrt{2}} = 2 + \sqrt{2}$ (just square both sides) and the proof is finished.

Problem for the reader: Prove the useful integral

$$\int_0^{2\pi} \ln(a + b\cos x)\,dx = 2\pi \ln \frac{a + \sqrt{a^2 - b^2}}{2} \quad (0 \le b < a)$$

(using the method of Chapter 2 will be helpful).

The trigonometric identities used above are very easy to prove. They both come from the double-angle formulas

$$\cos^2 t = \frac{1}{2}(1 + \cos 2t)$$

$$\sin^2 t = \frac{1}{2}(1 - \cos 2t)$$

after squaring both sides.

Example 5.4.13

In this example we will entertain the reader by proving the integral

$$\int_0^\infty \frac{\cos \lambda x}{x^4 + 1}\,dx = \frac{\pi}{2\sqrt{2}} e^{-\frac{\lambda}{\sqrt{2}}} \left(\cos \frac{\lambda}{\sqrt{2}} + \sin \frac{\lambda}{\sqrt{2}} \right), \quad \lambda \ge 0.$$

First some remarks.
For $b > 0$ by simple rescaling $x \to x / b$ we can write this in the form

$$\int_0^\infty \frac{\cos \lambda x}{x^4 + b^4}\,dx = \frac{\pi}{2b^3\sqrt{2}} e^{-\frac{\lambda b}{\sqrt{2}}} \left(\cos \frac{\lambda b}{\sqrt{2}} + \sin \frac{\lambda b}{\sqrt{2}} \right)$$

which is entry 3.727(1) in [25]. With $\lambda = 0$ we have

$$\int_0^\infty \frac{1}{x^4 + b^4} dx = \frac{\pi}{2b^3\sqrt{2}} \, .$$

Differentiation with respect to λ gives also entry 3.727(4) in [25]

$$\int_0^\infty \frac{x\sin\lambda x}{x^4 + b^4} dx = \frac{\pi}{2b^2} e^{-\frac{\lambda b}{\sqrt{2}}} \sin\frac{\lambda b}{\sqrt{2}}$$

while integration gives entry 3.734(1)

$$\int_0^\infty \frac{\sin\lambda x}{x(x^4 + b^4)} dx = \frac{\pi}{2b^4}\left(1 - e^{-\frac{\lambda b}{\sqrt{2}}} \cos\frac{\lambda b}{\sqrt{2}}\right).$$

Now we start working on the initial integral.
Well, our integral is a tough nut to crack! Such integrals are usually solved by contour integration and the residue theorem. We will solve the integral by using very simple facts about complex numbers. We start with the decomposition

$$\frac{1}{x^4 + 1} = \frac{1}{2i}\left(\frac{1}{x^2 - i} - \frac{1}{x^2 + i}\right)$$

and we also use some help from our old friend

$$\int_0^\infty \frac{\cos\lambda x}{x^2 + a^2} dx = \frac{\pi}{2a} e^{-\lambda a} \, .$$

This integral was evaluated in the book twice – in Chapter 2, Example 2.3.2, and also in Chapter 4, Section 4.3. Here $a > 0$. First we replace a by $\sqrt{a}$ to write

$$\int_0^\infty \frac{\cos\lambda x}{x^2 + a} dx = \frac{\pi}{2\sqrt{a}} e^{-\lambda\sqrt{a}} \, .$$

Now this equation can be extended by analytical continuation to the entire complex plane without the closed ray $(-\infty, 0]$. In particular, we can take $a = i = e^{i\frac{\pi}{2}}$ with

$$\sqrt{i} = e^{i\frac{\pi}{4}} = \cos\frac{\pi}{4} + i\sin\frac{\pi}{4} = \frac{1+i}{\sqrt{2}}, \quad \frac{1}{\sqrt{i}} = \frac{1-i}{\sqrt{2}}.$$

This way

$$\int_0^\infty \frac{\cos\lambda x}{x^2 + i} dx = \frac{\pi(1-i)}{2\sqrt{2}} e^{-\frac{\lambda}{\sqrt{2}}} \left(\cos\frac{\lambda}{\sqrt{2}} - i\sin\frac{\lambda}{\sqrt{2}} \right)$$

$$= \frac{\pi}{2\sqrt{2}} e^{-\frac{\lambda}{\sqrt{2}}} \left[\left(\cos\frac{\lambda}{\sqrt{2}} - \sin\frac{\lambda}{\sqrt{2}} \right) - i\left(\cos\frac{\lambda}{\sqrt{2}} + \sin\frac{\lambda}{\sqrt{2}} \right) \right].$$

Also, taking now $a = -i = e^{-i\frac{\pi}{2}}$ with

$$\sqrt{-i} = e^{-i\frac{\pi}{4}} = \frac{1-i}{\sqrt{2}}, \quad \frac{1}{\sqrt{-i}} = \frac{1+i}{\sqrt{2}}$$

we compute

$$\int_0^\infty \frac{\cos\lambda x}{x^2 - i} dx = \frac{\pi(1+i)}{2\sqrt{2}} e^{-\frac{\lambda}{\sqrt{2}}} \left(\cos\frac{\lambda}{\sqrt{2}} + i\sin\frac{\lambda}{\sqrt{2}} \right)$$

$$= \frac{\pi}{2\sqrt{2}} e^{-\frac{\lambda}{\sqrt{2}}} \left[\left(\cos\frac{\lambda}{\sqrt{2}} - \sin\frac{\lambda}{\sqrt{2}} \right) + i\left(\cos\frac{\lambda}{\sqrt{2}} + \sin\frac{\lambda}{\sqrt{2}} \right) \right].$$

Then

$$\int_0^\infty \frac{\cos\lambda x}{x^4 + 1} dx = \frac{1}{2i}\left(\int_0^\infty \frac{\cos\lambda x}{x^2 - i} dx - \int_0^\infty \frac{\cos\lambda x}{x^2 + i} dx \right)$$

$$= \frac{\pi}{2\sqrt{2}} e^{-\frac{\lambda}{\sqrt{2}}} \left(\cos\frac{\lambda}{\sqrt{2}} + \sin\frac{\lambda}{\sqrt{2}} \right).$$

Done!

Problem for the reader: Prove entry 3.734(1) in [25]

$$\int_0^\infty \frac{\sin \lambda x}{x(x^4 + b^4)} dx = \frac{\pi}{2b^4}\left(1 - e^{-\frac{\lambda b}{\sqrt{2}}} \cos\frac{\lambda b}{\sqrt{2}}\right).$$

Example 5.4.14

We will use again the integral

$$\int_0^\infty \frac{\cos \lambda x}{x^2 + a^2} dx = \frac{\pi}{2a} e^{-\lambda a} \quad (\lambda \geq 0,\ a > 0)$$

(Example 2.3.2) in order to prove the very unusual one

$$\int_0^\infty \frac{\sin(2n+1)x}{\sin x} \frac{1}{x^2 + a^2} dx = \frac{\pi}{a}\left(\frac{1}{2} + e^{-(n+1)a}\frac{\sinh na}{\sinh a}\right).$$

Here $n \geq 1$ is an integer and $a > 0$. The Dirichlet kernel $\sin(2n+1)x / \sin x$ is a bounded and continuous function on $[0, \infty)$ with period π and the integral is absolutely convergent.

To evaluate the integral we use the well-known representation

$$\frac{\sin(2n+1)x}{\sin x} = 1 + 2\sum_{k=1}^n \cos(2kx).$$

Thus we have

$$\int_0^\infty \frac{\sin(2n+1)x}{\sin x} \frac{1}{x^2 + a^2} dx = \int_0^\infty \frac{1}{x^2 + a^2}\left\{1 + 2\sum_{k=1}^n \cos(2kx)\right\} dx$$

$$= \frac{\pi}{2a} + 2\sum_{k=1}^n \left\{\int_0^\infty \frac{\cos(2kx)}{x^2 + a^2} dx\right\} = \frac{\pi}{2a} + \frac{\pi}{a}\sum_{k=1}^n e^{-2ka}$$

$$= \frac{\pi}{2a} + \frac{\pi}{a}e^{-2a}\frac{1 - e^{-2na}}{1 - e^{-2a}} = \frac{\pi}{a}\left(\frac{1}{2} + \frac{1 - e^{-2na}}{e^{2a} - 1}\right)$$

$$= \frac{\pi}{a}\left(\frac{1}{2} + \frac{e^{-na}(e^{na} - e^{-na})}{e^{a}(e^{a} - e^{-a})}\right) = \frac{\pi}{a}\left(\frac{1}{2} + e^{-(n+1)a}\frac{\sinh(na)}{\sinh a}\right)$$

as needed.

When n is not an integer, one needs to consider principal value integrals (see, for instance, the paper of the author "The integral formula of Poisson with principal value integrals..." at https://arxiv.org/abs/1503.01175v1).

Easy problem for the reader. Show that

$$\int_0^{\pi/2} \frac{\sin(2n+1)x}{\sin x}\, dx = \frac{\pi}{2}.$$

Three more problems, more challenging this time. Prove that

$$\int_0^{\pi/2} \frac{\cos(2n+1)x}{\cos x}\, dx = (-1)^n \frac{\pi}{2}$$

$$\int_0^{\pi/2} \frac{\sin(2nx)}{\sin x}\, dx = 2\sum_{k=1}^{n} \frac{(-1)^{k-1}}{2k-1} \quad (n \geq 1)$$

$$\int_0^{\pi/2} \frac{\sin(2nx)}{\cos x}\, dx = 2(-1)^n \sum_{k=1}^{n} \frac{(-1)^k}{2k-1} \quad (n \geq 1).$$

5.5 Catalan's Constant

Catalan's constant

$$G = \sum_{n=0}^{\infty} \frac{(-1)^n}{(2n+1)^2}$$

already appeared in previous chapters. In this section we will present some important integrals evaluated in terms of this constant. The most straightforward example is

(5.50) $$\int_0^1 \frac{\arctan x}{x}\, dx = G\,.$$

For the proof we expand the arctangent in Maclaurin series

$$\arctan x = \sum_{n=0}^{\infty} \frac{(-1)^n x^{2n+1}}{2n+1}$$

and integrating the expansion

$$\frac{\arctan x}{x} = \sum_{n=0}^{\infty} \frac{(-1)^n x^{2n}}{2n+1}$$

between 0 and 1 we come to (5.50). This is the most popular integral representation of G. Another popular representation is the integral

(5.51) $$G = -\int_0^1 \frac{\ln x}{1+x^2}\, dx$$

which is obtained from (5.50) using integration by parts. If we make the substitution $x = \tan\frac{t}{2}$ in (5.50) we get also

$$G = \frac{1}{4} \int_0^{\pi/2} \frac{t}{\tan\frac{t}{2}\cos^2\frac{t}{2}}\, dt = \frac{1}{4} \int_0^{\pi/2} \frac{t}{\sin\frac{t}{2}\cos\frac{t}{2}}\, dt = \frac{1}{2} \int_0^{\pi/2} \frac{t}{\sin t}\, dt$$

with the help from the double-angle formula $\sin 2\theta = 2\sin\theta\cos\theta$. Thus we have the very interesting evaluation

$$\int_0^{\pi/2} \frac{t}{\sin t}\, dt = 2G$$

(entry 2.5.4 (5) in [43]).
Another amazing similar evaluation is

$$\int_0^\infty \frac{t}{\cosh t}\,dt = 2G.$$

The proof now is very different. We write $2\cosh t = e^t + e^{-t}$ and then using geometric series we have

$$\int_0^\infty \frac{t}{2\cosh t}\,dt = \int_0^\infty \frac{t}{e^t + e^{-t}}\,dt = \int_0^\infty t\,\frac{e^{-t}}{1+e^{-2t}}\,dt$$

$$= \int_0^\infty t\left\{ e^{-t}\sum_{n=0}^\infty (-e^{-2t})^n \right\}dt = \int_0^\infty t\left\{ \sum_{n=0}^\infty (-1)^n e^{-(2n+1)t} \right\}dt$$

$$= \sum_{n=0}^\infty (-1)^n \left\{ \int_0^\infty t e^{-(2n+1)t}\,dt \right\} = \sum_{n=0}^\infty (-1)^n \frac{1}{(2n+1)^2} = G.$$

Next we prove the integral

(5.52) $$\int_0^1 \frac{\ln(1+x^2)}{1+x^2}\,dx = \frac{\pi}{2}\ln 2 - G$$

which will be used later in Example 5.6.4. With the substitution $x = \tan t$ we have

$$\int_0^1 \frac{\ln(1+x^2)}{1+x^2}\,dx = \int_0^{\pi/4} \ln \sec^2 t\,dt = -2\int_0^{\pi/4} \ln \cos t\,dt$$

$$= -2\left(-\frac{\pi}{4}\ln 2 + \frac{1}{2}G \right) = \frac{\pi}{2}\ln 2 - G$$

by using the results in Example 3.2.1 from Chapter 3. In a similar way

(5.53) $$\int_0^1 \frac{\ln(1-x^2)}{1+x^2}\,dx = \frac{\pi}{4}\ln 2 - G.$$

We use again the substitution $x = \tan t$ and also the trigonometric identity $\cos^2 t - \sin^2 t = \cos 2t$

$$\int_0^1 \frac{\ln(1-x^2)}{1+x^2} dx = \int_0^{\pi/4} \left(\ln(\cos^2 t - \sin^2 t) - \ln \cos^2 t \right) dt$$

$$= \int_0^{\pi/4} \ln \cos 2t \, dt - 2 \int_0^{\pi/4} \ln \cos t \, dt$$

$$= \frac{1}{2} \int_0^{\pi/2} \ln \cos t \, dt - 2\left(-\frac{\pi}{4}\ln 2 + \frac{G}{2} \right)$$

$$= \frac{1}{2}\left(-\frac{\pi}{2}\ln 2 \right) + \frac{\pi}{2}\ln 2 - G = \frac{\pi}{4}\ln 2 - G .$$

Next we present one really nice companion to (5.50)

(5.54) $$\int_0^1 \left(\frac{\arctan x}{x} \right)^2 dx = \frac{\pi}{4}\ln 2 - \frac{\pi^2}{16} + G .$$

We work this way: first we integrate by parts

$$\int_0^1 \frac{(\arctan x)^2}{x^2} dx = -\int_0^1 (\arctan x)^2 \, d\frac{1}{x}$$

$$= \frac{(\arctan x)^2}{x} \Big|_0^1 + 2\int_0^1 \frac{\arctan x}{x(1+x^2)} dx$$

$$= -\frac{\pi^2}{16} + 2\int_0^1 \left(\frac{1}{x} - \frac{x}{1+x^2} \right) \arctan x \, dx$$

$$= -\frac{\pi^2}{16} + 2G - 2\int_0^1 \frac{x \arctan x}{1+x^2} dx .$$

Then we evaluate the last integral

$$-2\int_0^1 \frac{x\arctan x}{1+x^2}\,dx = -\int_0^1 (\arctan x)\,d\ln(1+x^2)$$

$$= -\ln(1+x^2)\arctan x\Big|_0^1 + \int_0^1 \frac{\ln(1+x^2)}{1+x^2}\,dx$$

$$= -\frac{\pi}{4}\ln 2 + \frac{\pi}{2}\ln 2 - G = \frac{\pi}{4}\ln 2 - G.$$

Putting the pieces together we come to (5.54). Done!

Making the substitution $x = \tan t$ in (5.54) we come to another remarkable evaluation

(5.55)
$$\int_0^{\pi/4} \left(\frac{t}{\sin t}\right)^2\,dx = \frac{\pi}{4}\ln 2 - \frac{\pi^2}{16} + G$$

(entry 2.5.4 (2) in [43]).

And now we present a real jewel, an integral whose value is a combination of three important constants, Catalan's constant G, Apéry's constant $\zeta(3)$, and the Archimedes constant π.

$$\int_0^1 \frac{(\arctan t)^2}{t}\,dt = \frac{\pi}{2}G - \frac{7}{8}\zeta(3).$$

For the proof we use equation (2.46) from Chapter 2

$$\zeta(3) = \frac{8}{7}\int_0^1 \frac{\arctan t}{t}\arctan\frac{1}{t}\,dt$$

together with the identity

$$\arctan\frac{1}{t} = \frac{\pi}{2} - \arctan t.$$

This way

$$\frac{7}{8}\zeta(3) = \int_0^1 \frac{\arctan t}{t} \arctan\frac{1}{t}\, dt = \int_0^1 \frac{\arctan t}{t}\left(\frac{\pi}{2} - \arctan t\right) dt$$

$$= \frac{\pi}{2}\int_0^1 \frac{\arctan t}{t}\, dt - \int_0^1 \frac{(\arctan t)^2}{t}\, dt$$

$$= \frac{\pi}{2}G - \int_0^1 \frac{(\arctan t)^2}{t}\, dt$$

and the desired equation follows.

Problems for the reader: Show that

$$\int_0^{\pi/4} \left(\frac{t}{\cos t}\right)^2 dx = \frac{\pi}{4}\ln 2 + \frac{\pi^2}{16} - G$$

$$\int_0^{\pi/4} \frac{x}{\tan x}\, dx = \frac{G}{2} + \frac{\pi}{8}\ln 2$$

$$\int_0^{\pi/4} x\tan x\, dx = \frac{G}{2} - \frac{\pi}{8}\ln 2$$

$$\int_0^{\pi} x\tan x\, dx = -\pi\ln 2 \ .$$

The last one is a principal value integral

$$\int_0^{\pi} x\tan x\, dx = \lim_{\varepsilon\to 0}\left(\int_0^{\frac{\pi}{2}-\varepsilon} x\tan x\, dx + \int_{\frac{\pi}{2}+\varepsilon}^{\pi} x\tan x\, dx\right).$$

Enjoy!

Example 5.5.1

With some help from the above results we prove two interesting integrals

$$\int_0^1 \ln\left(\frac{1+x^2}{x}\right) \frac{dx}{1+x^2} = \frac{\pi}{2} \ln 2$$

$$\int_0^1 \ln\left(\frac{1-x^2}{x}\right) \frac{dx}{1+x^2} = \frac{\pi}{4} \ln 2$$

which are entries 4.2.98(8) and 4.298(11) in [25] correspondingly. Although not present in them, Catalan's constant plays a big role in their evaluation

$$\int_0^1 \ln\left(\frac{1+x^2}{x}\right) \frac{dx}{1+x^2} = \int_0^1 \frac{\ln(1+x^2) - \ln x}{1+x^2} dx$$

$$= \int_0^1 \frac{\ln(1+x^2)}{1+x^2} dx - \int_0^1 \frac{\ln x}{1+x^2} dx = \frac{\pi}{2} \ln 2 - G + G = \frac{\pi}{2} \ln 2 .$$

Likewise

$$\int_0^1 \ln\left(\frac{1-x^2}{x}\right) \frac{dx}{1+x^2} = \int_0^1 \frac{\ln(1-x^2) - \ln x}{1+x^2} dx$$

$$= \frac{\pi}{4} \ln 2 - G + G = \frac{\pi}{4} \ln 2 .$$

Example 5.5.2

The interesting and challenging integral

$$\int_0^1 \frac{1}{x^2} \ln(1-x^2) \arccos x \, dx = \frac{\pi^2}{4} - 4G$$

appears as entry 4.1.6(77) in Brychkov's handbook [20]. We will prove it now.

Starting with the substitution $x = \cos t$

$$\int_0^1 \frac{1}{x^2} \ln(1 - x^2) \arccos x \, dx = 2 \int_0^{\pi/2} \frac{t \sin t \ln(\sin t)}{\cos^2 t} dt$$

$$= -2 \int_0^{\pi/2} \frac{t \ln(\sin t)}{\cos^2 t} d\cos t = 2 \int_0^{\pi/2} t \ln(\sin t) d \frac{1}{\cos t}$$

$$= 2 \frac{t \ln(\sin t)}{\cos t} \Big|_0^{\frac{\pi}{2}} - 2 \int_0^{\pi/2} \frac{1}{\cos t} \left(\ln(\sin t) + \frac{t \cos t}{\sin t} \right) dt$$

$$= -2 \int_0^{\pi/2} \frac{\ln(\sin t)}{\cos t} dt - 2 \int_0^{\pi/2} \frac{t}{\sin t} dt$$

$$= \frac{\pi^2}{4} - 4G$$

by using a result from Example 3.2.19. We leave it to the reader to verify the limits

$$\lim_{t \to \pi/2} \frac{t \ln(\sin t)}{\cos t} = \lim_{t \to 0} \frac{t \ln(\sin t)}{\cos t} = 0 .$$

Next to this integral in [20] we see a very similar one, entry 4.1.6(76) whose value does not include Catalan's constant, but includes π^3.

$$\int_0^1 \frac{1}{x} \ln(1 - x^2) \arccos x \, dx = \frac{\pi}{2} \ln^2 2 - \frac{\pi^3}{24}$$

Possibly the reader is curious how such a small change in the integrand brings to a marked change in the value of the integral. In the next example we will solve a more general integral depending on a parameter α where the above integral corresponds to $\alpha = -1$.

Example 5.5.3

Here comes entry 4.1.6(75) from Brychkov's table [20]

$$\int_0^1 \frac{1}{x} \ln(1 + \alpha x^2) \arccos x\, dx$$

$$= \frac{\pi}{4} \left(\ln^2 \frac{1 + \sqrt{1 + \alpha}}{2} - 2\operatorname{Li}_2 \frac{1 - \sqrt{1 + \alpha}}{2} \right)$$

($|\alpha| \le 1$). We call this integral $J(\alpha)$ and apply the method from Chapter 2. For technical convenience during the work we assume that $0 < \alpha < 1$. Differentiating and integrating by parts we have

$$J'(\alpha) = \int_0^1 \frac{x \arccos x}{1 + \alpha x^2}\, dx = \frac{1}{2\alpha} \int_0^1 \frac{\arccos x}{1 + \alpha x^2}\, d(1 + \alpha x^2)$$

$$= \frac{1}{2\alpha} \int_0^1 \arccos x\, d\ln(1 + \alpha x^2) = \frac{1}{2\alpha} \arccos x \ln(1 + \alpha x^2) \Big|_0^1$$

$$+ \frac{1}{2\alpha} \int_0^1 \frac{\ln(1 + \alpha x^2)}{\sqrt{1 - x^2}}\, dx\,.$$

The term with the limits is zero and in the last integral we make the substitution $x = \sin t$ to get

$$J'(\alpha) = \frac{1}{2\alpha} \int_0^{\pi/2} \ln(1 + \alpha \sin^2 t)\, dt\,.$$

This integral was evaluated in Chapter 2, Example 2.2.15, equation (2.18).

$$J'(\alpha) = \frac{\pi}{2\alpha} \ln \frac{1 + \sqrt{1 + \alpha}}{2}\,.$$

Now we have to integrate this logarithm

$$J(\alpha) = \frac{\pi}{2} \int \frac{1}{\alpha} \ln \frac{1+\sqrt{1+\alpha}}{2} \, d\alpha .$$

Proceeding with the substitution $1+\alpha = \beta^2$, $\alpha = \beta^2 - 1$ we write

$$\frac{2}{\pi} J(\alpha) = \int \frac{2\beta}{\beta^2 - 1} \ln \frac{1+\beta}{2} \, d\beta$$

$$= \int \left(\frac{1}{1+\beta} + \frac{1}{\beta-1} \right) \ln \frac{1+\beta}{2} \, d\beta$$

$$= \int \left(\frac{1+\beta}{2} \right)^{-1} \ln \frac{1+\beta}{2} d\frac{1+\beta}{2} + \int \frac{1}{\beta-1} \ln \frac{1+\beta}{2} \, d\beta$$

$$= \int \ln \frac{1+\beta}{2} \, d \ln \frac{1+\beta}{2} + A$$

(calling the second integral A)

$$= \frac{1}{2} \ln^2 \frac{1+\beta}{2} + A .$$

For the evaluation of A we notice that

$$\frac{d}{d\beta} \mathrm{Li}_2 \frac{1-\beta}{2} = \frac{d}{d\beta} \sum_{n=1}^{\infty} \frac{(1-\beta)^n}{2^n n^2} = -\sum_{n=1}^{\infty} \frac{(1-\beta)^{n-1}}{2^n n}$$

$$= -\frac{1}{1-\beta} \sum_{n=1}^{\infty} \frac{1}{n} \left(\frac{1-\beta}{2} \right)^n = \frac{1}{1-\beta} \ln \frac{1+\beta}{2}$$

and therefore,

$$A = \int \frac{1}{\beta-1} \ln \frac{1+\beta}{2} \, d\beta = -\mathrm{Li}_2 \frac{1-\beta}{2}$$

This gives

$$\frac{2}{\pi}J(\alpha) = \frac{1}{2}\ln^2\frac{1+\beta}{2} - \text{Li}_2\frac{1-\beta}{2}$$

and returning to the variable α we finish the proof. The constant of integration is zero.

We can extend the result to the disk $|\alpha| \le 1$ by analytic continuation.

Example 5.5.4

We want to pay attention to the integral

$$\int_0^\infty \frac{\ln(1+x)}{1+x^2}\,dx = \frac{\pi}{4}\ln 2 + G$$

(entry 4.291(9) in [25]) where the integrand is the same as in Example 2.1.2, but the limits are different and the Catalan constant appears. To evaluate this integral we follow the same method as in Example 2.1.2. Set

$$F(\lambda) = \int_0^\infty \frac{\ln(1+\lambda x)}{1+x^2}\,dx$$

for $\lambda > 0$. Then

$$F'(\lambda) = \int_0^\infty \frac{x}{(1+\lambda x)(1+x^2)}\,dx$$

$$= \frac{1}{\lambda^2+1}\int_0^\infty \left\{\frac{x+\lambda}{1+x^2} - \frac{\lambda}{1+\lambda x}\right\}dx$$

$$= \frac{1}{\lambda^2+1}\left(\lambda\arctan x + \ln\frac{\sqrt{x^2+1}}{1+\lambda x}\right)\Bigg|_0^\infty$$

$$= \frac{1}{\lambda^2+1}\left\{\frac{\pi\lambda}{2} - \ln\lambda\right\}.$$

Therefore,

$$F(\lambda) = \frac{\pi}{4}\ln(1+\lambda^2) - \int_0^\lambda \frac{\ln t}{1+t^2}\,dt \ .$$

Integrating by parts we come to the representation

$$F(\lambda) = \frac{\pi}{4}\ln(1+\lambda^2) - \ln\lambda\arctan\lambda + \int_0^\lambda \frac{\arctan t}{t}\,dt$$

and from here

$$F(1) = \int_0^\infty \frac{\ln(1+x)}{1+x^2}\,dx = \frac{\pi}{4}\ln 2 + \int_0^1 \frac{\arctan t}{t}\,dt = \frac{\pi}{4}\ln 2 + G \ .$$

The function

$$\mathrm{Ti}_2(\lambda) = \int_0^\lambda \frac{\arctan t}{t}\,dt = \sum_{n=0}^\infty \frac{(-1)^n \lambda^{2n+1}}{(2n+1)^2}$$

is known as the inverse tangent integral. It was studied by Ramanujan and Levin [33]. In terms of the dilogarithm we have

$$\mathrm{Ti}_2(\lambda) = \frac{1}{2i}\left[\mathrm{Li}_2(i\lambda) - \mathrm{Li}_2(-i\lambda)\right].$$

Example 5.5.5

We again recall the integrals

$$\int_0^{\pi/2} \ln\cos t\,dt = -\frac{\pi}{2}\ln 2$$

$$\int_0^{\pi/4} \ln\cos t\,dt = -\frac{\pi}{4}\ln 2 + \frac{1}{2}G$$

from Example 3.2.1. With the help of these two integrals we will prove the interesting entry 4.227(10) in [25]

$$\int_0^{\pi/2} \ln(1 + \tan x)\,dx = \frac{\pi}{4}\ln 2 + G.$$

Indeed, we have

$$\int_0^{\pi/2} \ln(1 + \tan x)\,dx = \int_0^{\pi/2} \ln\left(\frac{\cos x + \sin x}{\cos x}\right)dx$$

$$= \int_0^{\pi/2} \ln(\cos x + \sin x)\,dx - \int_0^{\pi/2} \ln(\cos x)\,dx$$

$$\int_0^{\pi/2} \ln\left(\sqrt{2}\left(\cos\frac{\pi}{4}\cos x + \sin\frac{\pi}{4}\sin x\right)\right)dx + \frac{\pi}{2}\ln 2$$

$$= \frac{\pi}{2}\ln\sqrt{2} + \int_0^{\pi/2} \ln\cos\left(x - \frac{\pi}{4}\right)dx + \frac{\pi}{2}\ln 2$$

$$= \int_{-\pi/4}^{\pi/4} \ln\cos t\,dt + \frac{3\pi}{4}\ln 2 = 2\int_0^{\pi/4} \ln\cos t\,dt + \frac{3\pi}{4}\ln 2$$

$$= 2\left(-\frac{\pi}{4}\ln 2 + \frac{1}{2}G\right) + \frac{3\pi}{4}\ln 2 = \frac{\pi}{4}\ln 2 + G.$$

A problem for the reader: Use the same technique to show that

$$\int_0^{\pi/4} \ln(1 + \tan x)\,dx = \frac{\pi}{8}\ln 2.$$

This is entry 4.227(9) in [25]. During the computation Catalan's constant appears twice and cancels out.

Further problems for the reader: Prove entries 4.227(11), 4227(12), and 4.227(13) from [25]. Namely, in this order,

$$\int_0^{\pi/4} \ln(1 - \tan x)\,dx = \frac{\pi}{8}\ln 2 - G$$

$$\int_0^{\pi/2} \ln((1-\tan x)^2)\,dx = \frac{\pi}{2}\ln 2 - 2G$$

$$\int_0^{\pi/4} \ln(1+\cot x)\,dx = \frac{\pi}{8}\ln 2 + G \,.$$

Example 5.5.6

In this last example for the section we present one curious integral also related to Catalan's constant

$$J = \int_0^{\pi/2} \operatorname{arcsinh}(\cos x)\,dx = G \,.$$

The substitution $x = \dfrac{\pi}{2} - t$ shows that $\cos x$ here can be replaced by $\sin x$. According to equation (1.21) we can write

$$J = \int_0^{\pi/2} \ln(\cos x + \sqrt{1 + \cos^2 x}\,)\,dx \,.$$

Integrating by parts gives

$$J = \int_0^{\pi/2} \frac{x \sin x}{\sqrt{1 + \cos^2 x}}\,dx$$

(the intermediate term is zero). To evaluate this integral we will use the binomial series

$$(1 + \cos^2 x)^{-1/2} = \sum_{n=0}^{\infty} \binom{-1/2}{n} \cos^{2n} x \,.$$

Putting this series in the integral and switching the order of summation and integration we compute

$$J = \int_0^{\pi/2} x \sin x \left\{ \sum_{n=0}^{\infty} \binom{-1/2}{n} \cos^{2n} x \right\} dx$$

$$= \sum_{n=0}^{\infty} \binom{-1/2}{n} \left\{ \int_0^{\pi/2} x(\cos^{2n} x) \sin x \, dx \right\}$$

$$= \sum_{n=0}^{\infty} \binom{-1/2}{n} \left\{ -\frac{1}{2n+1} \int_0^{\pi/2} x \, d(\cos x)^{2n+1} \right\}.$$

After integration by parts where the intermediate term is zero

$$J = \sum_{n=0}^{\infty} \frac{1}{2n+1} \binom{-1/2}{n} \left\{ \int_0^{\pi/2} (\cos x)^{2n+1} \, dx \right\}.$$

Easy computation shows that

$$\binom{-1/2}{n} = \frac{(-1)^n}{4^n} \binom{2n}{n}$$

and the value of the Wallis integral in the braces is well-known

$$\int_0^{\pi/2} (\cos x)^{2n+1} \, dx = \frac{4^n}{(2n+1)} \binom{2n}{n}^{-1}.$$

The result is

$$J = \sum_{n=0}^{\infty} \frac{(-1)^n}{(2n+1)^2} = G \ !$$

5.6 Summation of Series by Using Integrals

In the previous sections we saw some examples of evaluating integrals by reducing them to series. Here we evaluate some challenging series by reducing them to integrals and then solving these integrals.

Example 5.6.1

Our first example is Problem 2063 from the *Mathematics Magazine* (March, 2019). Evaluate the series

$$S = \sum_{n=0}^{\infty} \sum_{k=1}^{\infty} \frac{(-1)^{n+k-1}}{(n+k)^2}.$$

The answer is $S = \ln 2$. Here is the proof.

Separating the sum with $n = 0$ and $n \geq 1$ we write

$$S = \sum_{k=0}^{\infty} \frac{(-1)^{k-1}}{k^2} + \sum_{n=1}^{\infty} \sum_{k=1}^{\infty} \frac{(-1)^{n+k-1}}{(n+k)^2} = S_1 + S_2$$

where S_1 is the first sum and S_2 is the double sum. We have

$$\frac{1}{(n+k)^2} = \int_0^{\infty} te^{-(n+k)t} dt = \int_0^{\infty} te^{-nt} e^{-kt} dt$$

and therefore,

$$S_2 = -\int_0^{\infty} t \left\{ \sum_{n=1}^{\infty} \sum_{k=1}^{\infty} e^{-nt} e^{-kt} \right\} dt = -\int_0^{\infty} t \left\{ \sum_{n=1}^{\infty} e^{-kt} \right\}^2 dt$$

$$= -\int_0^{\infty} t \left\{ \frac{-e^{-t}}{1 + e^{-t}} \right\}^2 dt = -\int_0^{\infty} \frac{te^{-2t}}{(1 + e^{-t})^2} dt$$

$$= \int_0^{\infty} \frac{te^{-t}}{(1 + e^{-t})^2} d(1 + e^{-t}) = -\int_0^{\infty} te^{-t} d \frac{1}{1 + e^{-t}}.$$

(Changing the order of summation and integration is easily justified.) Next we integrate by parts

$$S_2 = -\frac{te^{-t}}{1 + e^{-t}} \Bigg|_0^{\infty} + \int_0^{\infty} \frac{e^{-t} - te^{-t}}{1 + e^{-t}} dt$$

$$= \int_0^{\infty} \frac{e^{-t}}{1 + e^{-t}} dt - \int_0^{\infty} \frac{te^{-t}}{1 + e^{-t}} dt.$$

Here

$$-\int_0^\infty \frac{te^{-t}}{1+e^{-t}}dt = \int_0^\infty t\left\{\sum_{n=1}^\infty (-e^{-t})^n\right\}dt = \sum_{n=1}^\infty (-1)^n \int_0^\infty te^{-nt}dt$$

$$=\sum_{n=1}^\infty \frac{(-1)^n}{n^2} = -S_1$$

and also

$$\int_0^\infty \frac{e^{-t}}{1+e^{-t}}dt = -\ln(1+e^{-t})\Big|_0^\infty = \ln 2.$$

This way $S_2 = \ln 2 - S_1$ and $S = S_1 + S_2 = \ln 2$. Done!

Example 5.6.2

We present our solution to Problem 997 from the *College Mathematics Journal* (May 2013, p. 142). Evaluate the series

$$\sigma = \sum_{n=0}^\infty \left(\sum_{k=1}^n \frac{(-1)^{k-1}}{n+k}\right)^2.$$

We show that $\sigma = \ln 2$. A nice companion to the previous example!

For the solution we use the representation

$$\frac{1}{n+k} = \int_0^1 x^{n+k-1}dx$$

and we write

$$\sum_{k=1}^\infty \frac{(-1)^{k-1}}{n+k} = \int_0^1 x^n\left\{\sum_{k=1}^\infty (-1)^{k-1}x^{k-1}\right\}dx = \int_0^1 \frac{x^n}{1+x}dx$$

$$\sigma = \sum_{n=0}^\infty \left(\int_0^1 \frac{x^n dx}{1+x}\right)^2 = \sum_{n=0}^\infty \left(\int_0^1 \frac{x^n dx}{1+x}\right)\left(\int_0^1 \frac{y^n dy}{1+y}\right)$$

$$= \sum_{n=0}^{\infty} \int_0^1 \int_0^1 \frac{x^n y^n \, dx \, dy}{(1+x)(1+y)} = \int_0^1 \int_0^1 \left\{ \sum_{n=0}^{\infty} (xy)^n \right\} \frac{dx \, dy}{(1+x)(1+y)}$$

$$= \int_0^1 \int_0^1 \frac{dx \, dy}{(1-xy)(1+x)(1+y)} \, .$$

The integral is convergent, as will be seen in the process of computation. We set $y = \dfrac{u}{x}$ to get

$$\sigma = \int_0^1 \left(\int_0^x \frac{du}{(1-u)(u+x)} \right) \frac{dx}{(1+x)} \, .$$

Using partial fractions we can evaluate the inside integral. The result is

$$\sigma = \int_0^1 \ln\left(\frac{2}{1-x} \right) \frac{dx}{(1+x)^2} \, .$$

Next, the substitution $\dfrac{2}{1-x} = 1+t$, or $x = \dfrac{t-1}{1+t}$, brings this to the integral

$$\sigma = \frac{1}{2} \int_1^{\infty} \frac{\ln(1+t)}{t^2} \, dt$$

which is obviously convergent. It can be evaluated easily by parts,

$$\int_1^{\infty} \frac{\ln(1+t)}{t^2} \, dt = -\frac{\ln(1+t)}{t} \bigg|_1^{\infty} + \int_1^{\infty} \frac{dt}{t(1+t)}$$

$$= \ln 2 + \ln \frac{t}{1+t} \bigg|_1^{\infty} = \ln 2 + \ln 2 = 2 \ln 2 \, .$$

Thus $\sigma = \ln 2$. Done!

Example 5.6.3

Monthly Problem 11682 (December 2012, p. 881). This is an alternating version of the previous series by the same author (Ovidiu Furdui). Evaluate

$$\sigma = \sum_{n=0}^{\infty} (-1)^n \left(\sum_{k=1}^{n} \frac{(-1)^{k-1}}{n+k} \right)^2 .$$

This time $\sigma = \dfrac{\pi^2}{24}$.

Solution: Starting again from the representations

$$\frac{1}{n+k} = \int_0^1 x^{n+k-1} dx$$

$$\sum_{k=1}^{\infty} \frac{(-1)^{k-1}}{n+k} = \int_0^1 x^n \left\{ \sum_{k=1}^{\infty} (-1)^{k-1} x^{k-1} \right\} dx = \int_0^1 \frac{x^n}{1+x} dx$$

we continue this way

$$\sigma = \sum_{n=0}^{\infty} (-1)^n \left(\int_0^1 \frac{x^n dx}{1+x} \right)^2 = \sum_{n=0}^{\infty} (-1)^n \left(\int_0^1 \frac{x^n dx}{1+x} \right) \left(\int_0^1 \frac{y^n dy}{1+y} \right)$$

$$= \sum_{n=0}^{\infty} (-1)^n \int_0^1 \int_0^1 \frac{x^n y^n dx dy}{(1+x)(1+y)}$$

$$= \int_0^1 \int_0^1 \left\{ \sum_{n=0}^{\infty} (-xy)^n \right\} \frac{dx\,dy}{(1+x)(1+y)} = \int_0^1 \int_0^1 \frac{dx\,dy}{(1+xy)(1+x)(1+y)} .$$

Here we set $y = \dfrac{u}{x}$ to get

$$\sigma = \int_0^1 \left(\int_0^x \frac{du}{(1+u)(u+x)} \right) \frac{dx}{(1+x)} .$$

Using partial fractions we can evaluate the inside integral. The result is

$$\sigma = \int_0^1 \ln\left(\frac{2}{1+x}\right)\frac{dx}{1-x^2}.$$

The substitution $\dfrac{2}{1+x}=1+t$ or $x=\dfrac{1-t}{1+t}$ transforms this integral into a more transparent one

$$\sigma = \frac{1}{2}\int_0^1 \frac{\ln(1+t)}{t}\,dt$$

which is evaluated this way

$$\int_0^1 \frac{\ln(1+t)}{t}\,dt = \int_0^1\left\{\sum_{k=1}^{\infty}\frac{(-1)^{k-1}t^k}{k}\right\}\frac{dt}{t} = \sum_{k=1}^{\infty}\frac{(-1)^{k-1}}{k^2} = \frac{\pi^2}{12}.$$

Hence $\sigma = \dfrac{\pi^2}{24}$. The evaluation of the last series above is left to the reader.

Example 5.6.4

Ovidiu Furdui and Huizeng Qin proposed the following most interesting problem (Problem H-691, *Fibonacci Quarterly*, 47 (33), 2009/2010).

Evaluate the series

$$\sigma = \sum_{n=1}^{\infty}(-1)^n\left(\ln 2 - \frac{1}{n+1}-\frac{1}{n+2}-...-\frac{1}{2n}\right)^2.$$

We will evaluate this series and show that

(5.56)
$$\sigma = \frac{\pi^2}{48}-\frac{\pi}{8}\ln 2 -\frac{7}{8}(\ln 2)^2 +\frac{G}{2}$$

where G is Catalan's constant (see Section 5.5).

We start the solution by noticing that

$$\frac{1}{n+1}+\frac{1}{n+2}+\ldots+\frac{1}{2n}=\sum_{k=1}^{2n}\frac{(-1)^{k-1}}{k}$$

(an interesting known identity). At the same time

$$\ln 2 = \sum_{k=1}^{\infty}\frac{(-1)^{k-1}}{k}$$

so that

$$\ln 2 - \frac{1}{n+1}-\frac{1}{n+2}-\ldots-\frac{1}{2n}=\sum_{k=2n+1}^{\infty}\frac{(-1)^{k-1}}{k}=\int_0^1\frac{x^{2n}}{1+x}dx$$

(the last equality is proved by expanding $(1+x)^{-1}$ in power series and then integrating term by term). Next we have

$$\sigma = \sum_{n=1}^{\infty}(-1)^n\left(\int_0^1\frac{x^{2n}}{1+x}dx\right)^2=\sum_{n=1}^{\infty}(-1)^n\left(\int_0^1\frac{x^{2n}}{1+x}dx\right)\left(\int_0^1\frac{y^{2n}}{1+y}dy\right)$$

$$\sum_{n=1}^{\infty}(-1)^n\int_0^1\int_0^1\frac{x^{2n}y^{2n}}{(1+x)(1+y)}\,dxdy$$

$$=\int_0^1\int_0^1\left\{\sum_{n=1}^{\infty}(-x^2y^2)\right\}\frac{dx\,dy}{(1+x)(1+y)}$$

$$=-\int_0^1\int_0^1\frac{x^2y^2}{(1+x^2y^2)(1+x)(1+y)}dx\,dy\,.$$

Here we set $y = u/x$ to get

$$-\sigma = \int_0^1\left\{\int_0^x\frac{u^2\,du}{(1+u^2)(u+x)}\right\}\frac{dx}{1+x}$$

$$= \int_0^1 \left\{ \int_0^x \left\{ \frac{x^2}{1+x^2} \frac{1}{u+x} + \frac{1}{1+x^2} \frac{u}{1+u^2} - \frac{x}{1+x^2} \frac{1}{1+u^2} \right\} du \right\} \frac{dx}{1+x}$$

$$= \int_0^1 \left\{ \frac{x^2}{1+x^2} \ln 2 + \frac{\ln(1+x^2)}{2(1+x)} - \frac{x \arctan x}{1+x^2} \right\} \frac{dx}{1+x}$$

$$= \ln 2 \int_0^1 \frac{x^2}{(1+x^2)(1+x)} dx + \frac{1}{2} \int_0^1 \frac{\ln(1+x^2)}{(1+x^2)(1+x)} dx + \int_0^1 \frac{-x \arctan x}{(1+x^2)(1+x)} dx.$$

Let us call these thee integrals A, B and C so that

(5.57)
$$-\sigma = A \ln 2 + \frac{1}{2} B + C$$

and now we will evaluate these integrals one by one.

The first one is easy

$$A = \int_0^1 \left\{ \frac{1}{2(1+x)} + \frac{x}{2(1+x^2)} - \frac{1}{2(1+x^2)} \right\} dx$$

$$= \frac{1}{2} \ln 2 + \frac{1}{4} \ln 2 - \frac{\pi}{8} = \frac{3}{4} \ln 2 - \frac{\pi}{8}.$$

Next

$$B = \frac{1}{2} \left\{ \int_0^1 \frac{\ln(1+x^2)}{1+x} dx + \int_0^1 \frac{\ln(1+x^2)}{1+x^2} dx - \int_0^1 \frac{x \ln(1+x^2)}{1+x^2} dx \right\}.$$

We have

$$\int_0^1 \frac{x \ln(1+x^2)}{1+x^2} dx = \frac{1}{2} \int_0^1 \ln(1+x^2) d \ln(1+x^2)$$

$$= \frac{1}{4} (\ln(1+x^2))^2 \Big|_0^1 = \frac{(\ln 2)^2}{4}$$

and from (5.44)

$$\int_0^1 \frac{\ln(1+x^2)}{1+x} dx = \frac{3}{4}(\ln 2)^2 - \frac{\pi^2}{48}.$$

Now the integral in the middle needs attention. We know from Section 5.5 that

(5.58) $$\int_0^1 \frac{\ln(1+x^2)}{1+x^2} dx = \frac{\pi}{2}\ln 2 - G.$$

Now we can write

$$B = \frac{1}{2}\left(\frac{(\ln 2)^2}{2} + \frac{\pi \ln 2}{2} - \frac{\pi^2}{48} - G \right).$$

It remains to compute C.

$$C = \int_0^1 \frac{-x\arctan x}{(1+x^2)(1+x)} dx$$

$$= \frac{1}{2}\int_0^1 \left\{ \frac{1}{1+x} - \frac{x}{1+x^2} - \frac{1}{1+x^2} \right\} \arctan x \, dx$$

$$= \frac{1}{2}\left\{ \int_0^1 \frac{\arctan x}{1+x} dx - \int_0^1 \frac{x\arctan x}{1+x^2} dx - \int_0^1 \frac{\arctan x}{1+x^2} dx \right\}.$$

Here

$$\int_0^1 \frac{\arctan x}{1+x} dx = \frac{\pi}{8}\ln 2$$

(evaluated in Chapter 2, see equation (2.3)). Integrating by parts and using equation (5.58) we write

$$\int_0^1 \frac{x\arctan x}{1+x^2} dx = \frac{\pi}{8}\ln 2 - \frac{1}{2}\int_0^1 \frac{\ln(1+x^2)}{1+x^2} dx$$

$$= \frac{\pi}{8}\ln 2 - \frac{1}{2}\left(\frac{\pi}{2}\ln 2 - G\right) = \frac{G}{2} - \frac{\pi}{8}\ln 2 \ .$$

$$\int_0^1 \frac{\arctan x}{1+x^2}\,dx = \int_0^1 \arctan x\, d\arctan x = \frac{1}{2}(\arctan x)^2\Big|_0^1 = \frac{\pi^2}{32}\ .$$

Collecting these numbers we find

$$C = \frac{1}{2}\left(\frac{\pi}{4}\ln 2 - \frac{\pi^2}{32} - \frac{G}{2}\right).$$

Now we put these numbers in equation (5.57) and simple algebra brings to the desired evaluation (5.56).

Another challenging series similar to (5.56) is

$$\sum_{n=1}^{\infty} \frac{(-1)^{n-1}}{n}\left(\sum_{k=1}^{\infty} \frac{(-1)^{k-1}}{n+k}\right)^2 = \sum_{n=1}^{\infty} \frac{(-1)^{n-1}}{n}\left(\ln 2 - \sum_{k=1}^{n} \frac{(-1)^{k-1}}{k}\right)^2$$

$$= \frac{\pi^2}{12}\ln 2 + \frac{1}{3}(\ln 2)^3 - \frac{1}{2}\zeta(3)\ .$$

It has been evaluated in [12] by the same method - reducing the series to integrals.

Example 5.6.5

In line with the previous examples in this section we evaluate the series

$$\sigma = \sum_{n=1}^{\infty}\left(\left(\frac{1}{n^2} + \frac{1}{(n+2)^2} + \frac{1}{(n+4)^2} + \dots\right) - \frac{1}{2n}\right)$$

by using integrals. This is Monthly Problem 12250 (*The American Mathematical Monthly,* November 2020, p. 853).

Using the representation

$$\frac{1}{(n+2k)^2} = \int_0^\infty te^{-(n+2k)t}\,dt$$

we have

$$\frac{1}{n^2} + \frac{1}{(n+2)^2} + \frac{1}{(n+4)^2} + \ldots = \sum_{k=0}^\infty \frac{1}{(n+2k)^2}$$

$$= \sum_{k=0}^\infty \int_0^\infty te^{-(n+2k)t}\,dt = \int_0^\infty te^{-nt}\left\{\sum_{k=0}^\infty e^{-2kt}\right\}dt = \int_0^\infty \frac{te^{-nt}}{1-e^{-2t}}\,dt$$

by changing the order of summation and integration (this is easy to justify and is left to the reader). Also

$$\frac{1}{2n} = \frac{1}{2}\int_0^\infty e^{-nt}\,dt \,.$$

This way we can write

$$\sigma = \sum_{n=1}^\infty \left\{\int_0^\infty \frac{te^{-nt}}{1-e^{-2t}}\,dt - \frac{1}{2}\int_0^\infty e^{-nt}\,dt\right\}dt = \sum_{n=1}^\infty \left\{\int_0^\infty e^{-nt}\left(\frac{t}{1-e^{-2t}} - \frac{1}{2}\right)dt\right\}$$

and again by exchanging integration and summation

$$\sigma = \int_0^\infty \sum_{n=1}^\infty e^{-nt}\left(\frac{t}{1-e^{-2t}} - \frac{1}{2}\right)dt = \int_0^\infty \frac{e^{-t}}{1-e^{-t}}\left(\frac{t}{1-e^{-2t}} - \frac{1}{2}\right)dt$$

$$= \int_0^\infty \left(\frac{te^{-t}}{(1-e^{-2t})(1-e^{-t})} - \frac{e^{-t}}{2(1-e^{-t})}\right)dt$$

$$= \int_0^\infty \left(\frac{te^{2t}}{(e^{2t}-1)(e^t-1)} - \frac{1}{2(e^t-1)}\right)dt\,.$$

From here we use the technique of partial fractions to decompose the integrand

$$\sigma = \int\limits_0^\infty \left(\frac{1}{4}\frac{t}{e^t+1} + \frac{3}{4}\frac{t}{e^t-1} + \frac{1}{2}\frac{t}{(e^t-1)^2} - \frac{1}{2}\frac{1}{e^t-1} \right) dt$$

$$= \frac{1}{4}\int\limits_0^\infty \frac{t}{e^t+1}dt + \frac{3}{4}\int\limits_0^\infty \frac{t}{e^t-1}dt + \frac{1}{2}\int\limits_0^\infty \left(\frac{t}{(e^t-1)^2} - \frac{1}{e^t-1} \right) dt .$$

The first two integrals are easy to evaluate. Using the well-known representations of Riemann's zeta function for $\operatorname{Re} s > 1$

$$\int\limits_0^\infty \frac{t^{s-1}}{e^t-1}dt = \zeta(s)\Gamma(s); \quad \int\limits_0^\infty \frac{t^{s-1}}{e^t+1}dt = (1-2^{1-s})\zeta(s)\Gamma(s)$$

(see (4.35) and (4.36)) we have

$$\frac{1}{4}\int\limits_0^\infty \frac{t}{e^t+1}dt = \frac{1}{8}\zeta(2) = \frac{\pi^2}{48}; \quad \frac{3}{4}\int\limits_0^\infty \frac{t}{e^t-1}dt = \frac{3}{4}\zeta(2) = \frac{\pi^2}{8} .$$

We are left now with the integral

$$J = \frac{1}{2}\int\limits_0^\infty \left(\frac{t}{(e^t-1)^2} - \frac{1}{e^t-1} \right) dt .$$

The substitution $u = e^t - 1$ brings it to the form

$$J = \frac{1}{2}\int\limits_0^\infty \left(\frac{\ln(1+u)}{u^2} - \frac{1}{u} \right) \frac{du}{u+1} = \frac{1}{2}\int\limits_0^\infty \frac{\ln(1+u)-u}{u^2(1+u)} du .$$

Here we integrate by parts

$$2J = -\int\limits_0^\infty \frac{\ln(1+u)-u}{1+u}d\frac{1}{u} = -\frac{\ln(1+u)-u}{(1+u)u}\bigg|_0^\infty - \int\limits_0^\infty \frac{\ln(1+u)}{u(1+u)^2}du$$

$$= -\int\limits_0^\infty \frac{\ln(1+u)}{u(1+u)^2}du$$

using the fact that

$$\left(\frac{\ln(1+u)-u}{1+u}\right)' = -\frac{\ln(1+u)}{(1+u)^2}.$$

Next we make the substitution $x = \ln(1+u)$ to write

$$2J = -\int_0^\infty \frac{\ln(1+u)}{u(1+u)^2}du = -\int_0^\infty \frac{x}{(e^x-1)e^x}dx = -\int_0^\infty \frac{x(1-e^x+e^x)}{(e^x-1)e^x}dx$$

$$= \int_0^\infty xe^{-x}dx - \int_0^\infty \frac{x}{e^x-1}dx = 1 - \frac{\pi^2}{6}$$

(the first integral in the last line is $\Gamma(2) = 1$ and the second was evaluated above). We found that

$$J = \frac{1}{2} - \frac{\pi^2}{12}$$

and putting the pieces together we finally get

$$\sigma = \frac{\pi^2}{16} + \frac{1}{2}.$$

5.7 Generating Functions for Harmonic and Skew-Harmonic Numbers

In this section we present several power series where the coefficients include harmonic or skew harmonic numbers. The creation of these series is based on the evaluation of some interesting and challenging integrals involving the logarithmic and polylogarithmic functions.

The first subsection here is dedicated to harmonic numbers and the second subsection – to skew harmonic numbers.

5.7.1 *Harmonic numbers*

The harmonic numbers

$$H_n = 1 + \frac{1}{2} + \ldots + \frac{1}{n}, \quad H_0 = 0$$

already appeared in this book. Here we will use also the generalized harmonic numbers

$$H_n^{(p)} = 1 + \frac{1}{2^p} + \ldots + \frac{1}{n^p}, \quad H_0^{(p)} = 0, \quad p \geq 1$$

and the polylogarithm [33]

$$\mathrm{Li}_p(t) = \sum_{n=1}^{\infty} \frac{t^n}{n^p}.$$

In this section we produce various generating functions for these numbers by repeated integration. In the process we will encounter some challenging integrals.

First a simple lemma.

Lemma 5.2. *Let* $f(t) = a_0 + a_1 t + a_2 t^2 + \ldots$ *be a power series. Then*

(5.59)
$$\frac{1}{1 - \lambda t} \sum_{n=0}^{\infty} a_n t^n = \sum_{n=0}^{\infty} t^n \left\{ \sum_{k=0}^{n} a_k \lambda^{n-k} \right\}$$

for an appropriate parameter λ.

For the proof we expand $(1 - \lambda t)^{-1}$ in geometric series assuming that $|\lambda t| < 1$ and multiply the two series.

Starting from the expansion

$$-\log(1 - t) = \sum_{n=1}^{\infty} \frac{1}{n} t^n \quad (|t| < 1)$$

we obtain (using the lemma with $\lambda = 1$)

(5.60)
$$\sum_{n=1}^{\infty} H_n t^n = \frac{-\log(1 - t)}{1 - t}.$$

This is the well-known generating function for the harmonic numbers. In the same way, using the lemma we find

(5.61) $$\sum_{n=1}^{\infty} H_n^{(p)} t^n = \frac{\mathrm{Li}_p(t)}{1-t}.$$

Integrations in (5.60) gives

(5.62) $$\sum_{n=1}^{\infty} H_n \frac{t^{n+1}}{n+1} = \sum_{n=1}^{\infty} H_{n-1} \frac{t^n}{n} = \frac{1}{2} \log^2(1-t).$$

From here, since $H_n = H_{n-1} + \dfrac{1}{n}$ we can write

(5.63) $$\sum_{n=1}^{\infty} H_n \frac{t^n}{n} = \frac{1}{2} \log^2(1-t) + \mathrm{Li}_2(t).$$

Clearly,

$$\mathrm{Li}_2(-1) = \sum_{n=1}^{\infty} \frac{(-1)^n}{n^2} = -\frac{1}{2}\zeta(2) = -\frac{\pi^2}{12}$$

and with $t=-1$ in (5.63) we find the evaluation

$$\sum_{n=1}^{\infty} \frac{(-1)^n}{n} H_n = \frac{1}{2} \log^2 2 - \frac{\pi^2}{12}.$$

The dilogarithm $\mathrm{Li}_2(t)$ satisfies Landen's identity [33]

(5.64) $$\frac{1}{2} \log^2(1-t) + \mathrm{Li}_2(t) = -\mathrm{Li}_2\left(\frac{-t}{1-t}\right)$$

and so (5.63) can be written in the form

$$\sum_{n=1}^{\infty} H_n \frac{t^n}{n} = -\mathrm{Li}_2\left(\frac{-t}{1-t}\right).$$

We notice that from (5.64) with $t=-1$

$$\mathrm{Li}_2\left(\frac{1}{2}\right) = \frac{\pi^2}{12} - \frac{1}{2}\log^2 2.$$

Next, with the help of the identity

$$H_n^2 - H_{n-1}^2 = \frac{2}{n}H_{n-1} + \frac{1}{n^2}$$

equation (5.62) leads to another generating function

(5.65) $$\sum_{n=1}^{\infty}(H_n^2 - H_{n-1}^2)t^n = \log^2(1-t) + \mathrm{Li}_2(t).$$

Applying the lemma and simplifying the telescoping sum on the left-hand side we come to

(5.66) $$\sum_{n=1}^{\infty}H_n^2 t^n = \frac{\log^2(1-t) + \mathrm{Li}_2(t)}{1-t}.$$

Dividing here both sides by t and integrating gives

(5.67) $$\sum_{n=1}^{\infty}H_n^2\frac{t^n}{n} = \mathrm{Li}_3(t) - \log(1-t)\mathrm{Li}_2(t) - \frac{1}{3}\log^3(1-t).$$

For this integration we use the decomposition

$$\frac{\log^2(1-t) + \mathrm{Li}_2(t)}{t(1-t)} = \frac{\log^2(1-t)}{t} + \frac{\log^2(1-t)}{1-t} + \frac{\mathrm{Li}_2(t)}{t} + \frac{\mathrm{Li}_2(t)}{1-t}$$

and then

$$\sum_{n=1}^{\infty}H_n^2\frac{t^n}{n} = \int_0^t\frac{\log^2(1-x)}{x}dx + \int_0^t\frac{\log^2(1-x)}{1-x}dx$$

$$+ \int_0^t\frac{\mathrm{Li}_2(x)}{x}dx + \int_0^t\frac{\mathrm{Li}_2(x)}{1-x}dx.$$

The second and the third integrals are solved easily, since

$$\int_0^t \frac{\mathrm{Li}_p(x)}{x}\,dx = \mathrm{Li}_{p+1}(t)\,.$$

Integrating by parts the last integral we find

(5.68) $$\int_0^t \frac{\mathrm{Li}_2(x)}{1-x}\,dx = -\log(1-t)\,\mathrm{Li}_2(t) - \int_0^t \frac{\log^2(1-x)}{x}\,dx$$

and (5.67) follows.

The series in (5.67) converges absolutely in the disk $|t| < 1$. It does not converge for $t = 1$, but converges for $t = -1$ and we obtain

$$\sum_{n=1}^{\infty} \frac{(-1)^n}{n} H_n^2 = \mathrm{Li}_3(-1) - \log(2)\,\mathrm{Li}_2(-1) - \frac{1}{3}\log^3(2)\,.$$

We have

$$\mathrm{Li}_3(-1) = \sum_{n=1}^{\infty} \frac{(-1)^n}{n^3} = -\frac{3}{4}\zeta(3)$$

so that

$$\sum_{n=1}^{\infty} \frac{(-1)^n}{n} H_n^2 = -\frac{3}{4}\zeta(3) + \frac{\pi^2}{12}\log(2) - \frac{1}{3}\log^3(2)\,.$$

Dividing both sides in (5.67) by t and integrating again we come to yet another generating function

(5.69) $$\sum_{n=1}^{\infty} H_n^2 \frac{t^n}{n^2}$$

$$= \mathrm{Li}_4(t) + \frac{1}{2}\bigl(\mathrm{Li}_2(t)\bigr)^2 - \frac{1}{3}\log(t)\log^3(1-t) - \log^2(1-t)\,\mathrm{Li}_2(1-t)$$

$$+ 2\log(1-t)\,\mathrm{Li}_3(1-t) - 2\,\mathrm{Li}_4(1-t) + 2\zeta(4)$$

$$= t + \frac{9}{16} t^2 + \frac{121}{324} t^3 + \frac{625}{2304} t^4 + \frac{18769}{9000} t^5 + \dots.$$

Here is the proof. Let first $0 < t < 1$. We have

$$\sum_{n=1}^{\infty} H_n^2 \frac{t^n}{n^2} = -\frac{1}{3} \int \frac{\log^3(1-t)}{t} \, dt + \mathrm{Li}_4(t) + \int \frac{-\log(1-t)}{t} \mathrm{Li}_2(t) \, dt$$

$$= \mathrm{Li}_4(t) + \frac{1}{2} \left(\mathrm{Li}_2(t) \right)^2 - \frac{1}{3} \int \frac{\log^3(1-t)}{t} \, dt$$

since

$$\frac{-\log(1-t)}{t} = \frac{d}{dt} \mathrm{Li}_2(t).$$

Now we evaluate the last integral. Integrating by parts we find

$$\int \frac{\log^3(1-t)}{t} \, dt = \log(t) \log^3(1-t) + 3 \int \frac{\log(t) \log^2(1-t)}{1-t} \, dt$$

and then with the substitution $1 - t = x$

$$\int \frac{\log(t) \log^2(1-t)}{1-t} \, dt = \int \frac{-\log(1-x) \log^2(x)}{x} \, dx$$

$$= \sum_{n=1}^{\infty} \frac{1}{n} \int x^{n-1} \log^2(x) \, dx = \sum_{n=1}^{\infty} \frac{1}{n} \left[\frac{x^n}{n} \log^2(x) - 2 \frac{x^n}{n^2} \log(x) + 2 \frac{x^n}{n^3} \right]$$

$$= \log^2(x) \mathrm{Li}_2(x) - 2 \log(x) \mathrm{Li}_3(x) + 2 \mathrm{Li}_4(x) + C.$$

Returning to t and putting all pieces together we form the right-hand side in (5.69). With $t \to 0$ we compute

$$C = 2 \mathrm{Li}_4(1) = \sum_{n=1}^{\infty} \frac{1}{n^4} = \zeta(4).$$

This way we finally come to (5.69).

Some terms on the right-hand side in (5.69), like $\log(t)\log^3(1-t)$, are not analytic in a neighborhood of zero, but all terms together represent an analytic function. The right-hand side in (5.69) is well defined for $0 < t < 1$ and equation (5.69) provides its analytic continuation in the disk $|t| < 1$. Both sides are defined also for $t = 1$, so we have

$$\sum_{n=1}^{\infty} \frac{H_n^2}{n^2} = 3\zeta(4) + \frac{1}{2}\left(\frac{\pi^2}{6}\right)^2 = \frac{17\pi^4}{360}$$

which is a well-known result.

Adding equation (5.61) with $p = 2$ to equation (5.66) we find

(5.70) $$\sum_{n=1}^{\infty}\left(H_n^2 + H_n^{(2)}\right)t^n = \frac{\log^2(1-t) + 2\operatorname{Li}_2(t)}{1-t}$$

$$= -\frac{2}{1-t}\operatorname{Li}_2\left(\frac{-t}{1-t}\right).$$

At this point we use the fact that for any positive integer p

$$\frac{d}{dt}\operatorname{Li}_{p+1}(t) = \frac{1}{t}\operatorname{Li}_p(t)$$

and then setting $u = \dfrac{-t}{1-t}$ we apply the chain rule

$$\frac{d}{dt}\operatorname{Li}_{p+1}(u) = \frac{1}{u}\operatorname{Li}_p(u)\frac{du}{dt}$$

that is,

$$\frac{d}{dt}\operatorname{Li}_{p+1}\left(\frac{-t}{1-t}\right) = \frac{1}{t(1-t)}\operatorname{Li}_p\left(\frac{-t}{1-t}\right).$$

From here we conclude that

(5.71) $$\int \frac{1}{t(1-t)} \operatorname{Li}_p \left(\frac{-t}{1-t} \right) dt = \operatorname{Li}_{p+1} \left(\frac{-t}{1-t} \right) + C.$$

We will apply this result now. Dividing both sides in (5.70) by t and integrating we come to another nice generating function

(5.72) $$\sum_{n=1}^{\infty} \left(H_n^2 + H_n^{(2)} \right) \frac{t^n}{n} = -2 \operatorname{Li}_3 \left(\frac{-t}{1-t} \right).$$

Repeating this step again, dividing both sides by t and integrating we find also

(5.73) $$\sum_{n=1}^{\infty} \left(H_n^2 + H_n^{(2)} \right) \frac{t^n}{n^2}$$

$$= -2 \operatorname{Li}_4 \left(\frac{-t}{1-t} \right) + \left(\operatorname{Li}_2 \left(\frac{-t}{1-t} \right) \right)^2 - 2 \log(1-t) \operatorname{Li}_3 \left(\frac{-t}{1-t} \right).$$

Applying the lemma (5.59) to equation (5.72) gives

$$\sum_{n=1}^{\infty} \left\{ \sum_{k=1}^{n} \frac{1}{k} \left(H_k^2 + H_k^{(2)} \right) \right\} t^n = \frac{-2}{1-t} \operatorname{Li}_3 \left(\frac{-t}{1-t} \right).$$

Now we can use the identity (see [3, p. 129])

$$\sum_{k=1}^{n} \frac{1}{k} \left(H_k^2 + H_k^{(2)} \right) = \frac{1}{3} \left(H_k^3 + 3 H_n H_n^{(2)} + 2 H_n^{(3)} \right)$$

to write

(5.74) $$\sum_{n=0}^{\infty} \left(H_n^3 + 3 H_n H_n^{(2)} + 2 H_n^{(3)} \right) t^n = \frac{-6}{1-t} \operatorname{Li}_3 \left(\frac{-t}{1-t} \right).$$

Dividing by t and integrating this yields in view of (5.70)

(5.75) $$\sum_{n=0}^{\infty} \left(H_n^3 + 3 H_n H_n^{(2)} + 2 H_n^{(3)} \right) \frac{t^n}{n} = -6 \operatorname{Li}_4 \left(\frac{-t}{1-t} \right).$$

Adding the equation

$$-2\sum_{n=1}^{\infty} H_n^{(2)} t^n = \frac{-2\operatorname{Li}_2(t)}{1-t}$$

to equation (5.66) we come to

(5.76) $$\sum_{n=0}^{\infty}\left(H_n^2 - H_n^{(2)}\right) t^n = \frac{\log^2(1-t)}{1-t}$$

and integration gives

$$\sum_{n=0}^{\infty}\left(H_n^2 - H_n^{(2)}\right)\frac{t^{n+1}}{n+1} = -\frac{1}{3}\log^3(1-t).$$

Next we prove the representation

(5.77) $$\sum_{n=1}^{\infty} H_n \frac{t^n}{n^2}$$

$$= \frac{1}{2}\log(t)\log^2(1-t) + \log(1-t)\operatorname{Li}_2(1-t) - \operatorname{Li}_3(1-t) + \operatorname{Li}_3(t) + \zeta(3).$$

An equivalent form of this representation was computed by Ramanujan in his notebooks. Ramanujan's result can be found in Berndt's monograph [6, p. 251], namely,

$$\sum_{n=1}^{\infty}\frac{H_n t^{n+1}}{(n+1)^2}$$

$$= \frac{1}{2}\log(t)\log^2(1-t) + \log(1-t)\operatorname{Li}_2(1-t) - \operatorname{Li}_3(1-t) + \zeta(3)$$

(see also [33, p. 303]). With $t = 1$ these representations give

$$\sum_{n=1}^{\infty}\frac{H_n}{n^2} = 2\zeta(3), \quad \sum_{n=1}^{\infty}\frac{H_n}{(n+1)^2} = \zeta(3).$$

Here is the proof of (5.77): We divide both sides in (5.63) by t and integrate. The result is

(5.78)
$$\sum_{n=1}^{\infty} H_n \frac{t^n}{n^2} = \frac{1}{2}\int \frac{\log^2(1-t)}{t}dt + \mathrm{Li}_3(t).$$

The evaluation of this integral can be found in Levin's book [33, p. 159]. We will give this evaluation here for the convenience of the reader. Integrating by parts we write

$$\int \frac{\log^2(1-t)}{t}dt = \log(t)\log^2(1-t) + 2\int \frac{\log(t)\log(1-t)}{1-t}dt$$

and then in the last integral we use the substitution $1-t = x$

$$\int \frac{\log(t)\log(1-t)}{1-t}dt = -\int \frac{\log(1-x)\log(x)}{x}dx$$

$$= \int \ln x\, d\mathrm{Li}_2(x) = \ln x\, \mathrm{Li}_2(x) - \int \frac{\mathrm{Li}_2(x)}{x}dx$$

$$= \ln x\, \mathrm{Li}_2(x) - \mathrm{Li}_3(x) + C$$

$$= \log(1-t)\mathrm{Li}_2(1-t) - \mathrm{Li}_3(1-t) + C.$$

Assembling these results we come to (5.77). The constant of integration is computed to be $C = \zeta(3)$ by setting $t = 0$.

Another interesting manipulation is this: From (5.61) we have

$$\sum_{n=1}^{\infty} H_n^{(2)}\, t^{n-1} = \frac{\mathrm{Li}_2(t)}{t(1-t)} = \frac{\mathrm{Li}_2(t)}{t} + \frac{\mathrm{Li}_2(t)}{1-t}$$

and then by integration

$$\sum_{n=1}^{\infty} H_n^{(2)}\, \frac{t^n}{n} = \mathrm{Li}_3(t) + \int \frac{\mathrm{Li}_2(t)}{1-t}dt.$$

Now in view of (5.68) we write

$$\sum_{n=1}^{\infty} H_n^{(2)} \frac{t^n}{n} = \text{Li}_3(t) - \log(1-t)\text{Li}_2(t) - \int_0^t \frac{\log^2(1-x)}{x} dx .$$

At the same time from (5.63)

$$2\sum_{n=1}^{\infty} H_n \frac{t^n}{n^2} = \int_0^t \frac{\log^2(1-x)}{x} dx + 2\text{Li}_3(t) .$$

Adding these two equations we find the useful representation

$$\sum_{n=1}^{\infty} H_n^{(2)} \frac{t^n}{n} + 2\sum_{n=1}^{\infty} H_n \frac{t^n}{n^2} = 3\text{Li}_3(t) - \text{Li}_2(t)\log(1-t) .$$

Dividing both sides by t and integrating (integration is easy) we come to

(5.79) $$\sum_{n=1}^{\infty} H_n^{(2)} \frac{t^n}{n^2} + 2\sum_{n=1}^{\infty} H_n \frac{t^n}{n^3} = 3\text{Li}_4(t) + \frac{1}{2}\left(\text{Li}_2(t)\right)^2 .$$

In particular, with $t = 1$ we find

$$\sum_{n=1}^{\infty} \frac{H_n^{(2)}}{n^2} + 2\sum_{n=1}^{\infty} \frac{H_n}{n^3} = 3\zeta(4) + \frac{1}{2}\left(\zeta(2)\right)^2 = \frac{17\pi^4}{360} .$$

Subtracting (5.67) from (5.72) we come to the generating function

$$\sum_{n=1}^{\infty} H_n^{(2)} \frac{t^n}{n}$$

$$= -2\text{Li}_3\left(\frac{-t}{1-t}\right) - \text{Li}_3(t) + \text{Li}_2(t)\ln(1-t) + \frac{1}{3}\log^3(1-t) .$$

A project for the reader: Dividing by t and integrating show that

(5.80) $$\sum_{n=1}^{\infty} H_n^{(2)} \frac{t^n}{n^2} = -2\text{Li}_4\left(\frac{-t}{1-t}\right) - \text{Li}_4(t) + 2\text{Li}_4(1-t) - 2\zeta(4)$$

$$-2\log(1-t)\operatorname{Li}_3\left(\frac{-t}{1-t}\right)+\left(\operatorname{Li}_2\left(\frac{-t}{1-t}\right)\right)^2-\frac{1}{2}\left(\operatorname{Li}_2(t)\right)^2$$

$$+\frac{1}{3}\log t \,\log^3(1-t)+\log^2(1-t)\operatorname{Li}_2(1-t)-2\log(1-t)\operatorname{Li}_3(1-t).$$

Using Landen's identities for the dilogarithm (5.64) and the trilogarithm

$$\operatorname{Li}_3\left(\frac{-z}{1-z}\right)=\frac{1}{6}\log^3(1-z)-\operatorname{Li}_3(1-z)$$

$$-\frac{1}{2}\log(z)\log^2(1-z)+\frac{\pi^2}{6}\log(1-z)-\operatorname{Li}_3(z)+\zeta(3)$$

some terms above can be simplified. Thus (5.80) becomes

$$(5.81)\qquad \sum_{n=1}^{\infty}H_n^{(2)}\frac{t^n}{n^2}=-2\operatorname{Li}_4\left(\frac{-t}{1-t}\right)-\operatorname{Li}_4(t)+2\operatorname{Li}_4(1-t)-2\zeta(4)$$

$$+\frac{1}{2}\left(\operatorname{Li}_2(t)\right)^2+2\log(1-t)(\operatorname{Li}_3(t)-\zeta(3))+\frac{1}{3}\log t\,\log^3(1-t)$$

$$-\frac{1}{12}\log^4(1-t)-\zeta(2)\log^2(1-t).$$

Subtracting (5.80) from (5.78) and simplifying we come to one important generating function

$$\sum_{n=1}^{\infty}H_n\frac{t^n}{n^3}=\operatorname{Li}_4\left(\frac{-t}{1-t}\right)+\log(1-t)\operatorname{Li}_3\left(\frac{-t}{1-t}\right)-\frac{1}{2}\left(\operatorname{Li}_2\left(\frac{-t}{1-t}\right)\right)^2$$

$$+2\operatorname{Li}_4(t)+\frac{1}{2}\left(\operatorname{Li}_2(t)\right)^2-\frac{1}{6}\log t\,\log^3(1-t)-\frac{1}{2}\log^2(1-t)\operatorname{Li}_2(1-t)$$

$$+\log(1-t)\operatorname{Li}_3(1-t)-\operatorname{Li}_4(1-t)+\zeta(4).$$

In simplified form

(5.82) $$\sum_{n=1}^{\infty} H_n \frac{t^n}{n^3} = \text{Li}_4\left(\frac{-t}{1-t}\right) + 2\,\text{Li}_4(t) - \text{Li}_4(1-t) + \zeta(4)$$

$$-\log(1-t)\,\text{Li}_3(t) - \frac{1}{6}\log(t)\log^3(1-t) + \frac{1}{24}\log^4(1-t)$$

$$+\zeta(3)\log(1-t) + \frac{\pi^2}{12}\log^2(1-t).$$

Using the obvious relation

$$\sum_{n=0}^{\infty} H_n \frac{t^{n+1}}{(n+1)^3} = \sum_{n=1}^{\infty} H_n \frac{t^n}{n^3} - \text{Li}_4(t)$$

the above generating function can be written also in the form

(5.83) $$\sum_{n=0}^{\infty} H_n \frac{t^{n+1}}{(n+1)^3} = \text{Li}_4\left(\frac{-t}{1-t}\right) + \text{Li}_4(t) - \text{Li}_4(1-t) + \zeta(4)$$

$$-\log(1-t)\,\text{Li}_3(t) - \frac{1}{6}\log(t)\log^3(1-t) + \frac{1}{24}\log^4(1-t)$$

$$+\zeta(3)\log(1-t) + \frac{\pi^2}{12}\log^2(1-t).$$

Ramanujan worked on this problem more than hundred years ago. After he died on April 26, 1920, his wife gave his notebooks to the University of Madras. The notebooks were analyzed in depth much later. In Entry 10, Chapter 6 of his notebooks Ramanujan showed some properties of the generating function for the numbers $\dfrac{H_n}{(n+1)^3}$, but this function was not given explicitly (see Berndt [6, p. 253]).

Problems for the reader: Using the above techniques show that

(5.84) $$\sum_{n=1}^{\infty} H_n H_n^{(2)} t^n = \frac{1}{1-t}\left(\frac{1}{6}\log^3(1-t) - \text{Li}_3\left(\frac{-t}{1-t}\right)\right)$$

and also

$$(5.85) \quad \sum_{n=0}^{\infty} H_n^3 t^n = \frac{-1}{1-t} \left[3\operatorname{Li}_3\left(\frac{-t}{1-t}\right) + 2\operatorname{Li}_3(t) + \frac{1}{2}\log^3(1-t) \right].$$

The following evaluations are also left to the reader.

$$\sum_{n=1}^{\infty} (-1)^{n-1} \frac{H_n^{(2)}}{n} = 2\operatorname{Li}_3\left(\frac{1}{2}\right) + \operatorname{Li}_3(-1) - \operatorname{Li}_2(-1)\log 2 - \frac{1}{3}\log^3 2$$

$$= \zeta(3) - \frac{\pi^2}{12}\log 2$$

$$\sum_{n=1}^{\infty} \frac{H_n}{n^2 2^n} = \zeta(3) - \frac{\pi^2}{12}\log 2$$

$$\sum_{n=1}^{\infty} (-1)^{n-1} \frac{H_n}{n^2} = \frac{5}{8}\zeta(3); \quad \sum_{n=1}^{\infty} (-1)^{n-1} \frac{H_n}{(n+1)^2} = \frac{1}{8}\zeta(3)$$

$$\sum_{n=1}^{\infty} H_n^2 \frac{(-1)^{n-1}}{n} = \frac{3}{4}\zeta(3) - \frac{\pi^2 \log 2}{12} + \frac{(\log 2)^3}{3}$$

$$\sum_{n=1}^{\infty} \left(H_n^2 + H_n^{(2)}\right) \frac{(-1)^{n-1}}{n} = \frac{7}{4}\zeta(3) - \frac{\pi^2 \log 2}{6} + \frac{(\log 2)^3}{3}$$

$$\sum_{n=1}^{\infty} H_n^{(2)} \frac{(-1)^{n-1}}{n} + 2\sum_{n=1}^{\infty} H_n \frac{(-1)^{n-1}}{n^2} = \frac{9}{4}\zeta(3) - \frac{\pi^2}{12}\log 2 .$$

5.7.2 Skew-harmonic numbers

The skew-harmonic numbers already appeared at the end of Chapter 2, in Examples 2.5.3 and 2.5.5. We will take a closer look at these numbers now and generate various power series with skew-harmonic numbers in their coefficients.

The skew-harmonic numbers are defined by

$$H_n^- = 1 - \frac{1}{2} + \ldots + \frac{(-1)^{n-1}}{n} \quad (n = 1, 2, \ldots), \ H_0^- = 0 .$$

Their relation to the digamma function is given by

$$H_n^- = \log 2 + \frac{(-1)^n}{2} \left[\psi \left(\frac{n+1}{2} \right) - \psi \left(\frac{n+2}{2} \right) \right]$$

and because of this representation sometimes it is convenient to work with the expression $H_n^- - \log 2$.

Applying Lemma 5.2 to the series

$$\log(1 + t) = \sum_{n=1}^{\infty} \frac{(-1)^{n-1}}{n} t^n \quad (|t| < 1)$$

we find the generating function for the skew-harmonic numbers

(5.86) $$\frac{\log(1 + t)}{1 - t} = \sum_{n=1}^{\infty} H_n^- t^n .$$

Combining this with the series

$$\frac{\log 2}{1 - t} = \sum_{n=0}^{\infty} (\log 2) t^n$$

we write also

(5.87) $$\sum_{n=0}^{\infty} \left(H_n^- - \log 2 \right) t^n = \frac{1}{1 - t} \log \left(\frac{1 + t}{2} \right) .$$

This series is convergent in the disk $|t| < 1$ and also for $t = 1$. We have

$$H_n^- - \log 2 = (-1)^{n-1} \left(\frac{1}{n+1} - \frac{1}{n+2} + \frac{1}{n+3} - \frac{1}{n+4} + \ldots \right)$$

$$= (-1)^{n-1} \left(\frac{1}{(n+1)(n+2)} + \frac{1}{(n+3)(n+4)} + \ldots \right)$$

where

$$b_n = \frac{1}{(n+1)(n+2)} + \frac{1}{(n+3)(n+4)} + \ldots$$

$$\leq \frac{1}{(n+1)(n+2)} + \frac{1}{(n+2)(n+3)} + \ldots = \frac{1}{n+1}$$

so $b_n \to 0$ and b_n is also monotone decreasing. Therefore, the series in (5.87) for $t = 1$ is a convergent alternating series. Computing

$$\lim_{t \to 1} \frac{1}{1-t} \log\left(\frac{1+t}{2}\right) = -\frac{1}{2}$$

we find

$$\sum_{n=0}^{\infty} \left(H_n^- - \log 2\right) = -\frac{1}{2}.$$

Proposition 5.3. *For all* $|t| \leq 1, t \neq 1$

$$(5.88) \quad \sum_{n=1}^{\infty} H_n^- \frac{t^n}{n} = \mathrm{Li}_2\left(\frac{1-t}{2}\right) - \mathrm{Li}_2\left(\frac{1}{2}\right) - \mathrm{Li}_2(-t) - \log(1-t)\log 2$$

$$= \mathrm{Li}_2\left(\frac{1-t}{2}\right) - \mathrm{Li}_2(-t) - \log(1-t)\log 2 + \frac{1}{2}\log^2 2 - \frac{\pi^2}{12}.$$

(This representation appears in a different form as entry 5.5.27 in Hansen's table [26].)

Proof. We notice that

$$(5.89) \quad \frac{d}{dx}\mathrm{Li}_2\left(\frac{1-t}{2}\right) = \frac{1}{1-t}\log\left(\frac{1+t}{2}\right) = \frac{\log(1+t)}{1-t} - \frac{\log 2}{1-t}$$

and therefore,

$$\sum_{n=1}^{\infty} H_n^- \, t^n = \frac{\log(1+t)}{1-t} = \frac{d}{dt} \mathrm{Li}_2\left(\frac{1-t}{2}\right) + \frac{\log 2}{1-t} \, .$$

Integrating this we find

(5.90) $\qquad \displaystyle\sum_{n=1}^{\infty} H_n^- \frac{t^{n+1}}{n+1} = \mathrm{Li}_2\left(\frac{1-t}{2}\right) - \mathrm{Li}_2\left(\frac{1}{2}\right) - \log 2 \log(1-t)$

which reduces to (5.88) because

$$\sum_{n=1}^{\infty} H_n^- \frac{t^{n+1}}{n+1} = \sum_{n=1}^{\infty} H_n^- \frac{t^n}{n} + \mathrm{Li}_2(-t) \, .$$

Equation (5.88) can be written also in the form

(5.91) $\qquad \displaystyle\sum_{n=1}^{\infty} \left(H_n^- - \log 2 \right) \frac{t^n}{n} = \mathrm{Li}_2\left(\frac{1-t}{2}\right) - \mathrm{Li}_2\left(\frac{1}{2}\right) - \mathrm{Li}_2(-t)$

by moving the term $\log(1-t)\log 2$ to the left-hand side and expanding $\log(1-t)$. This equation is defined on the whole closed disk $|t| \le 1$ while the series in (5.88) is not convergent for $t = 1$. This way we have

$$\sum_{n=1}^{\infty} \left(H_n^- - \log 2 \right) \frac{1}{n} = -\mathrm{Li}_2\left(\frac{1}{2}\right) - \mathrm{Li}_2(-1) = \frac{1}{2}\log^2 2$$

by using the values

$$\mathrm{Li}_2(-1) = \frac{-\pi^2}{12}, \quad \mathrm{Li}_2\left(\frac{1}{2}\right) = \frac{\pi^2}{12} - \frac{1}{2}\log^2 2 \, .$$

With $t = -1$

$$\sum_{n=1}^{\infty} \left(H_n^- - \log 2 \right) \frac{(-1)^{n-1}}{n} = \frac{\pi^2}{12} - \frac{\log^2 2}{2}$$

or

$$\sum_{n=1}^{\infty} H_n^- \frac{(-1)^{n-1}}{n} = \frac{\pi^2}{12} + \frac{\log^2 2}{2}.$$

In view of (5.89) equation (5.87) can be written as

$$\sum_{n=0}^{\infty} \left(H_n^- - \log 2 \right) t^n = \frac{d}{dx} \operatorname{Li}_2\left(\frac{1-x}{2} \right)$$

and integrating this we come to

$$\sum_{n=0}^{\infty} \left(H_n^- - \log 2 \right) \frac{t^{n+1}}{n+1} = \operatorname{Li}_2\left(\frac{1-t}{2} \right) - \operatorname{Li}_2\left(\frac{1}{2} \right).$$

Proposition 5.4. *For* $|t| < 1$

(5.92)
$$\sum_{n=1}^{\infty} \left(H_n^- \right)^2 t^n$$

$$= \frac{1}{1-t} \left\{ \operatorname{Li}_2(t) + 2\log 2 \log(1+t) + 2\operatorname{Li}_2\left(\frac{1}{2} \right) - 2\operatorname{Li}_2\left(\frac{1+t}{2} \right) \right\}$$

and for $|t| \le 1$

(5.93)
$$\sum_{n=0}^{\infty} \left(H_n^- - \log 2 \right)^2 t^n$$

$$= \frac{1}{1-t} \left\{ \operatorname{Li}_2(t) + \log^2 2 - 2\left[\operatorname{Li}_2\left(\frac{1+t}{2} \right) - \operatorname{Li}_2\left(\frac{1}{2} \right) \right] \right\}$$

where the value at $t = 1$ *of the right-hand side is understood as the limit for* $t \to 1$.

For the proof of the proposition we need a lemma.

Lemma 5.5. *For every* $n = 1, 2, \ldots$ *we have*

(5.94)
$$2\sum_{k=1}^{n} \frac{(-1)^{k-1}}{k} H_k^- = \left(H_n^- \right)^2 + H_n^{(2)}.$$

(Recall that $H_n^{(2)} = 1 + \dfrac{1}{2^2} + \ldots + \dfrac{1}{n^2}$.)

Proof of the lemma by induction. For $n = 1$ the equality is true. Suppose now it is true for some $n \geq 1$. Then

$$2\sum_{k=1}^{n+1} \frac{(-1)^{k-1} H_k^-}{k} = 2\sum_{k=1}^{n} \frac{(-1)^{k-1} H_k^-}{k} + 2\frac{(-1)^n H_{n+1}^-}{n+1}$$

$$= \left(H_n^-\right)^2 + H_n^{(2)} + 2\frac{(-1)^n H_{n+1}^-}{n+1}$$

$$= \left(H_{n+1}^- - \frac{(-1)^n}{n+1}\right)^2 + H_{n+1}^{(2)} - \frac{1}{(n+1)^2} + 2\frac{(-1)^n H_{n+1}^-}{n+1}$$

$$= \left(H_{n+1}^-\right)^2 + H_{n+1}^{(2)}.$$

The proof is completed.

Now we turn to the proof of the proposition. Replacing t by $-t$ in (5.88) we have

$$\sum_{n=1}^{\infty} \frac{(-1)^n H_n^-}{n} t^n = \mathrm{Li}_2\left(\frac{1+t}{2}\right) - \mathrm{Li}_2\left(\frac{1}{2}\right) - \mathrm{Li}_2(t) - \log(1+t)\log 2$$

and with some help from Lemma 5.2 and Lemma 5.5 we write

$$\sum_{n=1}^{\infty} \left\{2\sum_{k=1}^{n} \frac{(-1)^{k-1} H_k^-}{k}\right\} t^n = \sum_{n=1}^{\infty} \left\{\left(H_n^-\right)^2 + H_n^{(2)}\right\} t^n$$

$$= \frac{2}{1-t}\left\{\mathrm{Li}_2(t) + \log(1+t)\log 2 - \mathrm{Li}_2\left(\frac{1+t}{2}\right) + \mathrm{Li}_2\left(\frac{1}{2}\right)\right\}.$$

At the same time

$$\sum_{n=1}^{\infty} H_n^{(2)} t^n = \frac{\mathrm{Li}_2(t)}{1-t}$$

which inserted in the equation above leads to (5.92).

After that we write

$$\left(H_n^- - \log 2\right)^2 = \left(H_n^-\right)^2 - (2\log 2)H_n^- + \log^2 2$$

and with summation starting from $n = 0$

$$\sum_{n=0}^{\infty}\left(H_n^- - \log 2\right)^2 t^n$$

$$= \sum_{n=0}^{\infty}\left(H_n^-\right)^2 t^n - 2\log 2 \sum_{n=0}^{\infty} H_n^- t^n + \sum_{n=0}^{\infty}(\log^2 2)t^n.$$

Now (5.93) follows from (5.92) and (5.86). The proposition is proved.

Remark. Setting $t = -1$ in (5.93) we compute

$$\sum_{n=0}^{\infty}(-1)^n\left(H_n^- - \log 2\right)^2 = \frac{1}{2}\left\{\mathrm{Li}_2(-1) + \log^2 2 + 2\mathrm{Li}_2\left(\frac{1}{2}\right)\right\} = \frac{\pi^2}{24}.$$

This provides an independent solution to problem 11682 in the *American Mathematical Monthly* (see Example 5.6.3). The series in (5.92) converges also for $t = 1$, as we have the estimate

$$\left(H_n^- - \log 2\right)^2 \le \frac{1}{(n+1)^2}.$$

Computing the limit of the right-hand side in (5.93) we find

$$\sum_{n=0}^{\infty}\left(H_n^- - \log 2\right)^2 = \log 2.$$

This equation repeats the result from Example 5.6.2.

Equation (5.93) presents an extension of Problem 997 from the *College Mathematic Journal*.

Proposition 5.6. *For all* $-\dfrac{1}{3} \le t \le 1$

(5.95)
$$\sum_{n=1}^{\infty} H_n^{-} \frac{t^{n+1}}{(n+1)^2}$$

$$= \text{Li}_3\left(\frac{2t}{1+t}\right) - \text{Li}_3\left(\frac{t}{1+t}\right) - \text{Li}_3\left(\frac{1+t}{2}\right) + \text{Li}_3\left(\frac{1}{2}\right) - \text{Li}_3(t)$$

$$+ \log(1+t)\left[\text{Li}_2(t) + \text{Li}_2\left(\frac{1}{2}\right) + \frac{1}{2}\log 2 \log(1+t)\right]$$

or, equivalently,

(5.96)
$$\sum_{n=1}^{\infty} H_n^{-} \frac{t^n}{n^2}$$

$$= \text{Li}_3\left(\frac{2t}{1+t}\right) - \text{Li}_3\left(\frac{t}{1+t}\right) - \text{Li}_3\left(\frac{1+t}{2}\right) + \text{Li}_3\left(\frac{1}{2}\right) - \text{Li}_3(t) - \text{Li}_3(-t)$$

$$+ \log(1+t)\left[\text{Li}_2(t) + \text{Li}_2\left(\frac{1}{2}\right) + \frac{1}{2}\log 2 \log(1+t)\right].$$

The representation (5.95) results from (5.90) by dividing both sides by t and integrating. The detailed proof can be found in [11].

Proposition 5.7. *For* $|t| < 1$

(5.97)
$$\sum_{n=0}^{\infty} \left(H_n^{-} - \log 2\right)^2 \frac{t^{n+1}}{n+1}$$

$$= \int_0^t \frac{\text{Li}_2(x)}{1-x} dx + \log(1-t)\left[2\text{Li}_2\left(\frac{1+t}{2}\right) - \frac{\pi^2}{6}\right]$$

$$+ 2\log(1+t)\left(\log^2(1-t) - \log^2 2\right) + 2\log 2\left[\text{Li}_2\left(\frac{1+t}{2}\right) - \text{Li}_2\left(\frac{1}{2}\right)\right]$$

$$-2\log 2\log^2(1-t) + 4\log(1-t)\,\mathrm{Li}_2\left(\frac{1-t}{2}\right) - 4\left[\mathrm{Li}_3\left(\frac{1-t}{2}\right) - \mathrm{Li}_3\left(\frac{1}{2}\right)\right],$$

where for $-1 < t < \dfrac{1}{2}$

$$\int_0^t \frac{\mathrm{Li}_2(x)}{1-x}\,dx$$

$$= -2\,\mathrm{Li}_3\left(\frac{-t}{1-t}\right) - 2\,\mathrm{Li}_3(t) + \log(1-t)\,\mathrm{Li}_2(t) + \frac{1}{3}\log^3(1-t)$$

and for $0 \le t < 1$ we have

$$\int_0^t \frac{\mathrm{Li}_2(x)}{1-x}\,dx = 2\left[\mathrm{Li}_3(1-t) - \zeta(3)\right] - \log(1-t)\left[\mathrm{Li}_2(1-t) + \frac{\pi^2}{6}\right].$$

For the proof we divide by t in (5.93) and integrate. Details can be found in [11].

5.7.3 Double integrals related to the above series

Some of the generating functions considered above can be represented by double integrals. Starting from

$$\frac{1}{n+k} = \int_0^1 x^{n+k-1}\,dx$$

we write

$$\log 2 - H_n^- = \sum_{k=1}^{\infty} \frac{(-1)^{k-1}}{n+k} = \int_0^1 x^n \left\{\sum_{k=1}^{\infty}(-1)^{k-1}x^{k-1}\right\}dx = \int_0^1 \frac{x^n}{1+x}\,dx$$

and then for $|z| < 1$

$$\sum_{n=0}^{\infty}\left(H_n^- - \log 2\right)^2 z^n = \sum_{n=0}^{\infty}\left(\int_0^1 \frac{x^n dx}{1+x}\right)^2 z^n$$

$$= \sum_{n=0}^{\infty}\left(\int_0^1 \frac{x^n dx}{1+x}\right)\left(\int_0^1 \frac{y^n dy}{1+y}\right)z^n = \sum_{n=0}^{\infty}\int_0^1\int_0^1 \frac{x^n y^n z^n dx dy}{(1+x)(1+y)}$$

$$= \int_0^1\int_0^1\left\{\sum_{n=0}^{\infty}(xyz)^n\right\}\frac{dx dy}{(1+x)(1+y)}$$

$$= \int_0^1\int_0^1 \frac{dx dy}{(1-xyz)(1+x)(1+y)}.$$

Therefore, from Propositions 5.4 and 5.7 we come to the following corollary.

Corollary 5.8. *For* $|z| < 1$

$$(5.98) \qquad g(z) \equiv \int_0^1\int_0^1 \frac{dx dy}{(1-xyz)(1+x)(1+y)}$$

$$= \frac{1}{1-z}\left\{\operatorname{Li}_2(z) + \log^2 2 - 2\left[\operatorname{Li}_2\left(\frac{1+z}{2}\right) - \operatorname{Li}_2\left(\frac{1}{2}\right)\right]\right\}$$

with $g(1) = \log 2$ *and* $g(-1) = \dfrac{\pi^2}{24}$. *Also, by integration*

$$(5.99) \qquad -\int_0^1\int_0^1 \frac{\log(1-xyz)\, dx dy}{xy(1+x)(1+y)}$$

$$= \int_0^z \frac{\operatorname{Li}_2(t)}{1-t}dt + \log(1-z)\left[2\operatorname{Li}_2\left(\frac{1+z}{2}\right) - \frac{\pi^2}{6}\right]$$

$$+ 2\log(1+z)\left(\log^2(1-z) - \log^2 2\right) + 2\log 2\left[\operatorname{Li}_2\left(\frac{1+z}{2}\right) - \operatorname{Li}_2\left(\frac{1}{2}\right)\right]$$

$$-2\log 2\log^2(1-z)+4\log(1-z)\operatorname{Li}_2\left(\frac{1-z}{2}\right)-4\left[\operatorname{Li}_3\left(\frac{1-z}{2}\right)-\operatorname{Li}_3\left(\frac{1}{2}\right)\right].$$

5.7.4 *Expansions of dilogarithms and trilogarithms*

Consider the series

$$-\log(1-\mu x)=\sum_{n=1}^{\infty}\frac{\mu^n}{n}x^n$$

with an appropriate parameter μ, where $-1\le\mu x<1$. Using Lemma 5.2 with $\lambda=-1$ we have

$$\frac{-\log(1-\mu x)}{1+x}=\sum_{n=1}^{\infty}\left\{\sum_{k=1}^{n}\frac{(-\mu)^k}{k}\right\}(-1)^n x^n.$$

We notice that

$$(5.100)\qquad \frac{d}{dx}\left[\operatorname{Li}_2\left(\frac{(1+\mu)x}{1+x}\right)-\operatorname{Li}_2\left(\frac{x}{1+x}\right)\right]$$

$$=\frac{-\log(1-\mu x)}{x(1+x)}=\sum_{n=1}^{\infty}\left\{\sum_{k=1}^{n}\frac{(-\mu)^k}{k}\right\}(-1)^n x^{n-1}.$$

Integrating this we come to the following result.

Proposition 5.9. *For any* $|\mu|<1$, $|x|<1$ *we have the expansion*

$$(5.101)\qquad \operatorname{Li}_2\left(\frac{(1+\mu)x}{1+x}\right)-\operatorname{Li}_2\left(\frac{x}{1+x}\right)=\sum_{n=1}^{\infty}\left\{\sum_{k=1}^{n}\frac{(-\mu)^k}{k}\right\}\frac{(-x)^n}{n}.$$

With $\mu=-1$ this turns into equation (5.63) in view of Landen's identity (5.64). When $\mu=1$ we have a new version of (5.88).

Equation (5.100) also says that

$$\frac{d}{dx}\left[\mathrm{Li}_2\left(\frac{(1+\mu)x}{1+x}\right)-\mathrm{Li}_2\left(\frac{x}{1+x}\right)\right]$$

$$=\frac{-\log(1-\mu x)}{x}+\frac{\log(1-\mu x)}{1+x}=\frac{d}{dx}\mathrm{Li}_2(\mu x)+\frac{\log(1-\mu x)}{1+x}.$$

Integrating this by using (5.100) we come to the known identity

(5.102) $\mathrm{Li}_2\left(\dfrac{(1+\mu)x}{1+x}\right)+\mathrm{Li}_2\left(\dfrac{\mu(1+x)}{1+\mu}\right)-\mathrm{Li}_2(\mu x)$

$$=\mathrm{Li}_2\left(\frac{x}{1+x}\right)+\mathrm{Li}_2\left(\frac{\mu}{1+\mu}\right)+\log(1+\mu)\log(1+x).$$

With $\mu=1$ this becomes

$$\mathrm{Li}_2\left(\frac{2x}{1+x}\right)+\mathrm{Li}_2\left(\frac{1+x}{2}\right)-\mathrm{Li}_2(x)$$

$$=\mathrm{Li}_2\left(\frac{x}{1+x}\right)+\mathrm{Li}_2\left(\frac{1}{2}\right)+\log2\log(1+x).$$

This identity shows that the function $\mathrm{Li}_2\left(\dfrac{2x}{1+x}\right)$ extends on the closed unit disk $|x|\le1$ except for $x=-1$. This function appears implicitly in the works of Ramanujan (see [6, Entry 8, p. 249]). Ramanujan considered the function

$$f(t)=\sum_{n=1}^{\infty}\left(1+\frac{1}{3}+\ldots+\frac{1}{2n-1}\right)\frac{t^{2n-1}}{2n-1}$$

and proved that

$$f\left(\frac{t}{2-t}\right)=\frac{1}{8}\log^2(1-t)+\mathrm{Li}_2(t).$$

With the substitution $x = \dfrac{t}{2-t}$ we can put Ramanujan's result in the form

$$\mathrm{Li}_2\left(\frac{2x}{1+x}\right) + \frac{1}{4}\log^2\frac{1-x}{1+x} = 2\sum_{n=1}^{\infty}\left(1+\frac{1}{3}+\ldots+\frac{1}{2n-1}\right)\frac{x^{2n-1}}{2n-1}.$$

Let now

$$\mathrm{Li}_m(x) = \sum_{n=1}^{\infty}\frac{x^n}{n^m}$$

be the polylogarithm. It is easy to check that for every integer $m \geq 2$ and for an appropriate parameter μ we have

$$\frac{d}{dx}\left[\mathrm{Li}_{m+1}\left(\frac{(1+\mu)x}{1+x}\right) - \mathrm{Li}_{m+1}\left(\frac{x}{1+x}\right)\right]$$

$$= \frac{1}{x(1+x)}\left[\mathrm{Li}_m\left(\frac{(1+\mu)x}{1+x}\right) - \mathrm{Li}_m\left(\frac{x}{1+x}\right)\right].$$

With $m = 2$ from Proposition 5.9 we have

$$\frac{d}{dx}\left[\mathrm{Li}_3\left(\frac{(1+\mu)x}{1+x}\right) - \mathrm{Li}_3\left(\frac{x}{1+x}\right)\right]$$

$$= \frac{1}{x(1+x)}\left[\mathrm{Li}_2\left(\frac{(1+\mu)x}{1+x}\right) - \mathrm{Li}_2\left(\frac{x}{1+x}\right)\right]$$

$$= \frac{1}{x(1+x)}\sum_{n=1}^{\infty}\left\{\frac{(-1)^n}{n}\sum_{k=1}^{n}\frac{(-\mu)^k}{k}\right\}x^n$$

and applying Lemma 5.2 this becomes

$$= \frac{1}{x}\sum_{n=1}^{\infty}\left\{\sum_{k=1}^{n}\left\{(-1)^{n-k}\frac{(-1)^k}{k}\sum_{j=1}^{k}\frac{(-\mu)^j}{j}\right\}\right\}x^n$$

$$= \sum_{n=1}^{\infty} \left\{ \sum_{k=1}^{n} \left\{ \frac{1}{k} \sum_{j=1}^{k} \frac{(-\mu)^j}{j} \right\} \right\} (-1)^n x^{n-1} .$$

After integration we come to the remarkable expansion:

Proposition 5.10. *For any* $|\mu| \leq 1$, $|x| < 1$

$$(5.103) \qquad \mathrm{Li}_3 \left(\frac{(1+\mu)x}{1+x} \right) - \mathrm{Li}_3 \left(\frac{x}{1+x} \right)$$

$$= \sum_{n=1}^{\infty} \left\{ \sum_{k=1}^{n} \left\{ \frac{1}{k} \sum_{j=1}^{k} \frac{(-\mu)^j}{j} \right\} \right\} \frac{(-1)^n}{n} x^n .$$

(with analytic extension of the left-hand side when needed).

With $\mu = 1$

$$\mathrm{Li}_3 \left(\frac{2x}{1+x} \right) - \mathrm{Li}_3 \left(\frac{x}{1+x} \right) = \sum_{n=1}^{\infty} \left\{ \sum_{k=1}^{n} \frac{H_k^-}{k} \right\} \frac{(-1)^{n-1}}{n} x^n .$$

With $\mu = -1$ we have

$$\mathrm{Li}_3 \left(\frac{x}{1+x} \right) = \sum_{n=1}^{\infty} \left\{ \sum_{k=1}^{n} \frac{H_k}{k} \right\} \frac{(-1)^{n-1}}{n} x^n$$

which is equation (5.72) in disguise because of the identity

$$2 \sum_{k=1}^{n} \frac{H_k}{k} = H_n^2 + H_n^{(2)} .$$

This nice identity is similar to the identity in (5.94) and can be proved the same way.

5.8 Fun with Lobachevsky

In this short section we look at some simple, but amusing integrals.

First, recall the classical integral solved at the beginning of Chapter 2

(5.104)
$$\int_0^\infty \frac{\sin x}{x}\,dx = \frac{\pi}{2}.$$

Can we replace here $\sin x$ by $\cos x$?

$$\int_0^\infty \frac{\cos x}{x}\,dx = ?$$

No! This integral is divergent, because the ratio $\dfrac{\cos x}{x}$ behaves like $\dfrac{1}{x}$ close to zero.

However, we can multiply by $\cos x$ the integrand in (5.104) to get

(5.105)
$$\int_0^\infty \frac{\cos x \sin x}{x}\,dx = \frac{\pi}{4}$$

(second case in entry 3.741(2) in [25]). This result is very easy to explain by using the trigonometric identity $\sin(2x) = 2\sin x \cos x$

$$\int_0^\infty \frac{\cos x \sin x}{x}\,dx = \int_0^\infty \frac{\sin 2x}{2x}\,dx = \frac{1}{2}\int_0^\infty \frac{\sin 2x}{2x}\,d2x = \frac{\pi}{4}.$$

Can we replace here $\cos x$ by $\sin x$? No, the integral becomes divergent

(5.106)
$$\int_0^\infty \frac{\sin x \sin x}{x}\,dx = \int_0^\infty \frac{\sin^2 x}{x}\,dx = \infty.$$

To see why this integral is divergent the reader can analyze the proof that (5.104) is convergent in Example 2.1.1 and see why it does not work for (5.106).

At the same time

(5.107)
$$\int_0^\infty \frac{\sin(ax)\sin(bx)}{x}\,dx = \frac{1}{2}\ln\left|\frac{a+b}{a-b}\right|$$

for $a > 0, b > 0$, $a \neq b$ (entry 3.741(1) in [25]). This result comes from equation (5.13) after writing the product $\sin(ax)\sin(bx)$ as the difference of two cosines

$$\sin(ax)\sin(bx) = \frac{1}{2}\big(\cos(a-b)x - \cos(a+b)x\big).$$

(See also equation (2.2) in Example 2.2.10.)

Representing the product $\sin(ax)\cos(bx)$ as a difference of two sines when $a > b > 0$ we find (first case in entry 3.741(2))

$$(5.108) \qquad \int_0^\infty \frac{\sin(ax)\cos(bx)}{x}\,dx = \frac{\pi}{2}.$$

This is somewhat counterintuitive in view of (5.105). (See also Example 2.2.12.)

Multiplying inside (5.105) by $\cos x$ we get

$$(5.109) \qquad \int_0^\infty \frac{\cos^2 x \sin x}{x}\,dx = \frac{\pi}{4}.$$

Interesting! The same value as in (5.105)! But how do we evaluate this integral? In fact, it is true that

$$\int_0^\infty \frac{\cos^2 x \sin x}{x}\,dx = \int_0^{\pi/2} \cos^2 x\,dx = \frac{\pi}{4}.$$

What we see here is a special case of the following.

Lobachevsky's Integral Formula. *Suppose* $f(x)$ *is a continuous function defined for* $x \geq 0$ *with the property* $f(x + \pi) = f(x)$ *and* $f(\pi - x) = f(x)$ *for any* $x \geq 0$. *Then*

$$(5.110) \qquad \int_0^\infty f(x)\frac{\sin x}{x}\,dx = \int_0^{\pi/2} f(x)\,dx$$

$$\int_0^\infty f(x)\left(\frac{\sin x}{x}\right)^2 dx = \int_0^{\pi/2} f(x)\,dx.$$

(Nikolai Ivanovich Lobachevsky (1792-1856) was a Russian mathematician who invented the hyperbolic geometry.) More details about this integral formula together with an interesting extensions can be found in [55].

For the cosine function we have

$$\cos(x-\pi) = \cos(\pi-x) = -\cos(x)$$

and we see that the functions

$$f(x) = (\cos x)^{2n}, \quad f(x) = |\cos x|$$

satisfy the conditions for Lobachevsky's formula. Clearly the same is true for the functions

$$f(x) = (\sin x)^{2n}, \quad f(x) = |\sin x|.$$

For our examples we will use the well-known Wallis integrals

$$W_n = \int_0^{\pi/2} \sin^n x\,dx = \int_0^{\pi/2} \cos^n x\,dx \quad (n = 0, 1, 2, \ldots)$$

where

(5.111) $$W_{2n} = \frac{(2n)!}{4^n (n!)^2} \frac{\pi}{2} = \frac{1}{4^n}\binom{2n}{n}\frac{\pi}{2}$$

It follows that

$$\int_0^\infty \frac{\cos^{2n} x \sin x}{x}\,dx = \int_0^\infty \cos^{2n} x \left(\frac{\sin x}{x}\right)^2 dx = \int_0^{\pi/2} \cos^{2n} x\,dx = W_{2n}$$

$$\int_0^\infty \frac{\sin^{2n+1} x}{x}\,dx = \int_0^\infty \sin^{2n} x \left(\frac{\sin x}{x}\right)^2 dx = W_{2n}$$

for any integer $n \geq 0$. The case $n = 1$ is equation (5.109). For $n = 2, 3$ we have correspondingly

$$\int_0^\infty \frac{\cos^4 x \sin x}{x} dx = \int_0^{\pi/2} \cos^4 x\, dx = \frac{3\pi}{16}$$

$$\int_0^\infty \frac{\cos^6 x \sin x}{x} dx = \int_0^{\pi/2} \cos^6 x\, dx = \frac{5\pi}{32}$$

etc. We also have

$$\int_0^\infty \frac{\sin^3 x}{x} dx = \int_0^{\pi/2} \sin^2 x\, dx = \frac{\pi}{4}$$

$$\int_0^\infty \frac{\sin^5 x}{x} dx = \int_0^{\pi/2} \sin^4 x\, dx = \frac{3\pi}{16}.$$

At the same time all integrals of the form

$$\int_0^\infty \frac{\sin^{2n} x}{x} dx$$

($n = 0, 1, 2, \ldots$) are divergent.

With $f(x) = |\sin x|$, $f(x) = |\cos x|$ we have

$$\int_0^\infty |\sin x| \frac{\sin x}{x} dx = \int_0^{\pi/2} |\sin x|\, dx = \int_0^{\pi/2} \sin x\, dx = 1.$$

$$\int_0^\infty |\cos x| \frac{\sin x}{x} dx = \int_0^{\pi/2} \cos x\, dx = 1.$$

Problems for the reader:

1. For any integer $p \geq 0$ evaluate

$$A_1 = \int_0^\infty \frac{(\cos x)^2 (\sin x)^{2p+1}}{x} dx$$

$$A_2 = \int_0^\infty \frac{\cos^{2p} x (\sin x)^3}{x} dx .$$

2. Show that for every $n \geq 2$

$$W_n = \frac{n-1}{n} W_{n-2}$$

and using that $W_0 = \frac{\pi}{2}$, $W_1 = 1$ prove the Wallis formulas (5.111) and

(5.112) $$W_{2n+1} = \frac{4^n (n!)^2}{(2n+1)!} = \frac{4^n}{(2n+1)} \binom{2n}{n}^{-1} .$$

5.9 More Special Functions

We have already seen integrals evaluated in terms of special functions (in Section 4.3, for example). Here we want to mention briefly that certain integral representations of some special functions can be nicely used for solving integrals.

Example 5.9.1

The second Binet formula for the log-gamma function was used in Example 3.2.11

$$\ln \Gamma(\lambda) = \left(\lambda - \frac{1}{2} \right) \ln \lambda - \lambda + \frac{\ln 2\pi}{2} + 2 \int_0^\infty \frac{\arctan(t / \lambda)}{e^{2\pi t} - 1} dt .$$

Here and throughout this example we assume $\operatorname{Re} \lambda > 0$. Differentiation gives an integral representation for the digamma function

$$\psi(\lambda) = \ln \lambda - \frac{1}{2\lambda} - 2 \int_0^\infty \frac{t}{(t^2 + \lambda^2)(e^{2\pi t} - 1)} dt .$$

Replacing λ by $\lambda / 2$ we write this in the form

(5.113) $$\psi\left(\frac{\lambda}{2}\right) = \ln\frac{\lambda}{2} - \frac{1}{\lambda} - 8\int_0^\infty \frac{t}{(4t^2 + \lambda^2)(e^{2\pi t} - 1)}\,dt\ .$$

From Section 3.3 we have also the representation

$$\psi\left(z + \frac{1}{2}\right) = \ln z + 2\int_0^\infty \frac{x}{(x^2 + z^2)(e^{2\pi x} + 1)}\,dx$$

which we write in the form

(5.114) $$\psi\left(\frac{\lambda+1}{2}\right) = \ln\frac{\lambda}{2} + 8\int_0^\infty \frac{x}{(x^2 + z^2)(e^{2\pi x} + 1)}\,dx\ .$$

Using these two integral representations we will prove two interesting Laplace transforms.

(5.115) $$\int_0^\infty e^{-\lambda x}\left(\frac{1}{x} - \coth x\right)dx = \psi\left(\frac{\lambda}{2}\right) - \ln\frac{\lambda}{2} + \frac{1}{\lambda}$$

(5.116) $$\int_0^\infty e^{-\lambda x}\left(\frac{1}{x} - \frac{1}{\sinh x}\right)dx = \psi\left(\frac{\lambda+1}{2}\right) - \ln\frac{\lambda}{2}\ .$$

(These are correspondingly entries 3.554(4) and 3.554(2) from [25].)

To prove the first one we use the important integral from Example 4.5.2

$$4\int_0^\infty \frac{\sin 2xt}{e^{2\pi t} - 1}\,dt = \coth x - \frac{1}{x}$$

(after a simple adjustment). Thus

$$\int_0^\infty e^{-\lambda x}\left(\frac{1}{x} - \coth x\right)dx = -4\int_0^\infty e^{-\lambda x}\left\{\int_0^\infty \frac{\sin 2xt}{e^{2\pi t} - 1}\,dt\right\}dx$$

$$= -4\int_0^\infty \frac{1}{e^{2\pi t} - 1}\left\{\int_0^\infty e^{-\lambda x}\sin(2xt)\,dx\right\}dt = -4\int_0^\infty \frac{2t}{(e^{2\pi t} - 1)(\lambda^2 + 4t^2)}\,dt$$

$$= \psi\left(\frac{\lambda}{2}\right) - \ln\frac{\lambda}{2} + \frac{1}{\lambda}$$

according to (5.113). This way (5.115) is proved. Likewise we use the integral (after adjustment)

$$4\int_0^\infty \frac{\sin 2xt}{e^{2\pi t} + 1}\, dt = \frac{1}{x} - \frac{1}{\sinh x}$$

from Example 4.5.2 for the proof of (5.116)

$$\int_0^\infty e^{-\lambda x}\left(\frac{1}{x} - \frac{1}{\sinh x}\right) dx = \int_0^\infty e^{-\lambda x}\left\{4\int_0^\infty \frac{\sin 2xt}{e^{2\pi t} + 1}\, dt\right\} dx$$

$$= 4\int_0^\infty \frac{1}{e^{2\pi t} + 1}\left\{\int_0^\infty e^{-\lambda x}\sin(2xt)\, dx\right\} dt$$

$$= 4\int_0^\infty \frac{2t}{(e^{2\pi t} + 1)(\lambda^2 + 4t^2)}\, dt = \psi\left(\frac{\lambda+1}{2}\right) - \ln\frac{\lambda}{2}$$

by using (5.114). Done!

Example 5.9.2

The reader may looks at the group of entries 3.554 in [25] from where the above two integrals (5.115) and (5.116) were taken. The first entry in this group will surely catch the attention

(5.117) $$\int_0^\infty e^{-\lambda x}\left(1 - \frac{1}{\cosh x}\right)\frac{dx}{x} = 2\ln\frac{\Gamma\left(\frac{\lambda+3}{4}\right)}{\Gamma\left(\frac{\lambda+1}{4}\right)} - \ln\frac{\lambda}{4}$$

(here again $\mathrm{Re}\,\lambda > 0$). This integral resembles (5.115) and (5.116), but requires a different approach. For the proof we recall an important formula from Section 4.5.2, namely,

(5.118) $$\int_0^\infty \frac{e^{-\lambda x}}{\cosh x} dx = \frac{1}{2}\left[\psi\left(\frac{\lambda+3}{4}\right) - \psi\left(\frac{\lambda+1}{4}\right)\right].$$

Therefore, we write

$$\int_0^\infty e^{-\lambda x}\left(1 - \frac{1}{\cosh x}\right)dx = \int_0^\infty e^{-\lambda x}dx - \int_0^\infty \frac{e^{-\lambda x}}{\cosh x}dx$$

$$\frac{1}{\lambda} - \frac{1}{2}\left[\psi\left(\frac{\lambda+3}{4}\right) - \psi\left(\frac{\lambda+1}{4}\right)\right].$$

Now we integrate this equation with respect to λ and remembering that $\psi(z) = \dfrac{d}{dz}\ln\Gamma(z)$ we find

$$-\int_0^\infty \frac{e^{-\lambda x}}{x}\left(1 - \frac{1}{\cosh x}\right)dx$$

$$= \ln\lambda - 2\left[\ln\Gamma\left(\frac{\lambda+3}{4}\right) - \ln\Gamma\left(\frac{\lambda+1}{4}\right)\right] + C$$

or,

$$\int_0^\infty e^{-\lambda x}\left(1 - \frac{1}{\cosh x}\right)\frac{dx}{x} = 2\ln\frac{\Gamma\left(\dfrac{\lambda+3}{4}\right)}{\Gamma\left(\dfrac{\lambda+1}{4}\right)} - \ln\lambda + C.$$

In order to compute the constant of integration we set $\lambda \to \infty$. The left-hand side approaches zero. Using the properties of the gamma function it is easy to compute that (with $\lambda = 4x$)

$$\lim_{x \to \infty} \frac{\Gamma\left(x + \dfrac{3}{4}\right)}{\Gamma\left(x + \dfrac{1}{4}\right)\sqrt{x}} = 1$$

so with $C = \ln 4$ the limit of the right-hand side is also zero. Thus (5.117) is proved.

Example 5.9.3

Using the properties of Euler's beta function $B(x, y)$ and the representation

$$B(x, y) = \frac{\Gamma(x)\Gamma(y)}{\Gamma(x + y)}$$

one can derive the formula (see entry 8.380(10) in [25])

$$\int_0^\infty \frac{\cosh(2yt)}{\cosh^{2x}(t)} dt = 2^{2x-2} \frac{\Gamma(x + y)\Gamma(x - y)}{\Gamma(2x)}$$

where $\operatorname{Re} x > 0, \operatorname{Re} x > |\operatorname{Re} y|$. From here with $x = 1$ and $|y| < 1$

$$\int_0^\infty \frac{\cosh(2yt)}{\cosh^2(t)} dt = \Gamma(1 + y)\Gamma(1 - y) = \frac{\pi y}{\sin \pi y}.$$

With $x = 2$

$$\int_0^\infty \frac{\cosh(2yt)}{\cosh^4(t)} dt = \frac{2}{4}\Gamma(2 + y)\Gamma(2 - y) = \frac{2\pi y(1 - y^2)}{3 \sin \pi y}.$$

Replacing y by iy we have also

$$\int_0^\infty \frac{\cos(2yt)}{\cosh^2(t)} dt = \frac{\pi y}{\sinh \pi y}$$

$$\int_0^\infty \frac{\cos(2yt)}{\cosh^4(t)} dt = \frac{2\pi y(1 + y^2)}{3 \sinh \pi y}$$

Other interesting integrals van be obtained from here by differentiation with respect to x or y.

Example 5.9.4

In Example 2.3.1, Chapter 2, we evaluated the integral

$$\int_0^\infty e^{-t^2}\cos(2xt)\,dt = \frac{\sqrt{\pi}}{2}e^{-x^2}.$$

Differentiating $2n$ times with respect to x we find

$$\int_0^\infty t^{2n}e^{-t^2}\cos(2xt)\,dt = \frac{(-1)^n\sqrt{\pi}}{2^{2n+1}}\left(\frac{d}{dx}\right)^{2n}e^{-x^2}.$$

At this point we turn to the Rodrigues formula for the Hermite polynomials

$$H_p(x) = (-1)^p e^{x^2}\left(\frac{d}{dx}\right)^p e^{-x^2}$$

($p = 0, 1, 2, \ldots$) where

$$H_0(x) = 1,\ H_1(x) = 2x,\ H_2(x) = 4x^2 - 2,$$

$$H_3(x) = 8x^3 - 12x,\ H_4(x) = 16x^4 - 48x^2 + 12, \ldots .$$

We conclude that for $n = 0, 1, 2, \ldots$

$$(5.119)\qquad \int_0^\infty t^{2n}e^{-t^2}\cos(2xt)\,dt = \frac{(-1)^n\sqrt{\pi}}{2^{2n+1}}e^{-x^2}H_{2n}(x).$$

In the same way we find

$$(5.120)\qquad \int_0^\infty t^{2n+1}e^{-t^2}\sin(2xt)\,dt = \frac{(-1)^n\sqrt{\pi}}{2^{2n+2}}e^{-x^2}H_{2n+1}(x)$$

($n = 0, 1, 2, \ldots$). These formulas can be used for solving integrals. For instance, with $n = 1$

$$\int_0^\infty t^2 e^{-t^2} \cos(2xt)\,dt = -\frac{\sqrt{\pi}}{8} e^{-x^2} (4x^2 - 2)$$

$$\int_0^\infty t^3 e^{-t^2} \sin(2xt)\,dt = -\frac{\sqrt{\pi}}{16} e^{-x^2} (8x^3 - 12x).$$

In the place of t^{2n} in (5.119) we can put any even polynomial and evaluate the integral in terms of Hermite polynomials. Likewise in the place of t^{2n+1} in (5.120) we can put any odd polynomial.

Example 5.9.5

Suppose we want to evaluate the integral

$$\int_0^\pi e^{\cos x}\,dx\,.$$

What can we do? The substitution $u = \cos x$ will not help, it brings to something worse

$$\int_{-1}^1 \frac{e^u}{\sqrt{1-u^2}}\,du\,.$$

Integration by parts does not help either.
We can introduce a parameter and try the method in Chapter 2, say,

(5.121) $$F(\lambda) = \int_0^\pi e^{\lambda \cos x}\,dx\,.$$

Then

$$F'(\lambda) = \int_0^\pi \cos x\, e^{\lambda \cos x}\,dx$$

and things become complicated.

Something natural will be to use the exponential series in (5.119), that is,

$$F(\lambda) = \int_0^\pi \left\{ \sum_{n=0}^\infty \frac{\lambda^n \cos^n x}{n!} \right\} dx = \sum_{n=0}^\infty \frac{\lambda^n}{n!} \left\{ \int_0^\pi \cos^n x \, dx \right\}.$$

Now

$$\int_0^\pi \cos^n x \, dx = 0 \quad (n \text{ odd})$$

because the substitution $u = x - \dfrac{\pi}{2}$ says that

$$\int_0^\pi \cos^n x \, dx = \int_{-\pi/2}^{\pi/2} \sin^n x \, dx = 0$$

as $\sin^n x$ is odd. For $n = 2p$ even we have from (5.111)

$$\int_0^\pi \cos^{2p} x \, dx = 2 \int_0^{\pi/2} \cos^{2p} x \, dx = 2W_{2p} = \frac{\pi (2p)!}{4^p (p!)^2}$$

and

$$F(\lambda) = \pi \sum_{p=0}^\infty \frac{1}{(p!)^2} \left(\frac{\lambda}{2} \right)^{2p}.$$

The series here is the modified Bessel function of the first kind $I_0(\lambda)$ of zero order, so that

(5.122)
$$\int_0^\pi e^{\lambda \cos x} \, dx = \pi I_0(\lambda)$$

and with $\lambda = 1$

$$\int_0^\pi e^{\cos x} \, dx = \pi \sum_{p=0}^\infty \frac{1}{4^p (p!)^2}.$$

More generally, we have

(5.123)
$$\int_0^\pi e^{\lambda \cos x} \cos(nx)dx = \pi I_n(\lambda)$$

with the modified Bessel function of the first kind $I_n(\lambda)$ (entry 2.5.40(3) in [43]). Here

$$I_n(\lambda) = \sum_{k=0}^{\infty} \frac{1}{k!\Gamma(k+n+1)} \left(\frac{\lambda}{2}\right)^{2k+n}$$

(see [40]).

In the same way one can use the integral representation

$$J_0(x) = \frac{1}{2\pi} \int_0^{2\pi} e^{ix\sin\theta}d\theta = \frac{1}{\pi}\int_0^\pi e^{ix\cos\theta}d\theta$$

where

$$J_0(x) = \sum_{n=0}^{\infty} \frac{(-1)^n}{2^{2n}(n!)^2} x^{2n}$$

is the Bessel function of zero order. Thus, for example,

$$\int_0^{\pi/2} \cos(x\sin\theta)d\theta = \int_0^{\pi/2} \cos(x\cos\theta)d\theta = \frac{\pi}{2}J_0(x).$$

Example 5.9.6

In this example we will show a very unusual method to prove two very unusual integrals

(5.124)
$$\int_0^\pi \frac{\ln(\cosh x + \sinh x \cos y)}{\cosh x + \sinh x \cos y}dy = \pi \ln\frac{2}{1+\cosh x}$$

where x is arbitrary. Also

$$(5.125) \quad \int_0^\pi \frac{\ln^2(\cosh x + \sinh x \cos y)}{\cosh x + \sinh x \cos y} dy = -2\pi \operatorname{Li}_2\left(\frac{1-\cosh x}{2}\right).$$

One strange feature of these integrals is that they are even functions of x although the integrands contain the odd function $\sinh(x)$.

To prove these integrals we use the Legendre function of the first kind $P_\lambda(z)$ which is a solution to the Legendre differential equation

$$(5.126) \quad\quad (1-z^2)y'' - 2zy' + \lambda(\lambda+1)y = 0.$$

When $\lambda = n$, a nonnegative integer, $P_n(z)$ is the Legendre polynomial of order n. Substituting $P_\lambda(z)$ in (5.126) and differentiating with respect to the parameter λ, Szmytkowski showed that

$$\frac{d}{d\lambda} P_\lambda(z)\bigg|_{\lambda=0} = \ln\left(\frac{1+z}{2}\right)$$

$$\frac{d^2}{d\lambda^2} P_\lambda(z)\bigg|_{\lambda=0} = -2\operatorname{Li}_2\left(\frac{1-z}{2}\right)$$

$$\frac{d^3}{d\lambda^3} P_\lambda(z)\bigg|_{\lambda=0} = 12\operatorname{Li}_3\left(\frac{1+z}{2}\right) - 6\ln\left(\frac{1+z}{2}\right)\operatorname{Li}_2\left(\frac{1+z}{2}\right)$$

$$+\pi^2 \ln\left(\frac{1+z}{2}\right) - 12\zeta(3)$$

(R. Szmytkowski, "The parameter derivatives ..." arXiv:1301.6586v1 (2013); the author attributes the second formula to G. P. Schramkowski.)

The Legendre function has several integral representations, one of which is

$$P_\lambda(\cosh x) = \frac{1}{\pi} \int_0^\pi \frac{1}{(\cosh x + \sinh x \cos y)^{\lambda+1}} dy$$

(see p. 203 in Nico Temme's nice book [44]). Differentiating here with respect to λ we find consecutively

(5.127) $\quad \dfrac{d}{d\lambda}P_\lambda(\cosh x)\bigg|_{\lambda=0} = -\dfrac{1}{\pi}\int\limits_0^\pi \dfrac{\ln(\cosh x + \sinh x \cos y)}{\cosh x + \sinh x \cos y}dy$

(5.128) $\quad \dfrac{d^2}{d\lambda^2}P_\lambda(\cosh x)\bigg|_{\lambda=0} = \dfrac{1}{\pi}\int\limits_0^\pi \dfrac{\ln^2(\cosh x + \sinh x \cos y)}{\cosh x + \sinh x \cos y}dy$

(5.129) $\quad \dfrac{d^3}{d\lambda^3}P_\lambda(\cosh x)\bigg|_{\lambda=0} = -\dfrac{1}{\pi}\int\limits_0^\pi \dfrac{\ln^3(\cosh x + \sinh x \cos y)}{\cosh x + \sinh x \cos y}dy \,.$

The two integrals (5.124) and (5.125) follow immediately from (5.127) and (5.128) in view of Szmytkowski's results. The reader can continue and write down the evaluation of the integral in (5.129).

Example 5.9.7

Here we solve problem 1184 from the *College Mathematics Journal* (September 2020, p. 306) by using one neat special function. Evaluate

$$J = \int\limits_0^\infty \int\limits_0^\infty \frac{\sin(x)\sin(x+y)}{x(x+y)}dxdy \,.$$

For the evaluation we will use the sine integral

$$\mathrm{Si}(z) = \int\limits_0^z \frac{\sin(t)}{t}dt$$

which has the obvious properties

$$\frac{d}{dz}\mathrm{Si}(z) = \frac{\sin z}{z}, \quad \lim_{z\to\infty}\mathrm{Si}(z) = \int\limits_0^\infty \frac{\sin(t)}{t}dt = \frac{\pi}{2}, \quad \mathrm{Si}(0) = 0 \,.$$

Using the substitution $x + y = u$ we write

$$J = \int\limits_0^\infty \frac{\sin x}{x}\left\{\int\limits_0^\infty \frac{\sin(x+y)}{x+y}dy\right\}dx = \int\limits_0^\infty \frac{\sin x}{x}\left\{\int\limits_x^\infty \frac{\sin(u)}{u}du\right\}dx$$

and changing the order of integration (very easy to justify)

$$J = \int_0^\infty \frac{\sin u}{u} \left\{ \int_0^u \frac{\sin(x)}{x} dx \right\} du = \int_0^\infty \frac{\sin u}{u} \operatorname{Si}(u) du$$

$$= \int_0^\infty \operatorname{Si}(u) \, d\operatorname{Si}(u) = \frac{1}{2} (\operatorname{Si}(u))^2 \Big|_0^\infty = \frac{1}{2} \left(\frac{\pi}{2} \right)^2 = \frac{\pi^2}{8}.$$

That's all!

(5.130) $$\int_0^\infty \int_0^\infty \frac{\sin(x)\sin(x+y)}{x(x+y)} dx dy = \frac{\pi^2}{8}.$$

Appendix A

List of Solved Integrals

The integrals here are listed in the order they appear in the book. Some of them are solutions to problems from mathematics journals and this is indicated in the next line. The following abbreviations are used:

American Mathematical Monthly (AMM)
College Mathematics Journal (CMJ)
Mathematics Magazine (MM)

(A.1) $\quad \displaystyle\int \frac{dx}{\sqrt{x^2+x+1}} = -\ln(2\sqrt{x^2+x+1}-1-2x)+C$

Example 1.2.1.

(A.2) $\quad \displaystyle\int \frac{dx}{(x+3)\sqrt{3x-x^2-2}} = \frac{-1}{\sqrt{5}}\arctan\left(\frac{2}{\sqrt{5}}\sqrt{\frac{2-x}{x-1}}\right)+C$

Example 1.2.2.

(A.3) $\qquad \displaystyle\int \frac{dx}{x\sqrt{x^2+6x+8}}$

$\displaystyle = \frac{1}{2\sqrt{2}}\ln\left|\frac{\sqrt{x^2+6x+8}-\sqrt{2}(x+2)}{\sqrt{x^2+6x+8}+\sqrt{2}(x+2)}\right|+C$

Example 1.2.3.

333

(A.4) $$\int_a^b \arccos \frac{x}{\sqrt{ax+bx-ab}}\,dx = \frac{\pi(b-a)^2}{4(b+a)}$$

Example 1.2.4. This is AMM Problem 11457. Here $0 \le a \le b$.

(A.5) $$\int \frac{dx}{x\sqrt{1-2x-x^2}} = \ln\left|\frac{x-1+\sqrt{1-2x-x^2}}{x}\right| + C$$

Example 1.2.5.

(A.6) $$\int \frac{dx}{(ax^2+bx+c)^{3/2}} = \frac{4ax+2b}{(4ac-b)\sqrt{ax^2+bx+c}} + C$$

Section 1.3.

(A.7) $$\int \frac{dx}{\sqrt{x}(1+\sqrt[3]{x})^2} = \frac{-3\sqrt[6]{x}}{1+\sqrt[3]{x}} + 3\arctan\sqrt[6]{x} + C$$

Section 1.4. Also there

(A.8) $$\int \frac{dx}{x\sqrt{1+x^5}} = \frac{1}{5}\ln\frac{\sqrt{1+x^5}+1}{\sqrt{1+x^5}-1} + C$$

(A.9) $$\int \frac{\sqrt{1+x^3}}{x}\,dx = \frac{2}{3}\sqrt{1+x^3} + \frac{1}{3}\ln\frac{\sqrt{1+x^3}-1}{\sqrt{1+x^3}+1} + C$$

(A.10) $$\int \frac{dx}{\sqrt{(1+x^2)^3}} = \frac{x}{\sqrt{1+x^2}} + C$$

(A.11) $$\int \frac{dx}{x^2\sqrt{x+x^4}} = \frac{-2}{3}\sqrt{x^{-3}+1} + C.$$

The next integral is from Section 5:

(A.12) $\qquad \displaystyle\int \frac{\sqrt{x^2-3}}{x^2}\,dx = \ln(x+\sqrt{x^2-3}) - \frac{\sqrt{x^2-3}}{x} + C$

Example 1.5.1.

(A.13) $\qquad \displaystyle\int \sqrt{x^2+1}\,dx = \frac{x}{2}\sqrt{x^2+1} + \frac{1}{2}\ln(x+\sqrt{x^2+1}) + C$

Example 1.5.2.

(A.14) $\qquad \displaystyle\int_0^\infty \frac{dx}{(x+\sqrt{1+x^2})^r} = \frac{r}{r^2-1}$

Example 1.5.3.

(A.15) $\qquad \displaystyle\int_0^\infty \frac{\cos t \sin\sqrt{t^2+1}}{\sqrt{t^2+1}}\,dt = \frac{\pi}{4}$

Example 1.5.4.

(A.16) $\qquad \displaystyle\int \frac{dx}{5+\sin x} = \frac{1}{\sqrt{6}}\arctan\left[\frac{1}{2\sqrt{6}}\left(5\tan\frac{x}{2}+1\right)\right] + C$

Example 1.6.1.

(A.17) $\qquad \displaystyle\int \frac{dx}{a+b\cos x} = \frac{1}{\sqrt{a^2-b^2}}\arctan\left(\sqrt{\frac{a-b}{a+b}}\tan\frac{x}{2}\right) + C$

Example 1.6.1.

(A.18) $\qquad \displaystyle\int \frac{dx}{2-\cos^2 x} = \frac{1}{\sqrt{2}}\arctan(\sqrt{2}\tan x) + C$

Example 1.6.2.

(A.19)
$$\int_0^{\pi/2} \frac{dx}{\sqrt{a^2 \cos^2 x + b^2 \sin^2 x}} = \frac{\pi}{2M(a,b)}$$

Section 1.7. Here $M(a,b)$ is the arithmetic-geometric mean for a,b.

(A.20)
$$\int_0^{\pi} \frac{x \sin x}{1 + \cos^2 x} dx = \frac{\pi^2}{4}.$$

Example 1.8.1.

(A.21)
$$\int_0^{\pi/2} \frac{(\cos x)^p}{(\cos x)^p + (\sin x)^p} dx = \frac{\pi}{4}$$

($p \geq 0$) Example 1.8.2.

(A.22)
$$\int_0^{\infty} \frac{\ln x}{x^2 + 1} dx = 0$$

Example 1.8.3.

(A.23)
$$\int_0^{\infty} (1 + x^{2\beta})^{-\frac{1}{\beta}} dx = \frac{1}{2\beta} \Gamma^2 \left(\frac{1}{2\beta} \right) \Gamma \left(\frac{1}{\beta} \right)^{-1}$$

($\beta > 0$) Example 1.8.4.

(A.24)
$$\int_0^{\infty} \frac{\arctan x}{\sqrt[\beta]{1 + x^{2\beta}}} dx = \frac{\pi}{8\beta} \Gamma^2 \left(\frac{1}{2\beta} \right) \Gamma \left(\frac{1}{\beta} \right)^{-1}$$

($\beta > 0$) Example 1.8.4. In particular,

(A.25)
$$\int_0^{\infty} \frac{\arctan x}{1 + x^2} dx = \frac{\pi^2}{8}$$

(A.26)
$$\int_0^{\infty} \frac{\arctan x}{\sqrt{1 + x^4}} dx = \frac{\sqrt{\pi}}{16} \Gamma^2 (1/4)$$

(A.27) $\displaystyle\int_0^\infty \frac{\arctan x}{\sqrt[3]{1+t^6}} dt = \frac{\pi}{24}\Gamma^2(1/6)\Gamma^{-1}(1/3)$.

(A.28) $\displaystyle\int \frac{dx}{\sqrt{(x-a)(b-x)}} = 2\arcsin\sqrt{\frac{x-a}{b-a}} + C \quad (a < b)$

Example 1.8.5. In particular,

(A.29) $\displaystyle\int_a^b \frac{dx}{\sqrt{(x-a)(b-x)}} = \pi, \quad \int_0^1 \frac{dx}{\sqrt{x(1-x)}} = \pi$.

(A.30) $\displaystyle\int \sqrt{x + \sqrt{x + \sqrt{x + \ldots}}}\, dx = \frac{x}{2} + \frac{1}{12}(1+4x)^{\frac{3}{2}} + C$

Example 1.8.6.

(A.31) $\displaystyle\int_0^\infty e^{-a^2x^2 - x^{-2}} dx = \frac{\sqrt{\pi}}{2a} e^{-2a} \quad (a > 0)$

Example 1.8.7.

(A.32) $\displaystyle\int_0^\infty e^{-\lambda x} \frac{\sin x}{x} dx = \frac{\pi}{2} - \arctan\lambda \quad (\lambda \geq 0)$

Example 2.1.1. In particular,

$$\int_0^\infty \frac{\sin x}{x} dx = \frac{\pi}{2}.$$

(A.33) $\displaystyle\int_0^\infty \frac{e^{-\lambda x} - e^{-\mu x}}{x} \sin x\, dx = \arctan\mu - \arctan\lambda$

($\lambda, \mu > 0$). Also in Example 2.1.1

(A.34) $\displaystyle\int_0^\infty e^{-\lambda x} \frac{\sinh x}{x} dx = \frac{1}{2}\ln\frac{\lambda+1}{\lambda-1}, \quad \lambda > 1$

(A.35)
$$\int_0^\infty \frac{\cos ax - \cos bx}{x^2}\,dx = \frac{\pi}{2}(b-a).$$

(A.36)
$$\int_0^1 \frac{\ln(1+x)}{1+x^2}\,dx = \frac{\pi}{8}\ln 2$$

Example 2.1.2. In the same place also

(A.37)
$$\int_0^1 \frac{\arctan x}{1+x}\,dx = \frac{\pi}{8}\ln 2.$$

(A.38)
$$\int_0^\infty \left(\frac{1-e^{-\lambda x}}{x}\right)^2 dx = \lambda \ln 4$$

Example 2.1.3.

(A.39)
$$\int_0^1 \frac{x^\alpha - x^\beta}{\ln x}\,dx = \ln \frac{1+\alpha}{1+\beta} \quad (\alpha, \beta \geq 0)$$

Example 2.2.1. In particular,

(A.40)
$$\int_0^1 \frac{x^\alpha - 1}{\ln x}\,dx = \ln(1+\alpha), \quad \alpha \geq 0.$$

(A.41)
$$\int_0^1 \frac{\arctan \lambda x}{x\sqrt{1-x^2}}\,dx = \frac{\pi}{2}\ln\left(\lambda + \sqrt{1+\lambda^2}\right)$$

Example 2.2.2.

(A.42)
$$\int_0^\infty \frac{\arctan \lambda x}{x(1+x^2)}\,dx = \frac{\pi}{2}\ln(1+\lambda)$$

Example 2.2.3.

(A.43)
$$\int_0^\infty \frac{\arctan(\lambda x)\arctan(\mu x)}{x^2}\,dx$$

$$= \frac{\pi}{2} \left[(\lambda + \mu) \ln(\lambda + \mu) - \lambda \ln \lambda - \mu \ln \mu \right] \quad (\lambda, \mu > 0)$$

Example 2.2.4. In particular, with $\lambda = \mu = 1$

(A.44)
$$\int_0^\infty \frac{(\arctan x)^2}{x^2} dx = \pi \ln 2 .$$

(A.45)
$$\int_0^\infty \frac{\ln(1 + \lambda^2 x^2)}{1 + x^2} dx = \pi \ln(1 + \lambda) \quad (\lambda \ge 0) .$$

Example 2.2.5. In particular, with $\lambda = 1$

(A.46)
$$\int_0^\infty \frac{\ln(1 + x^2)}{1 + x^2} dx = \pi \ln 2 .$$

(A.47)
$$\int_0^\infty e^{-\beta x} \frac{1 - \cos \lambda x}{x} dx = \frac{1}{2} \ln \left(1 + \frac{\lambda^2}{\beta^2} \right) \quad (\beta > 0)$$

Example 2.2.6. Also there

(A.48)
$$\int_0^\infty e^{-\beta x} \frac{\cos \lambda x - \cos \mu x}{x} dx = \frac{1}{2} \ln \left(\frac{\beta^2 + \mu^2}{\beta^2 + \lambda^2} \right)$$

(A.49)
$$\int_0^\infty \frac{\cos \lambda x - \cos \mu x}{x} dx = \ln \frac{\mu}{\lambda}$$

$(\lambda, \mu > 0)$.

(A.50)
$$\int_0^\infty \frac{e^{-\lambda x} - e^{-\mu x}}{x} \cos \beta x \, dx = \frac{1}{2} \ln \frac{\mu^2 + \beta^2}{\lambda^2 + \beta^2}$$

$(\lambda, \mu > 0)$ **Example 2.2.7.**

(A.51)
$$\int_0^\infty e^{-x^2} dx = \frac{\sqrt{\pi}}{2} , \quad \int_0^\infty e^{-x^p} dx = \frac{1}{p} \Gamma \left(\frac{1}{p} \right)$$

Example 2.2.8.

(A.52) $$\int_0^\infty \frac{1-e^{-\lambda x^2}}{x^2}\,dx = \sqrt{\lambda \pi} \quad (\lambda \geq 0)$$

Example 2.2.8.

(A.53) $$\int_0^\infty \frac{e^{-px}\cos qx - e^{-\lambda x}\cos \mu x}{x}\,dx = \frac{1}{2}\ln\frac{\lambda^2 + \mu^2}{p^2 + q^2}$$

$(\lambda, p > 0)$ Example 2.2.9.

(A.54) $$\int_0^\infty e^{-\lambda x}\frac{\sin(ax)\sin(bx)}{x}\,dx = \frac{1}{4}\ln\frac{\lambda^2 + (a+b)^2}{\lambda^2 + (a-b)^2}$$

$(a > b > 0)$ Example 2.2.10.

(A.55) $$\int_0^\infty \frac{\sin(ax)\sin(bx)}{x^2}\,dx = \frac{\pi b}{2} \quad (a > b > 0)$$

Example 2.2.11.

(A.56) $$\int_0^\infty e^{-\lambda x}\frac{\sin(ax)\cos(ax)}{x}\,dx = \frac{\pi}{4} - \frac{1}{2}\arctan\frac{\lambda}{2a}$$

$(a > b > 0, \lambda > 0)$ Example 2.2.12.

(A.57) $$\int_0^\infty e^{-\lambda x}\frac{\cos(ax) - \cos(bx)}{x^2}\,dx$$

$$= \frac{\lambda}{2}\ln\frac{\lambda^2 + a^2}{\lambda^2 + b^2} + b\arctan\frac{b}{\lambda} - a\arctan\frac{a}{\lambda} \quad (\lambda > 0)$$

Example 2.2.13.

(A.58) $$\int_0^\infty e^{-\lambda x}\frac{\sin^2(ax) - \sin^2(bx)}{x^2}\,dx$$

$$= \frac{\lambda}{4} \ln \frac{\lambda^2 + 4b^2}{\lambda^2 + 4a^2} + a \arctan \frac{2a}{\lambda} - b \arctan \frac{2b}{\lambda}$$

($\lambda > 0$) Example 2.2.13.

(A.59) $\displaystyle\int_0^{\pi/2} \ln(\alpha^2 - \cos^2 \theta) d\theta = \pi \ln \frac{\alpha + \sqrt{\alpha^2 - 1}}{2}$ ($\alpha > 1$)

Example 2.2.14.

(A.60) $\displaystyle\int_0^{\pi/2} \ln(1 - \beta^2 \cos^2 \theta) d\theta = \pi \ln \frac{1 + \sqrt{1 - \beta^2}}{2}$ ($0 \le \beta \le 1$)

Example 2.2.14. In particular, with $\beta = 1$

(A.61) $\displaystyle\int_0^{\pi/2} \ln(\sin \theta) d\theta = -\frac{\pi}{2} \ln 2 .$

(A.62) $\displaystyle\int_0^{\pi/2} \ln(1 + \alpha \sin^2 \theta) d\theta = \pi \ln \frac{1 + \sqrt{1 + \alpha}}{2}$ ($-1 \le \alpha$)

Example 2.2.15.

(A.63) $\displaystyle\int_0^{\pi} \frac{\ln(1 + \alpha \cos \theta)}{\cos \theta} d\theta = \pi \arcsin \alpha$ ($|\alpha| \le 1$)

Example 2.2.16. In particular,

(A.64) $\displaystyle\int_0^{\pi} \frac{\ln(1 \pm \cos \theta)}{\cos \theta} d\theta = \pm \frac{\pi^2}{2} .$

(A.65) $\displaystyle\int_0^{\pi} \ln(1 + \alpha \cos \theta) d\theta = \pi \ln \frac{1 + \sqrt{1 - \alpha^2}}{2}$ ($|\alpha| < 1$)

Example 2.2.17. With $\alpha \to \pm 1$ we have

(A.66) $\int_0^\pi \ln(1 \pm \cos\theta)\, d\theta = -\pi \ln 2.$

(A.67) $\int_0^\pi \ln(\beta^2 - 2\alpha\beta \cos x + \alpha^2)\, dx = 2\pi \ln|\alpha|$

$(|\beta| \le |\alpha|)$ Example 2.2.18.

(A.68) $\int_0^1 \frac{\ln(1 - \alpha^2 x^2)}{x^2 \sqrt{1-x^2}}\, dx = \pi\left(\sqrt{1-\alpha^2} - 1\right)$ $(|\alpha| < 1)$

Example 2.2.19.

(A.69) $\int_0^\infty \frac{a}{\sqrt{a^2 + x^2}} \arctan \frac{b}{\sqrt{a^2 + x^2}}\, dx$

$$= \frac{\pi a}{2} \ln \frac{b + \sqrt{b^2 + a^2}}{a} \quad (a, b > 0)$$

Example 2.2.20 (AMM Problem 11101).

(A.70) $\int_0^\infty \frac{1}{x^2}\left(1 - \frac{1}{x}\arctan x\right) dx = \frac{\pi}{4}$

Example 2.2.21.

(A.71) $\int_0^\infty e^{-t^2} \cos(2xt)\, dt = \frac{\sqrt{\pi}}{2} e^{-x^2}$

Example 2.3.1.

(A.72) $\int_{-\infty}^\infty e^{-at^2} \cos(xt)\, dt = \sqrt{\frac{\pi}{a}}\, e^{-x^2/4a}$ $(\mathrm{Re}\, a > 0)$

Example 2.3.1.

(A.73) $$\int_0^\infty \frac{\cos\sqrt{t}}{\sqrt{t}}\cos t\, dt = \sqrt{\frac{\pi}{2}}\left(\cos\frac{1}{4} + \sin\frac{1}{4}\right)$$

Example 2.3.1 (MM Problem 1896).

(A.74) $$\int_0^\infty \frac{\cos\lambda x}{a^2 + x^2}\, dx = \frac{\pi}{2a}e^{-a\lambda}$$

Example 2.3.2. Here and in the following three integrals $a > 0, \lambda \geq 0$.

(A.75) $$\int_0^\infty \frac{x\sin\lambda x}{a^2 + x^2}\, dx = \frac{\pi}{2}e^{-a\lambda}$$

Example 2.3.2. Also in this example

(A.76) $$\int_0^\infty \frac{\sin\lambda x}{x(a^2 + x^2)}\, dx = \frac{\pi}{2a^2}(1 - e^{-a\lambda})$$

(A.77) $$\int_0^\infty \frac{\sin\lambda x}{x(a^2 + x^2)^2}\, dx = \frac{\pi}{2a^4}(1 - e^{-a\lambda}) - \frac{\pi\lambda}{4a^3}e^{-a\lambda}.$$

(A.78) $$\int_0^\infty \frac{e^{-st}}{a^2 + t^2}\, dt = \frac{1}{a}\left[\mathrm{ci}(as)\sin(as) - \mathrm{si}(as)\cos(as)\right]$$

($a > 0$) Example 2.3.3. Also there

(A.79) $$\int_0^\infty \frac{te^{-st}}{a^2 + t^2}\, dt = -\mathrm{ci}(as)\cos(as) - \mathrm{si}(as)\sin(as).$$

(A.80) $$\int_0^\infty \frac{\sin ax}{x + b}\, dx = \mathrm{ci}(ab)\sin(ab) - \mathrm{si}(ab)\cos(ab)$$

($a, b > 0$) Example 2.3.3. Also there

(A.81) $$\int_0^\infty \frac{\cos ax}{x + b}\, dx = -\mathrm{ci}(ab)\cos(ab) - \mathrm{si}(ab)\sin(ab).$$

(A.82) $\displaystyle\int_0^\infty \exp\left(-x-\frac{\alpha}{x}\right)\frac{dx}{\sqrt{x}} = \sqrt{\pi}\,\exp(-2\sqrt{\alpha}),\ \ \alpha > 0$

Example 2.3.4. This integral is equivalent to (A.31). Also in this example

(A.83) $\displaystyle\int_0^\infty \exp\left(-x-\frac{\alpha}{x}\right)\frac{dx}{x\sqrt{x}} = \sqrt{\frac{\pi}{\alpha}}\,\exp(-2\sqrt{\alpha})$

(A.84) $\displaystyle\int_0^\infty \sqrt{x}\,\exp\left(-x-\frac{\alpha}{x}\right)dx = \sqrt{\pi}\left(\sqrt{\alpha}+\frac{1}{2}\right)\exp(-2\sqrt{\alpha}).$

(A.85a) $\displaystyle\int_0^\infty \exp(-x^2)\cos\left(\frac{\alpha^2}{x^2}\right)dx = \frac{\sqrt{\pi}}{2}e^{-\sqrt{2}\alpha}\cos\sqrt{2}\alpha$

($\alpha > 0$) Example 2.3.5. Also in this example

(A.85b) $\displaystyle\int_0^\infty \exp(-x^2)\sin\left(\frac{\alpha^2}{x^2}\right)dx = \frac{\sqrt{\pi}}{2}e^{-\sqrt{2}\alpha}\sin\sqrt{2}\alpha.$

(A.86) $\displaystyle\int_{\varphi(\alpha)}^1 \frac{\arctan(\alpha x)}{\sqrt{1-x^2}}dx$

$\displaystyle = \frac{-1}{8}\left(\ln\frac{\sqrt{\alpha^2+1}+1}{\sqrt{\alpha^2+1}-1}\right)^2 - \frac{1}{2}\left(\arctan\sqrt{\alpha^2-1}\right)^2 + \frac{\pi^2}{8}$

where $\alpha > 1$, $\varphi(\alpha) = \sqrt{1-\dfrac{1}{\alpha^2}}$.

Example 2.4.1. In particular,

(A.87) $\displaystyle\int_0^1 \frac{\arctan x}{\sqrt{1-x^2}}dx = \frac{\pi^2}{8} - \frac{1}{2}\left(\ln(1+\sqrt{2})\right)^2$

(A.88) $\displaystyle\int_{\sqrt{3}/2}^{1}\frac{\arctan(2x)}{\sqrt{1-x^2}}dx=\frac{\pi^2}{8}-\frac{1}{8}\left(\ln\frac{\sqrt{5}+1}{\sqrt{5}-1}\right)^2-\frac{1}{2}\left(\arctan\sqrt{3}\right)^2.$

(A.89) $\displaystyle\int_{0}^{1}\frac{\arcsin x}{1+x^2}dx=\frac{1}{2}\left(\ln(1+\sqrt{2})\right)^2$

Example 2.4.1. In the same place

(A.90) $\displaystyle\int_{1}^{\infty}\frac{\arctan t}{t\sqrt{t^2-1}}dt=\frac{\pi^2}{8}+\frac{1}{2}\left(\ln(1+\sqrt{2})\right)^2.$

(A.91) $\displaystyle\int_{1/\alpha}^{\infty}\frac{\ln(\alpha x+\sqrt{\alpha^2 x^2-1})}{x(1+x^2)}dx=\frac{1}{2}\ln^2(\sqrt{1+\alpha^2}+\alpha)$

($\alpha>0$) Example 2.4.2. In particular,

(A.92) $\displaystyle\int_{1}^{\infty}\frac{\ln(x+\sqrt{x^2-1})}{x(1+x^2)}dx=\frac{1}{2}\ln^2(\sqrt{2}+1)$

(A.93) $\displaystyle\int_{2}^{\infty}\frac{\ln(x+\sqrt{x^2-4})}{x(1+x^2)}dx=\frac{1}{2}\ln^2\frac{\sqrt{5}+1}{2}+\ln 2\ln\frac{\sqrt{5}}{2}.$

(A.94) $\displaystyle\int_{1}^{\infty}\frac{\ln(x+\sqrt{x^2-1})}{x\sqrt{x^2-1}}dx=2G$

Example 2.4.2.

(A.95) $\displaystyle\int_{0}^{1}\frac{\arcsin\alpha x}{\sqrt{1-x^2}}dx=\frac{1}{2}\left[\mathrm{Li}_2(\alpha)-\mathrm{Li}_2(-\alpha)\right]$ ($|\alpha|\le 1$)

Section 2.5.1.

(A.96) $\displaystyle 2\alpha\int_{0}^{1}\frac{\arccos x}{\sqrt{1-\alpha^2 x^2}}dx=\mathrm{Li}_2(\alpha)-\mathrm{Li}_2(-\alpha)$ ($|\alpha|\le 1$)

Section 2.5.1.

(A.97) $$\int_0^1 \frac{x \arccos x}{\sqrt{\alpha^2 - x^2}} dx = \frac{\pi \alpha}{2} - \alpha \, \text{E}\left(\frac{\pi}{2}, \frac{1}{\alpha}\right) \quad (\alpha \geq 1)$$

Example 2.5.1. Here $E(\mu, k) = \int_0^\mu \sqrt{1 - k^2 \sin^2 t} \, dt$. Also there

(A.98) $$\int_0^1 \frac{x \arcsin x}{\sqrt{\alpha^2 - x^2}} dx = -\frac{\pi \sqrt{\alpha^2 - 1}}{2} + \alpha \, \text{E}\left(\frac{\pi}{2}, \frac{1}{\alpha}\right).$$

(A.99) $$2\int_0^\infty \frac{\arctan \alpha x}{1 + x^2} dx = \ln \alpha \ln \frac{1 - \alpha}{1 + \alpha} + \text{Li}_2(\alpha) - \text{Li}_2(-\alpha)$$

$0 \leq \alpha \leq 1$. Example 2.5.2

(A.100) $$2\int_0^1 \frac{\arctan \alpha x}{1 + x^2} dx = \sum_{n=0}^\infty (\ln 2 - H_n^-) \frac{\alpha^{2n+1}}{2n+1}$$

$(|\alpha| \leq 1)$. Example 2.5.3.

(A.101) $$\int_0^\infty \frac{\ln(1 + \alpha x)}{x(1 + x)} dx = \ln \alpha \ln(1 - \alpha) + \text{Li}_2(\alpha)$$

$(0 \leq \alpha \leq 1)$ Example 2.5.4. In particular,

(A.102) $$\int_0^\infty \frac{\ln(1 + x)}{x(1 + x)} dx = \frac{\pi^2}{6}.$$

(A.103) $$\int_0^1 \frac{\ln(1 + \alpha x)}{x(1 + x)} dx = \text{Li}_2\left(\frac{1}{2}\right) - \text{Li}_2\left(\frac{1 - \alpha}{2}\right)$$

$(-1 \leq \alpha \leq 1)$ Example 2.5.5. In particular,

(A.104) $$\int_0^1 \frac{\ln(1 + x)}{x(1 + x)} dx = \frac{\pi^2}{12} - \frac{1}{2}\ln^2 2.$$

(A.105)
$$\int_0^1 \frac{1}{x}(\log x)^p \log\frac{1-\beta x}{1+\beta x}\,dx$$

$$=(-1)^{p+1}\Gamma(p+1)\left\{\mathrm{Li}_{p+2}(\beta)-\mathrm{Li}_{p+2}(-\beta)\right\}$$

$(|\beta|\le 1,\ p\ge 0)$. Example 2.5.6. Also there

(A.106)
$$\int_0^1 \frac{1}{x}(\log x)^p \log(1-\beta x)\,dx$$

$$=(-1)^{p+1}\Gamma(p+1)\mathrm{Li}_{p+2}(\beta)$$

(A.107)
$$\zeta(3)=\frac{4}{7}\int_0^\infty \frac{1}{t}\arctan t\ \arctan\frac{1}{t}\,dt$$

(A.108)
$$\zeta(3)=\frac{8}{7}\int_0^1 \frac{1}{t}\arctan t\ \arctan\frac{1}{t}\,dt$$

(A.109)
$$\zeta(3)=\frac{1}{2}\int_0^\infty \frac{1}{t}\log(1+t)\log\left(1+\frac{1}{t}\right)dt$$

(A.110)
$$\zeta(3)=\int_0^1 \frac{1}{t}\log(1+t)\log\left(1+\frac{1}{t}\right)dt\,.$$

(A.111)
$$\int_0^x \frac{1}{1-t}\ln\left(\frac{1+t}{2}\right)dt=\mathrm{Li}_2\left(\frac{1-x}{2}\right)-\mathrm{Li}_2\left(\frac{1}{2}\right)$$

Example 2.5.7. This integral can be put also in the form

(A.112)
$$\int_0^x \frac{\ln(1+t)}{1-t}\,dt=\mathrm{Li}_2\left(\frac{1-x}{2}\right)-\mathrm{Li}_2\left(\frac{1}{2}\right)-\ln 2\ln(1-x)\,.$$

(A.113)
$$\int_0^x \ln(1+t)\ln(1-t)\,dt$$

$$= \text{Li}_2\left(\frac{1-x}{2}\right) - \text{Li}_2\left(\frac{1+x}{2}\right) + (1 - \ln 2)\ln\frac{1-x}{1+x}$$

$$+ x\ln(1+x)\ln(1-x) - x\ln(1-x^2) + 2x.$$

Here $|x| < 1$. In particular, with $x \to 1$

$$\int_0^1 \ln(1+t)\ln(1-t)\,dt = 2 - \frac{\pi^2}{6} + (\ln 2)^2 - 2\ln 2$$

Example 2.5.7.

(A.114) $$\int_0^{\pi/2} \ln(\sin t)\,dt = \int_0^{\pi/2} \ln(\cos t)\,dt = \frac{-\pi}{2}\ln 2$$

Example 3.2.1. Also there

(A.115) $$\int_0^{\pi} \ln(\sin t)\,dt = -\pi \ln 2$$

(A.116) $$\int_0^{\pi/4} \ln(\sin t)\,dt = \frac{-\pi}{4}\ln 2 - \frac{1}{2}G$$

(A.117) $$\int_0^{\pi/4} \ln(\cos t)\,dt = \frac{-\pi}{4}\ln 2 + \frac{1}{2}G$$

(A.118) $$\int_0^{\pi/4} t\ln(\sin t)\,dt = \frac{-\pi^2}{32}\ln 2 - \frac{\pi}{8}G + \frac{35}{128}\zeta(3)$$

(A.119) $$\int_0^{\pi/2} t\ln(\sin t)\,dt = \frac{-\pi^2}{8}\ln 2 + \frac{7}{16}\zeta(3)$$

(A.120) $$\int_0^{\pi} t\ln(\sin t)\,dt = \frac{-\pi^2}{2}\ln 2$$

(A.121) $$\int\limits_0^{\pi/2} \sin t \ln(\sin t)\,dt = \ln 2 - 1$$

(A.122) $$\int\limits_0^1 \frac{x \ln x}{\sqrt{1-x^2}}\,dx = \ln 2 - 1$$

(A.123) $$\int\limits_0^1 \frac{\ln x}{\sqrt{4-x^2}}\,dx = -\frac{1}{2}\mathrm{Cl}_2\left(\frac{\pi}{3}\right).$$

(A.124) $$\int\limits_0^\pi \cos x \ln \cot \frac{x}{2}\,dx = \pi$$

Example 3.2.2. Also there

(A.125) $$\int\limits_0^{\pi/2} \cos x \ln \cot \frac{x}{2}\,dx = \frac{\pi}{2}$$

(A.126) $$\int\limits_0^{\pi/4} \cos x \ln \cot \frac{x}{2}\,dx = \frac{\pi}{4} + \frac{\sqrt{2}}{2}\ln(1+\sqrt{2}).$$

(A.127) $$\int\limits_0^1 \arctan(x^2)\,dx = \frac{\pi(1-\sqrt{2})}{4} + \frac{\sqrt{2}}{2}\ln(1+\sqrt{2})$$

Example 3.2.3. Also there

(A.128) $$\int\limits_0^1 \frac{x^2}{1+x^4}\,dx = \frac{\pi\sqrt{2}}{8} - \frac{\sqrt{2}}{4}\ln(1+\sqrt{2})$$

(A.129) $$\int\limits_0^{\pi/4} \sqrt{\tan\theta}\,d\theta = \frac{\pi\sqrt{2}}{4} - \frac{\sqrt{2}}{2}\ln(1+\sqrt{2}).$$

(A.130) $$\int\limits_0^{\pi p} \log(1 - 2\alpha \cos x + \alpha^2)\,dx = 0$$

($|\alpha| < 1$, p any integer). Example 3.2.4. Also there

(A.131) $$\int_0^{\pi p} \ln(1 - \cos x)\,dx = -\pi p \ln 2$$

(A.132) $$\int_0^{\pi/2} \log(1 - 2\alpha \cos x + \alpha^2)\,dx = i\left(\mathrm{Li}_2(i\alpha) - \mathrm{Li}_2(-i\alpha)\right)$$

(A.133) $$\int_0^{\pi/2} \ln(1 - \cos x)\,dx = -\frac{\pi}{2}\ln 2 - 2G$$

(A.134) $$\int_0^{\pi p} \ln(1 - 2\alpha \cos x + \alpha^2)\,dx = 2\pi p \ln|\alpha|$$

$(|\alpha| > 1)$. Also

(A.135) $$\int_0^{\pi p} \ln(\beta^2 - 2\alpha\beta \cos x + \alpha^2)\,dx = 2\pi p \ln|\alpha|$$

$(|\beta| \le |\alpha|,\ p$ any integer).

(A.136) $$\int_0^{\pi p} \cos mx \log(1 - 2\alpha \cos x + \alpha^2)\,dx = \frac{-\pi p}{m}\alpha^m$$

$(m > 0,\ p$ integers). Example 3.2.5.

(A.137) $$\int_0^{\pi p} \ln(\cosh \beta \pm \cos x)\,dx = (\beta - \ln 2)p\pi$$

$(\beta > 0,\ p$ arbitrary integer) Example 3.2.6.

(A.138) $$\int_0^\infty \frac{\log(1 - 2\alpha \cos x + \alpha^2)}{x^2 + b^2}\,dx = \frac{\pi}{b}\log(1 - \alpha e^{-b})$$

$(|\alpha| < 1, \mathrm{Re}\,b > 0)$ Example 3.2.7.

(A.139) $$\int_0^\infty x^p \log(1 - ae^{-\lambda x})\,dx = -\frac{\Gamma(p+1)}{\lambda^{p+1}}\mathrm{Li}_{p+2}(a)$$

Example 3.2.8. Also there with $\lambda > 0, b \neq 0$

(A.140a)
$$\int_0^\infty \sin bx \log(1 - e^{-\lambda x}) \, dx$$

$$= \frac{-1}{2b}\left[\psi\left(1 + \frac{ib}{\lambda}\right) + \psi\left(1 - \frac{ib}{\lambda}\right) + 2\gamma\right]$$

(A.140b)
$$\int_0^\infty \sin bx \log(1 + e^{-\lambda x}) \, dx$$

$$= \frac{-1}{4b}\left[4\ln 2 - \psi\left(1 + \frac{ib}{2\lambda}\right) - \psi\left(1 - \frac{ib}{2\lambda}\right) + \psi\left(\frac{\lambda + ib}{2\lambda}\right) + \psi\left(\frac{\lambda - ib}{2\lambda}\right)\right]$$

(A.141)
$$\int_0^\infty \cos bx \log(1 - e^{-\lambda x}) \, dx = \frac{\lambda}{2b^2} - \frac{\pi}{2b}\coth\left(\frac{\pi b}{\lambda}\right).$$

(A.142)
$$\int_0^\pi e^{at} \ln(\sin t) \, dt$$

$$= \frac{1 - e^{a\pi}}{a}\left\{\ln 2 + \gamma + \frac{1}{2}\left[\psi\left(1 + \frac{ia}{\lambda}\right) + \psi\left(1 - \frac{ia}{\lambda}\right)\right]\right\}$$

($a \neq 0$) Example 3.2.9. Also there

(A.143)
$$\int_{-\pi/2}^{\pi/2} e^{at} \ln(\cos t) \, dt$$

$$= \frac{-1}{a}\sinh\frac{\pi a}{2}\left\{2\ln 2 + 2\gamma + \psi\left(1 + \frac{ia}{\lambda}\right) + \psi\left(1 - \frac{ia}{\lambda}\right)\right\}$$

(A.144)
$$\int_0^{\pi/2} e^{at} \ln(\cos t) \, dt$$

$$= \frac{1 - e^{a\pi/2}}{a}\ln 2 - \frac{e^{a\pi/2}}{2a}\left[\psi\left(1 + \frac{ia}{\lambda}\right) + \psi\left(1 - \frac{ia}{\lambda}\right) + 2\gamma\right]$$

$$+\frac{1}{2a}\left[\beta\left(1+\frac{ia}{2}\right)+\beta\left(1-\frac{ia}{2}\right)-2\beta(1)\right]$$

(A.145)
$$\int_0^{\pi/2} e^{at}\ln(\sin t)\,dt$$

$$=\frac{1-e^{a\pi/2}}{a}\ln 2-\frac{e^{a\pi/2}}{2a}\left[\beta\left(1+\frac{ia}{\lambda}\right)+\beta\left(1-\frac{ia}{\lambda}\right)-2\beta(1)\right]$$

$$+\frac{1}{2a}\left[\psi\left(1+\frac{ia}{\lambda}\right)+\psi\left(1-\frac{ia}{\lambda}\right)+2\gamma\right]$$

(A.146)
$$\int_0^1 \frac{\ln(1+x+x^2)}{x}=\frac{\pi^2}{9}$$

Example 3.2.10.

(A.147) $$\int_0^\infty \frac{\ln(1-e^{-2\pi\lambda x})}{1+x^2}\,dx=\pi\left[\ln\Gamma(\lambda)-\left(\lambda-\frac{1}{2}\right)\ln\lambda+\lambda-\frac{\ln 2\pi}{2}\right]$$

($\operatorname{Re}\lambda>0$) Example 3.2.11. Also there

(A.148) $$2\lambda\int_0^\infty \frac{\arctan(x)}{e^{2\pi\lambda x}-1}\,dx=\ln\Gamma(\lambda)-\left(\lambda-\frac{1}{2}\right)\ln\lambda+\lambda-\frac{\ln 2\pi}{2}$$

and in particular,

(A.149) $$\int_0^\infty \frac{\arctan(x)}{e^{2\pi x}-1}\,dx=\frac{1}{2}-\frac{\ln 2\pi}{4}.$$

(A.150) $$\int_0^\infty \left(\frac{1}{\sinh t}-\frac{1}{t}\right)\frac{\cos\mu t}{t}\,dt=-\ln(1+e^{-\pi\mu})$$

($\mu>0$) Example 3.2.12. In particular

(A.151) $$\int_0^\infty \left(\frac{1}{\sinh t} - \frac{1}{t} \right) \frac{dt}{t} = -\ln 2$$

and also there

(A.152) $$\int_0^\infty \left(\frac{1}{\sinh t} - \frac{1}{t} \right) \sin \mu t \, dt = \frac{-\pi e^{-\pi\mu}}{1 + e^{-\pi\mu}} = \frac{-\pi}{e^{\pi\mu} + 1}.$$

(A.153) $$\int_0^1 \ln \Gamma(x) dx = \frac{\ln \pi}{2} + \frac{\ln 2}{2} = \frac{1}{2} \ln 2\pi$$

Example 3.2.13. Also there

(A.154) $$\int_0^{1/2} \ln \Gamma(x) \, dx = \frac{\gamma + 3\ln \pi}{8} + \frac{\ln 2}{3} - \frac{3}{4\pi^2} \zeta'(2)$$

(A.155) $$\int_0^{1/4} \ln \Gamma(x) \, dx = \frac{3\gamma + 7\ln 2\pi}{32} + \frac{G}{4\pi} - \frac{9\zeta'(2)}{16\pi^2}.$$

(A.156) $$\int_0^1 \ln \Gamma(x+y) dx = \int_y^{y+1} \ln \Gamma(u) \, du = y \ln y - y + \ln \sqrt{2\pi}$$

Example 3.2.14. Also there

(A.157) $$\int_0^1 \int_0^1 \ln \Gamma(x+y) \, dx dy = \ln \sqrt{2\pi} - \frac{3}{4}$$

(Problem 904 from CMJ).

(A.158) $$\int_0^\infty 2^{-x} \Gamma(x) dx = 2 \int_0^1 2^{-x} \Gamma(x) dx - \frac{\gamma + \ln \ln 2}{\ln 2}$$

(A.159) $$\int_0^\infty x 2^{-x} \Gamma(x) dx$$

$$= 2\int_0^1 (x+1)2^{-x}\Gamma(x)\,dx - \frac{(\gamma + \ln\ln 2)(1 + 2\ln 2) - 1}{\ln^2 2}$$

Example 3.2.15. This is Problem 11329 from AMM.

(A.160) $$\int_0^1 \ln(\arcsin x)\,dx = \ln\frac{\pi}{2} - \mathrm{Si}\!\left(\frac{\pi}{2}\right)$$

Example 3.2.16. Also there

(A.161) $$\int_0^1 \ln x(\arcsin x)\,dx = 2 - \frac{\pi}{2} - \ln 2$$

(A.162) $$\int_0^1 \ln x(\arccos x)\,dx = \ln 2 - 2\,.$$

(A.163) $$\int_0^1 \ln x(\arctan x)\,dx = \frac{\pi^2}{48} + \frac{\ln 2}{2} - \frac{\pi}{4}$$

Example 3.2.17.

(A.164) $$\int_0^\pi \ln^2(2\sin t)\,dt = \frac{\pi^3}{12}$$

Example 3.2.18. Also there

(A.165) $$\int_0^{\pi/2} \ln^2(2\sin t)\,dt = \frac{\pi^3}{24}$$

This is Problem 11639 from AMM.

(A.166) $$\int_0^1 \ln x\ln(1-\lambda x)\,dx = \frac{1-\lambda}{\lambda}\ln(1-\lambda) - \frac{1}{\lambda}\mathrm{Li}_2(\lambda) + 2$$

($|\lambda| < 1$) **Example 3.2.19.** With $\lambda \to 1$ and $\lambda \to -1$

(A.167) $$\int_0^1 \ln x \ln(1-x)\, dx = 2 - \frac{\pi^2}{6}$$

(A.168) $$\int_0^1 \ln x \ln(1+x)\, dx = -2\ln 2 + 2 - \frac{\pi^2}{12}.$$

(A.169) $$\int_0^{\pi/2} \frac{\ln(\sin t)}{\cos t}\, dt = -\frac{\pi^2}{8}$$

Example 3.2.20.

(A.170) $$\ln \Gamma\left(z + \frac{1}{2}\right) = z \ln z - z + \ln\sqrt{2\pi} - 2\int_0^\infty \frac{\arctan(x/z)}{e^{2\pi x} + 1}\, dx$$

Section 3.3. Also there

(A.171) $$\psi\left(z + \frac{1}{2}\right) = \ln z + 2\int_0^\infty \frac{x}{(x^2 + z^2)(e^{2\pi x} + 1)}\, dx$$

(A.172) $$\int_0^\infty \frac{\arctan(x)}{e^{2\pi x} + 1}\, dx = \frac{3}{4}\ln 2 - \frac{1}{2}.$$

(A.173) $$\int_0^\infty \frac{x - \sin x}{x^3}\, dx = \frac{\pi}{4}$$

Example 4.2.1.

(A.174) $$\int_0^\infty \left(\frac{\sin x}{x}\right)^3 dx = \frac{3\pi}{8}$$

Example 4.2.2.

(A.175) $$\int_0^\infty \frac{t^{u-1}}{1+t}\, dt = \frac{\pi}{\sin \pi u} \quad (0 < u < 1)$$

Example 4.2.3. Also there

(A.176) $\displaystyle\int_0^\infty \frac{x^{-q}}{s^2+x^2}dx = \frac{\pi}{2s^{q+1}}\sec\frac{\pi q}{2},\quad -1<q<1$

(A.177) $\displaystyle\int_0^\infty x^{-p}\cos(xt)dx = \frac{\pi t^{p-1}}{2\Gamma(p)}\sec\frac{\pi p}{2}$

(A.178) $\displaystyle\int_0^\infty x^{-p}\sin(xt)dx = \frac{\pi t^{p-1}}{2\Gamma(p)}\csc\frac{\pi p}{2}$

$(0<p<1)$. For $\alpha>1$

(A.179) $\displaystyle\int_0^\infty \cos(z^\alpha)dz = \frac{1}{\alpha}\Gamma\left(\frac{1}{\alpha}\right)\cos\frac{\pi}{2\alpha}$

(A.180) $\displaystyle\int_0^\infty \sin(z^\alpha)dz = \frac{1}{\alpha}\Gamma\left(\frac{1}{\alpha}\right)\sin\frac{\pi}{2\alpha}$

(A.181) $\displaystyle\int_0^\infty x^s\cos xt\,dx = -\frac{\Gamma(s+1)}{t^{s+1}}\sin\frac{\pi s}{2}$

$(t>0,\ -1<\operatorname{Re}s<0)$.

(A.182) $\displaystyle\int_0^\infty\int_x^\infty e^{-(x-y)^2}\sin^2(x^2+y^2)\frac{x^2-y^2}{(x^2+y^2)^2}dydx$

$$=\frac{1}{16}\log\frac{1}{5}-\frac{1}{4}\arctan\frac{1}{2}$$

Example 4.2.4. This is Problem 11650 from AMM. Also there

(A.183) $\displaystyle\int_0^\infty e^{-st}\left(\frac{\sin t}{t}\right)^2 dt = \frac{s}{4}\log\frac{s^2}{s^2+4}-\arctan\frac{s}{2}+\frac{\pi}{2}.$

(A.184a) $\displaystyle\int_0^\infty t^{x-1}e^{-at}\cos bt\,dt = \frac{\Gamma(x)}{(a^2+b^2)^{\frac{x}{2}}}\cos\left(x\arctan\frac{b}{a}\right)$

(A.184b) $\displaystyle\int_0^\infty t^{x-1}e^{-at}\sin bt\,dt = \dfrac{\Gamma(x)}{(a^2+b^2)^{\frac{x}{2}}}\sin\left(x\arctan\dfrac{b}{a}\right)$

$(a,b,x>0)$ Example 4.2.5. Also in this example

(A.185a) $\displaystyle\int_0^\infty e^{-ax}\dfrac{\cos bx}{\sqrt{x}}\,dx = \sqrt{\dfrac{\pi}{2}}\,\dfrac{\sqrt{\sqrt{b^2+a^2}+a}}{\sqrt{b^2+a^2}}$

(A.185b) $\displaystyle\int_0^\infty e^{-ax}\dfrac{\sin bx}{\sqrt{x}}\,dx = \sqrt{\dfrac{\pi}{2}}\,\dfrac{\sqrt{\sqrt{b^2+a^2}-a}}{\sqrt{b^2+a^2}}$.

(A.186) $\displaystyle\int_0^\infty \dfrac{\sin(xt)}{x^2+1}\,dx$

$\displaystyle = \dfrac{1}{2}\left[e^t\,\mathrm{Ein}(t)-e^{-t}\,\mathrm{Ein}(-t)\right]-(\ln t+\gamma)\sinh t$

$\displaystyle = -(\ln t+\gamma)\sinh t + \sum_{n=0}^{\infty}\dfrac{t^{2n+1}}{(2n+1)}H_{2n+1}$.

(A.187) $\displaystyle\int_0^\infty \dfrac{\sin^{2n}(xt)}{x^2+1}\,dx$

$\displaystyle = \dfrac{\pi}{2^{2n+1}}\left\{\sum_{k=0}^{n-1}\binom{2n}{k}(-1)^{n-k}2e^{-2(n-k)t}+\binom{2n}{n}\right\}$

and in particular, for $n=1$

$\displaystyle\int_0^\infty \dfrac{\sin^2(xt)}{x^2+1}\,dx = \dfrac{\pi}{4}(1-e^{-2t})$

(A.188) $\displaystyle\int_0^\infty \dfrac{\cos^{2n}(xt)}{x^2+1}\,dx = \dfrac{\pi}{2^{2n+1}}\left\{\sum_{k=0}^{n-1}\binom{2n}{k}2e^{-2(n-k)t}+\binom{2n}{n}\right\}$

(A189)
$$\int_0^\infty \frac{\cos^{2n-1}(xt)}{x^2+1}dx = \frac{\pi}{2^{2n-1}}\sum_{k=0}^{n-1}\binom{2n-1}{k}e^{-(2n-2k-1)t}$$

Section 4.3.

(A.190)
$$\int_0^\infty \ln\left(1+\frac{a^2}{x^2}\right)\frac{dx}{x^2+b^2} = \frac{\pi}{b}\ln\frac{a+b}{b} \quad (a,b>0)$$

Example 4.4.1.

(A.191)
$$\int_0^\infty \frac{e^{-y^2}}{y^2+s^2}dy = \frac{\pi e^{s^2}}{2s}(1-\mathrm{erf}(s))$$

and in particular, with $s=1$

$$\int_0^\infty \frac{e^{-y^2}}{y^2+1}dy = \sqrt{\pi}e\int_1^\infty e^{-u^2}du$$

Example 4.4.2.

(A.192)
$$\int_0^\infty \left(\frac{1}{y^2}-\frac{\pi^2}{\sinh^2 \pi y}\right)y^{-\alpha}dy = \pi\alpha\sec\frac{\pi\alpha}{2}\zeta(1+\alpha)$$

($0<\alpha<1$). Example 4.4.3. In particular,

(A.193)
$$\int_0^\infty \left(\frac{1}{y^2}-\frac{\pi^2}{\sinh^2 \pi y}\right)dy = \pi .$$

(A.194)
$$\int_0^\infty x^{-\beta}\ln\left(1+\frac{a^2}{x^2}\right)dx = \frac{\pi}{a^\beta(1-\beta)}\sec\frac{\pi\beta}{2}$$

($a>0, 0<\beta<1$) Example 4.4.4.

(A.195)
$$\int_0^\infty \mathrm{sech}\left(\frac{\pi x}{2}\right)\frac{dx}{x^2+b^2} = \frac{1}{b}\beta\left(\frac{b+1}{2}\right)$$

Example 4.4.6.

(A.196)
$$\int_0^\infty \frac{\sin xt}{\sinh \dfrac{\pi t}{2}} dt = \tanh x$$

Example 4.5.1.

(A.197)
$$\int_0^\infty \frac{\sin xt}{e^t - 1} dt = \frac{\pi}{2} + \frac{\pi}{e^{2\pi x} - 1} - \frac{1}{2x}$$

$(x > 0)$. Also in the form

(A.198)
$$\int_0^\infty \frac{\sin xt}{e^t - 1} dt = \frac{\pi}{2} \coth \pi x - \frac{1}{2x}$$

Example 4.5.2. In the same place also

(A.199)
$$\int_0^\infty \frac{\sin xt}{e^t + 1} dt = \frac{1}{2x} - \frac{\pi}{2} \operatorname{csch} \frac{\pi x}{2}.$$

(A.200)
$$\int_0^\infty \frac{e^{-by}}{\cosh y} dy = \beta \left(\frac{b+1}{2} \right)$$

$$= \frac{1}{2} \left[\psi \left(\frac{b+3}{4} \right) - \psi \left(\frac{b+1}{4} \right) \right]$$

$$= 2 \int_0^{\pi/4} \tan^b x \, dx.$$

Section 4.5.2. Also in this section

(A.201)
$$\int_0^\infty \frac{\cos(xy)}{\cosh x} dx = \frac{\pi}{2} \operatorname{sech} \left(\frac{\pi y}{2} \right)$$

(A.202)
$$\int_0^\infty \frac{\sin(xy)}{x \cosh x} dx = 2 \arctan \left(e^{\frac{\pi y}{2}} \right) - \frac{\pi}{2}$$

(A.203)
$$\int_0^\infty \frac{\sin(xy)}{\cosh x}\,dx = \operatorname{Im}\beta\!\left(\frac{1-iy}{2}\right).$$

(A.204)
$$\int_{-\infty}^\infty e^{-i\lambda t}\,\Gamma(1+it)\,dt = 2\pi e^{-e^{\lambda}+\lambda}$$

($\lambda \in \mathbb{R}$) Example 4.6.1. Also there

(A.205)
$$\int_{-\infty}^\infty e^{-i\lambda t} t^n \,\Gamma(1+it)\,dt = -2\pi\, i^n \varphi_{n+1}(-e^{-\lambda})\,e^{-e^{\lambda}}$$

(A.206) $$\int_{-\infty}^\infty e^{-i\lambda t} t^n \,\Gamma(a+it)\,dt = 2\pi i^n e^{a\lambda} e^{-e^{\lambda}} \sum_{k=0}^n \binom{n}{k}\varphi_k(-e^{\lambda})\,a^{n-k}$$

($a>0,\, n=0,1,2,\dots$). Here

$$\varphi_n(x) = \sum_{k=0}^n \binom{n}{k} S(n,k)\,x^k$$

and $S(n,k)$ are the Stirling numbers of the second kind.

(A.207) $$\int_{-\infty}^\infty e^{-i\lambda t} e^{i\mu t}\,\Gamma(b+it)\,dt = 2\pi e^{b\lambda} e^{-b\mu} e^{-e^{\lambda}e^{-\mu}}$$

($b>0,\ \lambda,\mu \in \mathbb{R}$) Example 4.6.2. Also there

(A.208)
$$\int_{-\infty}^\infty e^{-i\mu t} t^n \,\Gamma(a+it)\,\Gamma(b-it)\,dt$$

$$= 2\pi i^n e^{-b\mu} \sum_{k=0}^n \binom{n}{k} a^{n-k} \sum_{j=0}^k S(k,j)\,(-1)^j \frac{\Gamma(a+b+j)}{(1+e^{-\mu})^{a+b+j}}$$

($a,b>0$). In particular,

(A.209)
$$\int_{-\infty}^\infty t^n \,\Gamma(a+it)\,\Gamma(b-it)\,dt$$

$$= \pi i^n \sum_{k=0}^{n} \binom{n}{k} a^{n-k} \sum_{j=0}^{k} S(k,j)(-1)^j \frac{\Gamma(a+b+j)}{2^{a+b+j-1}}$$

(A.210) $\quad \int_{-\infty}^{\infty} e^{-i\mu t} \Gamma(a+it)\Gamma(b-it)dt = 2\pi\,\Gamma(a+b)\dfrac{e^{-b\mu}}{(1+e^{-\mu})^{a+b}}$.

(A.211) $\qquad\qquad \int_{-\infty}^{\infty} e^{-i\mu t} \dfrac{\Gamma(b+it)}{a^2+t^2}dt$

$$= \frac{2\pi^2}{a}\left[e^{-a\mu}\gamma(a+b,e^\mu)+e^{a\mu}\Gamma(b-a,e^\mu)\right]$$

Example 4.6.3. In particular,

(A.212) $\qquad\qquad \int_{-\infty}^{\infty} e^{-i\mu t} \dfrac{\Gamma(a+it)}{a^2+t^2}dt$

$$= \frac{2\pi^2}{a}\left[e^{-a\mu}\gamma(2a,e^\mu)-e^{a\mu}\,\mathrm{Ei}(-e^\mu)\right]$$

(A.213) $\quad \int_{-\infty}^{\infty} \dfrac{\Gamma(b+it)}{a^2+t^2}\,dt = \dfrac{2\pi^2}{a}\left(\gamma(a+b,1)+\Gamma(b-a,1)\right)$.

(A.214) $\qquad\qquad \int_{0}^{\infty} \dfrac{x^{s-1}}{e^{bx}-1}dx = b^{-s}\Gamma(s)\zeta(s)$

Section 4.7. Here and further $b > 0$. Also in this section

(A.215) $\qquad\qquad \int_{0}^{\infty} \dfrac{x^{s-1}}{e^{bx}+1}dx = b^{-s}\Gamma(s)\eta(s)$

and in particular, for $s = 1$

$$\int_{0}^{\infty} \frac{1}{e^{bx}+1}dx = \frac{\ln 2}{b}$$

(A.216) $$\int_0^\infty \frac{x^{s-1}}{\sinh bx}\,dx = 2b^{-s}(1-2^{-s})\Gamma(s)\zeta(s)$$

(A.217) $$\int_0^\infty x^{s-1}(1-\tanh bx)\,dx = 2(2b)^{-s}\Gamma(s)\eta(s)$$

in particular,

$$\int_0^\infty (1-\tanh bx)\,dx = \frac{\ln 2}{b}$$

(A.218) $$\int_0^\infty x^{s-1}(\coth bx - 1)\,dx = 2(2b)^{-s}\Gamma(s)\zeta(s)$$

(A.219) $$\int_0^\infty \frac{x^s}{\cosh^2 bx}\,dx = \frac{4}{(2b)^{s+1}}\Gamma(s+1)\eta(s).$$

From this, with $s = 1, 2$ we have

$$\int_0^\infty \frac{x}{\cosh^2 bx}\,dx = \frac{\ln 2}{b^2}$$

$$\int_0^\infty \left(\frac{x}{\cosh bx}\right)^2 dx = \frac{\pi^2}{12b^3}$$

(A.220) $$\int_0^\infty \frac{x^s}{\sinh^2 bx}\,dx = \frac{4}{(2b)^{s+1}}\Gamma(s+1)\zeta(s)$$

(A.221) $$\int_0^\infty \frac{t\cos xt}{\sinh \dfrac{\pi t}{2}}\,dt = \frac{1}{\cosh^2 x}.$$

(A.222) $$\int_0^\infty \frac{\ln x}{e^{bx}+1}\,dx = -\frac{\ln 2}{2b}(\ln 2 + 2\ln b)$$

Example 4.7.1.

(A.223) $\quad \int\limits_0^\infty \ln x \, (1 - \tanh bx) \, dx = -\dfrac{\ln 2}{2b}(3\ln 2 + 2\ln b)$.

(A.224) $\quad \int\limits_0^\infty \dfrac{x \ln x}{e^{bx} - 1} \, dx = \dfrac{\pi^2}{6b^2}\left(\ln \dfrac{2\pi}{b} + 1 - 12 \ln A \right)$.

In particular, with $b = 2\pi$

(A.225) $\quad \int\limits_0^\infty \dfrac{x \ln x}{e^{2\pi x} - 1} \, dx = \dfrac{1}{2}\zeta'(-1)$

Example 4.7.2.

(A.226) $\quad \int\limits_0^\infty \dfrac{x^{s-1}}{\cosh x} \, dx = 2\Gamma(s)L(s)$

Section 4.8.

(A.227) $\quad \int\limits_{\pi/4}^{\pi/2} \ln(\ln \tan x) \, dx = \dfrac{\pi}{2}\ln\left(\dfrac{\Gamma(3/4)}{\Gamma(1/4)}\sqrt{2\pi} \right) \equiv J$

Example 4.9.1. Also there, with the same value of J

(A.228) $\quad \int\limits_1^\infty \dfrac{\ln(\ln u)}{1 + u^2} \, du = J$

(A.229) $\quad \int\limits_0^1 \dfrac{1}{1 + u^2}\ln\left(\ln \dfrac{1}{u} \right) du = \int\limits_0^1 \dfrac{\ln(-\ln u)}{1 + u^2} \, du = J$

(A.230) $\quad \int\limits_0^\infty \dfrac{\ln u}{\cosh u} \, du = 2J$.

(A.231) $\quad \int\limits_0^1 \left(\ln \dfrac{1}{x} \right)^{\mu-1} \ln\left(\ln \dfrac{1}{x} \right) dx = \psi(\mu)\Gamma(\mu) \quad (\mathrm{Re}\,\mu > 0)$

Example 4.9.2. In particular,

(A.232)

$$\int_0^1 \ln\left(\ln\frac{1}{x}\right) dx = \psi(1) = -\gamma$$

(A.233)

$$\int_0^1 \left(\ln\frac{1}{x}\right)^{-1/2} \ln\left(\ln\frac{1}{x}\right) dx = -(\gamma + 2\ln 2)\sqrt{\pi}.$$

(A.234)

$$\int_0^\infty \frac{x^{2n}}{\cosh x} dx = (-1)^n \left(\frac{\pi}{2}\right)^{2n+1} E_{2n}$$

Example 4.9.4. (E_k, $k = 0, 1, 2,...$ are the Euler numbers). Also there

(A.235)

$$\int_0^\infty \frac{\sqrt{x}}{\cosh x} dx = \sqrt{\pi} \sum_{n=0}^\infty \frac{(-1)^n}{(2n+1)^{3/2}}$$

(A.236)

$$\int_0^\infty \frac{1}{\sqrt{x}\cosh x} dx = 2\sqrt{\pi} \sum_{n=0}^\infty \frac{(-1)^n}{\sqrt{2n+1}}.$$

(A.237)

$$\int_{-\infty}^\infty \frac{\cos at}{x^2 + (y-t)^2} dt = \frac{\pi}{x} e^{-ax} \cos ay$$

(A.238)

$$\int_{-\infty}^\infty \frac{\sin at}{x^2 + (y-t)^2} dt = \frac{\pi}{x} e^{-ax} \sin ay$$

($a, x > 0$) Example 5.1.1.

(A.239)

$$\int_0^\infty \frac{\cos bt \cos at}{x^2 + t^2} dt = \frac{\pi}{2x} e^{-bx} \cosh ax$$

($0 < a \le b, \ x > 0$) Example 5.1.2.

(A.240)

$$\int_0^\infty \frac{\arctan \alpha t}{t} \frac{dt}{x^2 + t^2} = \frac{\pi}{2} \frac{\log(1 + \alpha x)}{x^2} \quad (\alpha, x > 0)$$

Example 5.1.3.

(A.241) $$\int_0^\infty \frac{\sin\alpha t \cos\beta t}{t} \frac{dt}{x^2+t^2} = \frac{\pi}{2x^2} e^{-\beta x} \sinh\alpha x$$

$(0 < \alpha < \beta,\ x > 0)$ Example 5.1.4.

(A.242) $$\int_0^\infty \frac{\cos bt}{b^2+t^2} \frac{dt}{x^2+t^2} = \frac{\pi}{2bx} \frac{be^{-ax} - xe^{-ab}}{b^2 - x^2}$$

$(x > 0,\ a > 0,\ b > 0)$ Example 5.1.5. Also there

(A.243) $$\int_0^\infty \frac{\cos bt}{(b^2+t^2)^2} dt = \frac{\pi e^{-ab}(ab+1)}{4b^3}.$$

(A.244) $$\int_0^\infty \frac{e^{-\lambda ax}\cos^p(ax) - e^{-\lambda bx}\cos^p(bx)}{x} dx = \ln\frac{b}{a}$$

$(a, b, \lambda > 0,\ p - \text{any positive integer})$ Example 5.2.1. Also there

(A.245) $$\int_0^\infty \frac{e^{-\lambda ax}\sin^p(ax) - e^{-\lambda bx}\sin^p(bx)}{x} dx = 0.$$

(A.246) $$\int_0^\infty \frac{\cos^{2n}(ax) - \cos^{2n}(bx)}{x} dx = \frac{2}{4^n} \ln\frac{b}{a} \sum_{k=0}^{n-1} \binom{2n}{k}$$

$(a, b > 0)$ Example 5.2.2. Also there

(A.247) $$\int_0^\infty \frac{\cos^{2n-1}(ax) - \cos^{2n-1}(bx)}{x} dx = \ln\frac{b}{a}.$$

(A.248) $$\int_0^\infty \frac{\arctan^p(ax) - \arctan^p(bx)}{x} dx = \left(\frac{\pi}{2}\right)^p \ln\frac{a}{b}$$

$(a, b > 0,\ p - \text{any positive integer})$ Example 5.2.3.

(A.249) $$\int_0^\infty \frac{\operatorname{sech}^p(ax) - \operatorname{sech}^p(bx)}{x} dx = \ln\frac{b}{a}$$

$(a, b > 0, \ p -$ any positive integer) Example 5.2.4.

(A.250) $\qquad \int_0^\infty \dfrac{e^{-\lambda a x} \ln^p(1+ax) - e^{-\lambda b x} \ln^p(1+bx)}{x}\, dx = 0$

$(a, b, p, \lambda > 0)$ Example 5.2.5.

(A.251) $\qquad \int_0^\infty \dfrac{e^{-a x^p} - e^{-b x^p}}{x}\, dx = \dfrac{1}{p}\ln\dfrac{b}{a}$

$(a, b, p > 0)$ Example 5.2.6.

(A.252) $\qquad \int_0^\infty \dfrac{\ln(1+x)}{(1+x)^p\, x}\, dx = \zeta(2, p) = \sum_{n=0}^\infty \dfrac{1}{(n+p)^2}$

$(p > 0)$ Example 5.3.1.

(A.253) $\qquad \int_0^\infty \left(\dfrac{1-e^{-x}}{x} - e^{-x}\right)\dfrac{dx}{x} = 1$

Example 5.3.2.

(A.254) $\qquad \int_0^\infty \left(\dfrac{\sin\sqrt{x}}{\sqrt{x}} - \dfrac{1}{1+qx}\right)\dfrac{dx}{x} = 2(1-\gamma) + \ln q$

$(q > 0)$ Example 5.3.3. Also there, for $p > 0$ is arbitrary,

(A.255) $\qquad \int_0^\infty \left(\dfrac{\sin\sqrt{x}}{\sqrt{x}} - \dfrac{1}{1+x^p}\right)\dfrac{dx}{x} = 2(1-\gamma)$

or

$$\int_0^\infty \left(\dfrac{\sin t}{t} - \dfrac{1}{1+t^p}\right)\dfrac{dt}{t} = 1 - \gamma$$

(A.256)
$$\int_0^\infty \left(\frac{1}{1+x} - \frac{1}{1+x^p} \right) \frac{dx}{x} = 0 \, .$$

(A.257)
$$\int_0^\infty \left(\cos \sqrt{x} - \frac{1}{1+qx} \right) \frac{dx}{x} = -2\gamma + \ln q \quad (q > 0)$$

Example 5.3.4. Also there

(A.258)
$$\int_0^\infty \left(\frac{1}{1+t^p} - \cos t \right) \frac{dt}{t} = \gamma \quad (p > 0) \, .$$

(A.259)
$$\int_0^\infty \left(\frac{1 - \cos t}{t^2} - \frac{1}{2(1+t^p)} \right) \frac{dt}{t} = \frac{3}{4} - \frac{\gamma}{2} \quad (p > 0)$$

Example 5.3.5.

(A.260)
$$\int_0^\infty \left(\frac{\arctan \sqrt{x}}{\sqrt{x}} - \frac{1}{1+qx} \right) \frac{dx}{x} = 2 + \ln q \quad (q > 0)$$

Example 5.3.6.

(A.261)
$$\int_0^\infty \left(e^{-px} - \frac{1}{1+qx} \right) \frac{dx}{x} = \ln \frac{q}{p} - \gamma$$

$(p, q > 0)$ Example 5.3.7. Also there

(A.262)
$$\int_0^\infty \left(e^{-px} - \frac{1}{1+x^r} \right) \frac{dx}{x} = -\ln p - \gamma \quad (p, r > 0) \, .$$

(A.263)
$$\int_0^\infty \left(\beta(x+1) - \frac{\ln 2}{1+qx} \right) \frac{dx}{x} = \left(\frac{1}{2} \ln 2 - \gamma + \ln q \right) \ln 2$$

$(q > 0, \beta(s)$ - Nielsen's beta function). Example 5.3.8. Also there

(A.264)
$$\int_0^\infty \left(\beta(x+1) - \frac{\ln 2}{1+x^p} \right) \frac{dx}{x} = \left(\frac{1}{2} \ln 2 - \gamma \right) \ln 2 \quad (p > 0)$$

(A.265) $$\int_0^\infty \left(\frac{\psi(x+1)+\gamma}{x} - \frac{\zeta(2)}{1+qx} \right) \frac{dx}{x} = \zeta(2)\ln q - \zeta'(2)$$

($q > 0$) Example 5.3.9.

(A.266) $$\int_0^\infty \ln \frac{x^2+b^2}{x^2+a^2}\, dx = \pi(b-a) \quad (a,b>0)$$

Example 5.4.1.

(A.267) $$\int_0^\infty \frac{\sin(2n+1)x}{\sin x} e^{-\alpha x} x^{m-1}\, dx = (m-1)! \sum_{k=-n}^{n} \frac{1}{(\alpha-2ik)^m}$$

$$= (m-1)! \left(\frac{1}{\alpha^m} + 2\sum_{k=1}^{n} \frac{\cos(m\arctan(2k/\alpha))}{(\alpha^2+4k^2)^{m/2}} \right)$$

Example 5.4.2. This is Problem 11796 from AMM. Here $\alpha > 0$, and $m \geq 1, n \geq 0$ are integers.

(A.268) $$\int_0^1 \left\{ \frac{1+(p-1)\ln x}{1-x} + \frac{x\ln x}{(1-x)^2} \right\} x^{p-1} dx = \psi'(p) - 1$$

($p > 0$) Example 5.4.3.

(A.269) $$\sin\lambda\pi \int_0^\infty \frac{x^\lambda}{x+a} \frac{b^\mu - x^\mu}{b-x}\, dx = \sin\mu\pi \int_0^\infty \frac{x^\mu}{x+b} \frac{a^\lambda - x^\lambda}{a-x}\, dx$$

$$= \frac{\sin\lambda\pi \sin\mu\pi}{\pi} \int_0^\infty \int_0^\infty \frac{x^\mu t^\lambda}{(x+b)(t+a)(t+x)}\, dx\, dt$$

Example 5.4.4. Here $|\operatorname{Re}\lambda| < 1, |\operatorname{Re}\mu| < 1, |\operatorname{Re}(\lambda+\mu)| < 1$. This extends Problem 11506 from AMM.

(A.270) $$\int_0^\infty \frac{\ln x\, dx}{\sqrt{(a^2+x^2)(b^2+x^2)}} = \frac{\ln ab}{2a} K\left(\frac{\sqrt{a^2-b^2}}{a} \right)$$

Example 5.4.5. Here $0 < b < a$ and

$$K(k) = \int\limits_0^{\pi/2} \frac{1}{\sqrt{1 - k^2 \sin^2 \theta}} d\theta \,.$$

(A.271) $$\int\limits_0^1 \frac{x \ln(1+x)}{1+x^2} dx = \frac{\pi^2}{96} + \frac{(\ln 2)^2}{8}$$

Example 5.4.6. Problem 11966 (AMM). Also there

(A.272) $$\int\limits_0^1 \frac{\ln(1+x^2)}{1+x} dx = \frac{3}{4}(\ln 2)^2 - \frac{\pi^2}{48} \,.$$

(A.273) $$\int\limits_0^1 \left(\frac{\pi^2}{18} - 2 \arcsin^2 \frac{x}{2} \right) \frac{dx}{1-x}$$

$$= \frac{17}{9} \zeta(3) + \frac{4\pi^3}{27\sqrt{3}} - \frac{2\pi}{\sqrt{3}} \sum_{n=0}^{\infty} \frac{1}{(3n+1)^2}$$

Example 5.4.7.

(A.274) $$\int\limits_0^1 \frac{(\ln(1-t))^n}{t} dt = (-1)^n n! \zeta(n+1)$$

Example 5.4.8. Problem 1139 from CMJ.

(A.275) $$\int\limits_0^1 \left(\frac{\ln(1-t)}{t} \right)^n dt = n \sum_{k=0}^{n-1} (-1)^{k-1} s(n-1,k) \zeta(n+1-k)$$

Example 5.4.9. Problem 1117 (CMJ). Here $s(n,k)$ are the Stirling numbers of the first kind.

(A.276) $$\int\limits_{-\infty}^{\infty} \frac{x^2 \operatorname{sech}^2 x}{a - \tanh x} dx = \frac{1}{12} \ln \frac{a+1}{a-1} \left(\pi^2 + \ln^2 \frac{a+1}{a-1} \right)$$

($a > 1$). Example 5.4.10. Problem 11418 (AMM).

(A.277)
$$\int_0^1 \left(\frac{\text{Li}_2(1) - \text{Li}_2(x)}{1-x} \right)^2 dx = \frac{\pi^4}{15}$$

Example 5.4.11. This is Problem 12127 from AMM.

(A.278)
$$\int_1^\infty \frac{\ln(x^4 - 2x^2 + 2)}{x\sqrt{x^2 - 1}} dx = \pi \ln(2 + \sqrt{2})$$

Example 5.4.12. Problem 12184 (AMM).

(A.279)
$$\int_0^\infty \frac{\cos \lambda x}{x^4 + b^4} dx = \frac{\pi}{2b^3\sqrt{2}} e^{-\frac{\lambda b}{\sqrt{2}}} \left(\cos\frac{\lambda b}{\sqrt{2}} + \sin\frac{\lambda b}{\sqrt{2}} \right)$$

where $\lambda \geq 0$, $b > 0$. In particular, with $\lambda = 0$

$$\int_0^\infty \frac{1}{x^4 + b^4} dx = \frac{\pi}{2b^3\sqrt{2}}$$

Example 5.4.13. Also in this example

(A.280)
$$\int_0^\infty \frac{x\sin \lambda x}{x^4 + b^4} dx = \frac{\pi}{2b^2} e^{-\frac{\lambda b}{\sqrt{2}}} \sin\frac{\lambda b}{\sqrt{2}}.$$

(A.281)
$$\int_0^\infty \frac{\sin(2n+1)x}{\sin x} \frac{1}{x^2 + a^2} dx = \frac{\pi}{a}\left(\frac{1}{2} + e^{-(n+1)a} \frac{\sinh na}{\sinh a} \right)$$

Example 5.4.14. Here $a > 0$ and $n \geq 1$ is an integer.

(A.282)
$$\int_0^1 \frac{\arctan x}{x} dx = G$$

Section 5.5. (G is Catalan's constant). Also there

(A.283)
$$\int_0^1 \frac{\ln x}{1 + x^2} dx = -G$$

(A.284)
$$\int_0^{\pi/2} \frac{t}{\sin t} dt = 2G$$

(A.285)
$$\int_0^\infty \frac{t}{\cosh t} dt = 2G$$

(A.286a)
$$\int_0^1 \frac{\ln(1+x^2)}{1+x^2} dx = \frac{\pi}{2} \ln 2 - G$$

(A.286b)
$$\int_0^1 \frac{\ln(1-x^2)}{1+x^2} dx = \frac{\pi}{4} \ln 2 - G$$

(A.287)
$$\int_0^1 \left(\frac{\arctan x}{x}\right)^2 dx = \frac{\pi}{4} \ln 2 - \frac{\pi^2}{16} + G$$

(A.288)
$$\int_0^{\pi/4} \left(\frac{t}{\sin t}\right)^2 dx = \frac{\pi}{4} \ln 2 - \frac{\pi^2}{16} + G$$

(A.289)
$$\int_0^1 \frac{(\arctan t)^2}{t} dt = \frac{\pi}{2} G - \frac{7}{8} \zeta(3).$$

(A.290a)
$$\int_0^1 \ln\left(\frac{1+x^2}{x}\right) \frac{dx}{1+x^2} = \frac{\pi}{2} \ln 2$$

Example 5.5.1. Also there

(A.290b)
$$\int_0^1 \ln\left(\frac{1-x^2}{x}\right) \frac{dx}{1+x^2} = \frac{\pi}{4} \ln 2.$$

(A.291)
$$\int_0^1 \frac{1}{x^2} \ln(1-x^2) \arccos x \, dx = \frac{\pi^2}{4} - 4G$$

Example 5.5.2.

(A.292)
$$\int_0^1 \frac{1}{x} \ln(1 + \alpha x^2) \arccos x \, dx$$

$$= \frac{\pi}{4} \left(\ln^2 \frac{1 + \sqrt{1 + \alpha}}{2} - 2 \operatorname{Li}_2 \frac{1 - \sqrt{1 + \alpha}}{2} \right)$$

Example 5.5.3. Here $|\alpha| \le 1$. In particular, with $\alpha = -1$

(A.293)
$$\int_0^1 \frac{1}{x} \ln(1 - x^2) \arccos x \, dx = \frac{\pi}{2} \ln^2 2 - \frac{\pi^3}{24}.$$

(A.294)
$$\int_0^\infty \frac{\ln(1 + \lambda x)}{1 + x^2} dx = \frac{\pi}{4} \ln(1 + \lambda^2) - \ln \lambda \arctan \lambda + \operatorname{Ti}_2(\lambda)$$

($\lambda > 0$). In particular, with $\lambda = 1$

(A.295)
$$\int_0^\infty \frac{\ln(1 + x)}{1 + x^2} dx = \frac{\pi}{4} \ln 2 + G$$

Example 5.5.4.

(A.296)
$$\int_0^{\pi/2} \ln(1 + \tan x) \, dx = \frac{\pi}{4} \ln 2 + G$$

Example 5.5.5. Also there

(A.297)
$$\int_0^{\pi/4} \ln(1 + \tan x) \, dx = \frac{\pi}{8} \ln 2.$$

(A.298)
$$\int_0^{\pi/2} \operatorname{arcsinh}(\cos x) \, dx = G$$

Example 5.5.6. Also there

(A.299)
$$\int_0^{\pi/2} \ln(\cos x + \sqrt{1 + \cos^2 x}) \, dx = G.$$

(A.300)
$$\int_0^1\int_0^1 \frac{dx\,dy}{(1-xy)(1+x)(1+y)} = \ln 2$$

Example 5.6.2. Also there

(A.301)
$$\int_0^1 \ln\left(\frac{2}{1-x}\right)\frac{dx}{(1+x)^2} = \ln 2 \,.$$

(A.302)
$$\int_1^\infty \frac{\ln(1+t)}{t^2}\,dt = 2\ln 2 \,.$$

(A.303)
$$\int_0^1\int_0^1 \frac{dx\,dy}{(1+xy)(1+x)(1+y)} = \frac{\pi^2}{24}$$

Example 5.6.3. Also there

(A.304)
$$\int_0^1 \ln\left(\frac{2}{1+x}\right)\frac{dx}{1-x^2} = \frac{\pi^2}{24}$$

(A.305)
$$\int_0^1 \frac{\ln(1+t)}{t}\,dt = \frac{\pi^2}{12} \,.$$

(A.306)
$$\int_0^1\int_0^1 \frac{x^2 y^2}{(1+x^2 y^2)(1+x)(1+y)}\,dx\,dy$$
$$= -\frac{\pi^2}{48} + \frac{\pi}{8}\ln 2 + \frac{7}{8}(\ln 2)^2 - \frac{G}{2} \,.$$

Example 5.6.4.

(A.307)
$$\int \frac{\log(t)\log^2(1-t)}{1-t}\,dt$$
$$= \log^2(1-t)\,\mathrm{Li}_2(1-t) - 2\log(1-t)\,\mathrm{Li}_3(1-t) + 2\,\mathrm{Li}_4(1-t) + C$$

Section 5.7.1. Also there

(A.308) $$\int \frac{1}{t(1-t)} \mathrm{Li}_p \left(\frac{-t}{1-t} \right) dt = \mathrm{Li}_{p+1} \left(\frac{-t}{1-t} \right) + C$$

(A.309) $$\int \frac{\log(t)\log(1-t)}{1-t} dt$$

$$= \log(1-t) \mathrm{Li}_2(1-t) - \mathrm{Li}_3(1-t) + C .$$

(A.310) $$\int_0^t \frac{\mathrm{Li}_2(x)}{1-x} dx$$

$$= 2\left[\mathrm{Li}_3(1-t) - \zeta(3) \right] - \log(1-t)\left[\mathrm{Li}_2(1-t) + \frac{\pi^2}{6} \right]$$

$(0 < t < 1)$. Section 5.7.2.

(A.311) $$\int_0^1 \int_0^1 \frac{dx\,dy}{(1 - xyz)(1+x)(1+y)}$$

$$= \frac{1}{1-z} \left\{ \mathrm{Li}_2(z) + \log^2 2 - 2\left[\mathrm{Li}_2\left(\frac{1+z}{2} \right) - \mathrm{Li}_2\left(\frac{1}{2} \right) \right] \right\} .$$

Section 5.7.3.

(A.312) $$\int_0^\infty \frac{\cos x \sin x}{x} dx = \frac{\pi}{4}$$

Section 5.8. Also there

(A.313) $$\int_0^\infty \frac{\sin(ax)\sin(bx)}{x} dx = \frac{1}{2} \ln \left| \frac{a+b}{a-b} \right|$$

$(a, b > 0,\ a \neq b)$

(A.314) $$\int_0^\infty \frac{\sin(ax)\cos(bx)}{x} dx = \frac{\pi}{2}$$

$(a > b > 0)$

(A.315) $$\int_0^\infty \frac{\cos^2 x \sin x}{x} dx = \frac{\pi}{4}$$

(A.316) $$\int_0^\infty \frac{\cos^{2n} x \sin x}{x} dx = \int_0^\infty \cos^{2n} x \left(\frac{\sin x}{x}\right)^2 dx$$

$$= \frac{1}{4^n} \binom{2n}{n} \frac{\pi}{2}$$

(A.317) $$\int_0^\infty \frac{\sin^{2n+1} x}{x} dx = \int_0^\infty \sin^{2n} x \left(\frac{\sin x}{x}\right)^2 dx = \frac{1}{4^n} \binom{2n}{n} \frac{\pi}{2}$$

(A.318) $$\int_0^\infty |\sin x| \frac{\sin x}{x} dx = \int_0^\infty |\cos x| \frac{\sin x}{x} dx = 1.$$

(A.319) $$\int_0^\infty e^{-\lambda x} \left(\frac{1}{x} - \coth x\right) dx = \psi\left(\frac{\lambda}{2}\right) - \ln\frac{\lambda}{2} + \frac{1}{\lambda}$$

Example 5.9.1. Here $\operatorname{Re}\lambda > 0$. Also in this example

(A.320) $$\int_0^\infty e^{-\lambda x} \left(\frac{1}{x} - \frac{1}{\sinh x}\right) dx = \psi\left(\frac{\lambda+1}{2}\right) - \ln\frac{\lambda}{2}.$$

(A.321) $$\int_0^\infty e^{-\lambda x} \left(1 - \frac{1}{\cosh x}\right) \frac{dx}{x} = 2\ln\frac{\Gamma\left(\frac{\lambda+3}{4}\right)}{\Gamma\left(\frac{\lambda+1}{4}\right)} - \ln\frac{\lambda}{4}$$

Example 5.9.2.

(A.322) $$\int_0^\infty \frac{\cosh(2yt)}{\cosh^{2x}(t)} dt = 2^{2x-2} \frac{\Gamma(x+y)\Gamma(x-y)}{\Gamma(2x)}$$

$(\operatorname{Re}x > 0, \operatorname{Re}x > |\operatorname{Re}y|)$ Example 5.9.3. Also there

(A.323) $$\int_0^\infty \frac{\cosh(2yt)}{\cosh^2(t)}\,dt = \frac{\pi y}{\sin \pi y}$$

(A.324) $$\int_0^\infty \frac{\cos(2yt)}{\cosh^2(t)}\,dt = \frac{\pi y}{\sinh \pi y}$$

(A.325) $$\int_0^\infty \frac{\cosh(2yt)}{\cosh^4(t)}\,dt = \frac{2\pi y(1-y^2)}{3\sin \pi y}$$

(A.326) $$\int_0^\infty \frac{\cos(2yt)}{\cosh^4(t)}\,dt = \frac{2\pi y(1+y^2)}{3\sinh \pi y}$$

(A.327) $$\int_0^\infty t^{2n}e^{-t^2}\cos(2xt)\,dt = \frac{(-1)^n\sqrt{\pi}}{2^{2n+1}}e^{-x^2}H_{2n}(x)$$

Example 5.9.4. Here $H_k(x)$ are the Hermite polynomials. Also there

(A.328) $$\int_0^\infty t^{2n+1}e^{-t^2}\sin(2xt)\,dt = \frac{(-1)^n\sqrt{\pi}}{2^{2n+2}}e^{-x^2}H_{2n+1}(x).$$

(A.329) $$\int_0^\pi e^{\lambda \cos x}\cos(nx)dx = \pi I_n(\lambda)$$

Example 5.9.5. Here $I_n(\lambda)$ $(n = 0, 1, 2, ...)$ are the modified Bessel functions of the first kind. In particular

(A.330) $$\int_0^\pi e^{\lambda \cos x}dx = \pi I_0(\lambda)$$

(A.331) $$\int_0^\pi e^{\cos x}dx = \pi \sum_{p=0}^\infty \frac{1}{4^p(p!)^2}.$$

(A.332) $$\int_0^\pi \frac{\ln(\cosh x + \sinh x \cos y)}{\cosh x + \sinh x \cos y}dy = \pi \ln \frac{2}{1+\cosh x}$$

Example 5.9.6. Also in this example

(A.333) $\displaystyle\int_0^\pi \frac{\ln^2(\cosh x + \sinh x \cos y)}{\cosh x + \sinh x \cos y}\,dy = -2\pi \operatorname{Li}_2\left(\frac{1-\cosh x}{2}\right).$

(A.334) $\displaystyle\int_0^\infty\int_0^\infty \frac{\sin(x)\sin(x+y)}{x(x+y)}\,dx\,dy = \frac{\pi^2}{8}$

Example 5.9.7.

(A.335) $\displaystyle\int_0^\infty \frac{\sin^2 x - x\sin x}{x^3}\,dx = \frac{1}{2} - \ln 2.$

Example 4.2.6.

References

[1]   Naum I. Akhiezer, *Elements of the Theory of Elliptic Functions*, Amer. Math. Soc. (1990).

[2]   Jacob Ablinger, Discovering and proving infinite binomial sums identities, *Experimental Mathematics*, 26 (1) 2017.

[3]   Victor Adamchik, On Stirling numbers and Euler sums, *J. Comput. Appl. Math.*, 79 (1) (1997), 119-130.

[4]   Matthew Albano, Tewodros Amdeberhan, Erin Beyerstedt and Victor H. Moll, The integrals in Gradshteyn and Ryzhik. Part 15: Frullani integrals, *SCIENTIA, Series A: Mathematical Sciences,* Vol. 19 (2010), 113–119.

[5]   Harry Bateman, Artur Erdelyi, *Tables of Integral Transforms*, Volumes I & II], McGraw Hill 1954.

[6]   Bruce C. Berndt, *Ramanujan's notebooks, Part 1.* Springer 1983.

[7]   George Boros, Victor H. Moll, *Irresistable Integrals*, Cambridge University Press, 2004.

[8]   Khristo N. Boyadzhiev, *Notes on the Binomial Transform*, World Scientific, 2018.

[9]   Khristo N. Boyadzhiev, Evaluation of series with binomial sums, *Anal. Math.,* 40 (1) (2014), 13-23.

[10]   Khristo N. Boyadzhiev, Some integrals related to the Basel problem, *SCIENTIA. Series A: Mathematical Sciences,* 26 (2015), 1-13. Also arXiv: 1611.03571 [math.NT]

[11]   Khristo N. Boyadzhiev, Power series with skew-harmonic numbers, dilogarithms, and double integrals, *Tatra Mountain Math. Publications,* 56 (2013), 93-108.

[12]   Khristo N. Boyadzhiev, Evaluation of one exotic Furdui type series, *Far East J. Math. Sci.,* 75 (2) (2013), 359-367.

[13]   **Khristo N. Boyadzhiev**, Series Transformation Formulas of Euler Type, Hadamard Product of Functions, and Harmonic Number Identities, *Indian J. Pure Appl. Math.*, 42 (2011), 371-387.

[14]   **Khristo N. Boyadzhiev**, Exponential polynomials, Stirling numbers, and evaluation of some Gamma integrals, *Abstract Appl. Anal.*, Volume 2009, Article ID 168672 (electronic).

[15]   **Khristo Boyadzhiev, Larry Glasser**, Solution to problem 883, *The College Math. Journal*, 40 (4) (2009), 297-299.

[16]   **Khristo N. Boyadzhiev, Louis Medina and Victor H. Moll**, The integrals in Gradshteyn and Ryzhik, Part 11: The incomplete beta function, *SCIENTIA: Series A: Mathematical Sciences*, 18 (2009), 61-75.

[17]   **Khristo N. Boyadzhiev, Victor H. Moll**, The integrals in Gradshteyn and Ryzhik, Part 21: Hyperbolic functions, *SCIENTIA: Series A: Mathematical Sciences*, v. 22 (2012), 109-127.

[18]   **Khristo Boyadzhiev, Hans Kappus**, Solution to problem E 3140, *Amer. Math. Monthly*, 95 (1), (1988), 57-59.

[19]   **Sergio Bravo, Ivan Gonzalez, Karen Kohl, and Victor H. Moll**, Integrals of Frullani type and the method of brackets, Open Math. 15 (1) (2017), 1-12.

[20]   **Yuri A. Brychkov**, *Handbook of Special Functions*, CRC Press, 2008.

[21]   **Hongwey Chen**, Parametric differentiation and integration, *Internat. J. Math. Ed. Sci. Tech.* 40 (4) (2009), 559-579.

[22]   **Lokenath Debnath, Dambaru Bhatta**, *Integral Transforms and Their Applications*, Chapman and Hall/CRC, 2007.

[23]   **Grigorii M. Fikhtengol'ts**, *A Course of Differential and Integral Calculus* (Russian), Vol. 1, Vol. 2, Vol. 3, Nauka, Moscow, 1966. (Abbreviated version in English: *The Fundamentals of Mathematical analysis*, Vol. 2, Pergamon Press, 1965.)

[24]   **M. Laurence Glasser**, A remarkable property of definite integrals, *Math. Comp.*, 40 (162) (1983), 561-563

[25] **Izrail S. Gradshteyn and Iosif M. Ryzhik**, *Tables of Integrals, Series, and Products*, 8ᵗʰ edition, Academic Press, 2014.

[26] **Eldon R. Hansen**, *A Table of Series and Products*, Prentice Hall, 1975.

[27] **Godfrey H. Hardy**, On the Frullanian integral (formula), *Quarterly J. Math.* 33 (1902) 113-144.

[28] **Godfrey H. Hardy**, On differentiation and integration under the integral sign, *Quarterly J. Pure and Applied Math*, 32 (1901), 66-140. (Continued as *Quarterly J. Math.*)

[29] **Omar Hijab**, Introduction to Calculus and Classical Analysis, Springer, 1997.

[30] **Paul Koosis**, *Introduction to H_p spaces*, Cambridge University Press, 1980.

[31] **Dan Kalman and Mark McKinzie**, Another way to sum a series: generating functions, Euler, and the dilog function, *Amer. Math. Monthly*, 119 (1) (2012), 42-51.

[32] **Derrick H. Lehmer**, Interesting series involving the central binomial coefficient, *Amer. Math. Monthly*, 92 (7) (1985), 449-457.

[33] **Leonard Lewin**, *Polylogarithms and associated functions*, North Holland, 1981.

[34] **Constantin C. Maican**, Integral Evaluations Using the Gamma and Beta Functions and Elliptic Integrals in Engineering, International Press, 2005.

[35] **Victor H. Moll**, Special Integrals of Gradshteyn and Ryzhik: the Proofs - Volumes 1 and 2, CRC, 2016.

[36] **Habib Bin Muzaffar**, A New Proof of a Classical Formula, *Amer. Math. Monthly*, 120, No. 4 (2013), 355-358.

[37] **Paul J. Nahin**, *Inside Interesting Integrals*, Springer, 2015.

[38] **Niels Nielsen**, *Die Gammafunktion*, Chelsea, 1965.

[39] **Niels Nielsen**, *Theorie des Integrallogarithmus*, Teubner, Leipzig, 1906.

[40] **Frank W. Olver** et al (editors), *NIST Handbook of Mathematical Functions*, Cambridge Univ. Press, 2010.

[41] **Nikolai S. Piskunov**, *Differential and Integral Calculus*, Groningen: P. Noordhoff, 1969.

[42] **Alexander D. Poularikas**, The Transformations and Applications *Handbook*, CDC Press, 1996

[43] **Anatoli P. Prudnikov, Yuri. A. Brychkov, Oleg. I. Marichev**, *Integrals and Series, vol. 1 Elementary Functions*, Gordon and Breach 1986.

[44] **Nico M. Temme**, *Special Functions*, John Wiley, 1996.

[45] **Edward Charles Titchmarsh**, *The Theory of the Riemann Zeta-function*, (second edition), Clarendon Press, Oxford, 2001

[46] **Georgi P. Tolstov**, *Fourier Series*, Dover, 1976.

[47] **J. Trainin**, Integrating expressions of the form $\sin^n x / x^m$ and others, *Math. Gazette*, 94 (2010), 216-223.

[48] **Edmund Taylor Whittaker, George Neville Watson**, *A Course of Modern Analysis*, (fourth edition), Cambridge University Press, 1992.

[49] **Joseph Wiener**, Differentiation with respect to a parameter, *College Mathematics Journal*, 32, No. 3. (2001), pp. 180-184.

[50] **Frederick S. Woods**, Advanced Calculus: A Course Arranged with Special Reference to the Needs of Students of Applied Mathematics. New Edition, Ginn and Co., Boston, MA, 1934.

[51] **Aurel J. Zajta, Sudhir K. Goel**, Parametric integration techniques, *Math. Magazine*, 62 (5) (1989), 318-322.

[52] **Don Zagier**, The Dilogarithn Function, *Frontiers in Number Theory*, Physics, and Geometry II, Springer, 2007, 3-65.

[53] 66th Annual William Lowell Putnam Mathematical Competition, *Math. Magazine*, 79 (2006), 76-79.

[54] **Ilan Vardi**, Integrals, an Introduction to Analytic Number Theory, *Amer. Math. Monthly*, 95 (4), (1988), 308-315.

[55] **Hassan Jolany**, An extension of Lobachevsky formula, *Elemente der Mathematik*, 73 (3) (2018), 89-94.

Index

Printed in the United States
by Baker & Taylor Publisher Services